Topics in Applied Physics   Volume 13

Topics in Applied Physics   Founded by Helmut K.V. Lotsch

High-Resolution Laser Spectroscopy

Edited by K. Shimoda

With Contributions by
N. Bloembergen   V. P. Chebotayev   J. L. Hall
S. Haroche   P. Jacquinot   V. S. Letokhov
M. D. Levenson   J. A. Magyar   K. Shimoda

With 132 Figures

Springer-Verlag Berlin Heidelberg GmbH 1976

Professor Dr. KOICHI SHIMODA

Department of Physics, University of Tokyo
7–3–1 Hongo, Bunkyo-ku, Tokyo 113, Japan

ISBN 978-3-662-31249-0   ISBN 978-3-540-38237-9 (eBook)
DOI 10.1007/978-3-540-38237-9

Library of Congress Cataloging in Publication Data. Main entry under title: High-resolution laser spectroscopy. (Topics in applied physics; v. 13). Includes bibliographies. 1. Laser spectroscopy. 1. Shimoda, Koichi, 1920—. II. Bloembergen, N. QC454.L3H3. 535.5'8. 76–10235

© by Springer-Verlag Berlin Heidelberg 1976
Originally published by Springer-Verlag Berlin Heidelberg New York in 1976
Softcover reprint of the hardcover 1st edition 1976

Preface

Immediately after the laser materialized in 1960, quite a few researchers conceived of optical analogues of radio and microwave spectroscopy by using the laser. The resolution and sensitivity of laser spectrometers have now been verified to be many orders of magnitude higher than otherwise assumed. The laser is not only an ideal light source for a well-known method of spectroscopy, but it provides unprecedented means of spectroscopic techniques which are called coherent spectroscopy, nonlinear spectroscopy, sub-Doppler spectroscopy etc.

Before the advent of laser spectroscopy, the resolving power of a spectrometer was usually limited by instrumental factors. If the instrumental resolution is improved by using, for example, a high-quality grating of a very large dimension, one can find the resolving power that is limited by the Doppler effect of atoms and molecules moving in a gas. Not only has this limit been reached practically by laser spectroscopy, but also very recently novel methods which furnish much higher resolution have been developed. The very rapid progress of laser spectroscopy has necessitated the publishing of another volume of *Topics in Applied Physics* soon after the volume 2, entitled "Laser Spectroscopy of Atoms and Molecules", was edited by Professor H. WALTHER.

The purpose of this volume is to provide comprehensive information and understanding of this exciting field of spectroscopy both for the beginners and the research workers. Main important achievements of high-resolution laser spectroscopy during the last few years are reviewed. The resolution of a laser spectrometer, in contrast to that of classical spectrometers, is primarily determined by the physical processes between the matter and the laser radiation involved in the method of observation. The main objective of this book is, therefore, to make the readers familiar with the basic ideas of laser spectroscopy. A brief historical review of laser spectroscopy in Chapter 1 is followed by Chapter 2 which gives basic principles and concepts that are relevant to high-resolution laser spectroscopy described below. Chapters 3 through 8 describe the methods of atomic and molecular beams, quantum beats, Lamb dip or saturated absorption, coupled transitions and two-photon transitions. Authors of these chapters are active researchers in the respective field of investigation, and they present physical insight and suggestive information in their expositions.

Although the application of high-resolution laser spectroscopy has been restricted up to the present within narrow ranges of the available lasers, the recent progress in tunable lasers promises wider application of those techniques in the years to come. The methods described in this book will be extensively used for high-resolution spectroscopy in all regions from the ultraviolet to the far infrared. They will also be applied to studies of laser-induced chemistry, selective excitation of any desired quantum states, precision measurements of various quantities, and many others.

This book represents an international cooperation among authors of different countries and disciplines. The editor made every effort in coordinating chapters so as to avoid any inconsistencies, overlaps, and errors. Unfortunately, however, no internationally accepted nomenclatures and symbols exist for this rapidly growing field of laser spectroscopy. Some of them have not been unified throughout this book. Thereby we have preserved coherence of terminology between each chapter and its closely related references.

The editor is grateful to other authors of this book who have written their excellent contributions in time and have generously accepted modifications and suggestions of the editor. Finally, he would like to thank Dr. LOTSCH of Springer-Verlag for his help, advice and encouragement, without which this book might have never appeared.

Tokyo, March 1976   KOICHI SHIMODA

Contents

3. Atomic Beam Spectroscopy. By P. JACQUINOT
(With 20 Figures)

4. Saturation Spectroscopy. By V. S. LETOKHOV (With 38 Figures)

Contributors

BLOEMBERGEN, NICOLAAS
Harvard University, Cambridge, MA 02138, USA

CHEBOTAYEV, VIENNA P.
Institute of Semiconductor Physics, Siberian Branch of the Academy of Sciences USSR, SU-630090 Novosibirsk, USSR

HALL, JOHN L.
Joint Institute for Laboratory Astrophysics, University of Colorado, Boulder, CO 80302, USA

HAROCHE, SERGE
Ecole Normale Supérieure, 24, rue Lhomond, F-75005 Paris

JACQUINOT, PIERRE
CNRS, Laboratoire Aimé Cotton, Campus d'Orsay, F-91405 Orsay

LETOKHOV, VLADILEN S.
Institute of Spectroscopy, Academy of Sciences USSR Moscow, Podolskii Rayon, SU-142092 Academgorodok, USSR

LEVENSON, MARC DAVID
University of Southern California, Department of Physics, Los Angeles, CA 90007, USA

MAGYAR, JOHN A.
National Measurement Laboratory, C.S.I.R.O. University Grounds, City Road, Chippendale, N.S.W. Australia 2008

SHIMODA, KOICHI
Dept. of Physics, University of Tokyo, 7-3-1 Hongo, Bunkyo-ku, Tokyo 113, Japan

1. Introduction

K. SHIMODA

The advent of the laser in 1960 revolutionized spectroscopy. The laser has now become a powerful tool of spectroscopy over a range extending from the submillimeter waves to the vacuum ultraviolet region. Laser spectroscopy has achieved improvements in resolution, accuracy, and sensitivity of many orders of magnitude compared with the traditional methods of spectroscopy. It took more than 10 years before the laser matured for such use in various methods of spectroscopy as described in Section 1.2.

The recent progress in high resolution laser spectroscopy will be extended and find more universal applications for chemical analysis and the study of molecules. Atomic processes and chemical reactions of molecules in their particular states identified by the laser will be the subjects of intensive research in the years to come.

1.1 General Features of the Laser

Characteristics and properties of a variety of lasers are outside the scope of this article. It is the purpose of this section to consider fundamental properties of the laser which are pertinent to laser spectroscopy.

1.1.1 Temporal Coherence and Monochromaticity

It is well known that the laser is an extremely monochromatic source of light, and its temporal coherence is very good. The laser is more like a microwave oscillator rather than an ordinary light source that emits many statistically independent wave trains, each having a duration of about 10^{-9} s. A good laser shows only small fluctuations in amplitude and frequency. Even a relatively noisy laser has a narrower spectral width than a sharp linewidth of a low pressure gas. The resolving power of laser spectroscopy is therefore determined ordinarily by the width of the spectral line of the sample under study rather than the linewidth of the laser output.

The resolution of a conventional spectrometer is, on the contrary, instrumentally limited. The resolving power of a grating spectrometer with its narrowest possible slit is ordinarily expressed by the product of the order of interference m and the number of grooves N. The resolution in wave number $\delta\tilde{v} = \tilde{v}/mN$ may be rewritten approximately as $\delta\tilde{v} \simeq 1/L$, where L is the size of the grating. The resolution of Fourier transform spectroscopy may likewise be given by $\delta\tilde{v} \simeq 1/D$, where D is the magnitude of translation of the mirror. Hence the high resolution in conventional spectroscopy is of the order of 0.01 cm^{-1}, corresponding to the practical limit in size of about 1 m. The resolving power of a good conventional infrared spectrometer is 10^4–10^5.

The Doppler broadening of the spectral line of atoms and molecules in a gas is about 10^{-6} times the frequency. Thus laser spectroscopy can easily attain a resolving power of 10^6 and a resolved linewidth of the order of 100 MHz in the middle infrared region.

As described everywhere in this book, various techniques of laser spectroscopy having higher resolution than the Doppler width have been developed. Because the linewidth obtained in these methods of high resolution laser spectroscopy is 10^{-3}–10^{-4} times the Doppler width, the laser must be stable to better than $\pm(10$–$100)$ kHz.

Quantum noise of the laser is a result of spontaneous emission, while the coherent oscillation is produced by stimulated emission. The quantum fluctuation δv_{las} in frequency v of the laser oscillator[1] is given by [1.1]

$$\delta v_{\text{las}} = [2\pi h v (\Delta v)^2/P] N_2/(N_2 - N_1) \tag{1.1}$$

where Δv is approximated by the half resonance width of the optical resonator Δv_{cav}, N_1 is the lower state population and N_2 is the upper state population of the active medium. The power of stimulated emission in the optical resonator is denoted by P, the output power being a fraction of P.

As a typical example of a gas laser at 1 μm, take $v = 3 \times 10^{14}$ Hz, $\Delta v = 3 \times 10^6$ Hz, $P = 1$ mW, and $N_2 = 2N_1$; then the quantum noise of (1.1) gives a small value of $\delta v_{\text{las}} = 0.02$ Hz.

A practical laser shows much larger frequency variation, because of instability of its optical resonator. It is easy to show in the first-order approximation[2] that the frequency of laser oscillation is given by [1.2]

$$v = (v_0 \Delta v_{\text{cav}} + v_{\text{cav}} \Delta v_0)/(\Delta v_{\text{cav}} + \Delta v_0) \tag{1.2}$$

---

[1] The laser consists of a nearly homogeneous active medium and an optical resonator with two concave mirrors or one concave and one flat mirror.

[2] The expression is valid to higher orders, when the laser line is only homogeneously broadened to have a Lorentzian shape.

where v_0 is the center frequency of the laser transition, Δv_0 is the half width of fluorescence of the laser transition, and v_{cav} is the resonant frequency of the optical resonator. Because the cavity resonance is normally sharper than the laser line so that $\Delta v_{cav} \ll \Delta v_0$, a rough estimate may be $v - v_0 = \delta v_{cav}$. Thus thermal expansion of the optical resonator by 10^{-8}, for example, gives rise to a decrease in laser frequency by 10^{-8}. Acoustic noises and mechanical disturbances are very often major sources of laser frequency fluctuations. It is from this reason that good acoustic and mechanical isolation as well as good temperature control are required in lasers for high resolution spectroscopy.

1.1.2 Spatial Coherence

The laser exhibits good spatial coherence: light waves emitted from different parts of a laser have definite phase relations with each other. The spatial coherence is perfect in the single-mode laser. An evident example of spatial coherence is the very directive beam of light emitted from the laser.

The beam divergence of the laser output is close to the theoretical diffraction limit of coherent waves passing through a finite aperture. The beam divergence $\delta\theta$ is approximately given by

$$\delta\theta = \lambda/D \tag{1.3}$$

where λ is the wavelength and D is the diameter of the beam. Diffraction of light having a constant amplitude over a circular aperture is not normaly applicable to the laser beam. The amplitude of the laser light shows a gaussian distribution about the beam axis when the laser is operated on the lowest transverse mode. When the optical field across the beam is given by

$$E(r) = E_0 \exp(-r^2/w^2)$$

with a constant phase (planar wave front) on the surface at the origin (e.g., output mirror), the angular distribution at a distance becomes

$$E(\theta) = A \exp[-(\theta/\delta\theta)^2], \qquad \delta\theta = \lambda/\pi w. \tag{1.4}$$

The parameter w is called the beam radius (at the beam waist), and $\delta\theta$ is the beam divergence.

The laser beam can be focused to a small spot. The spot size is given by $f\,\delta\theta$, where f is the focal length. Since $10^{-4} - 10^{-3}$ radian is a typical value of the beam divergence, the laser beam can be focused to a size of a few times the wavelength, depending on the F-number (f/D) of focusing.

This feature allows laser spectroscopy of high spatial resolution and concentration of light energy. Metallurgical, biological and other applications of laser spectroscopy with a spatial resolution of a few micrometers have been achieved. Concentration of the laser beam gives rise to the strong optical field, which exhibits a large variety of nonlinear optic effects.

1.1.3 Tuning and Modulation

The laser is now available at thousands of different wavelengths from the vacuum ultraviolet to the submillimeter wave region. The frequency of each laser can be tuned within the rather narrow fluorescence linewidth around the proper frequency by adjusting the resonant frequency of its optical resonator. Wider tuning may be obtained either by tuning the frequency of the laser transition with an appropriate method or by widening the bandwidth of the active medium, while suppressing the multimode-multifrequency operation.

The former method of tuning has been materialized in Zeeman tuning of gas lasers [1.3], pressure tuning of semiconductor diode lasers [1.4], temperature tuning of solid-state lasers [1.5], etc. The latter method is restricted to the high gain laser, because widening of the bandwidth results in the reduction in gain. A promising example of the latter method is the high-pressure molecular gas laser, in which over-lapping of closely spaced vibrational-rotational lines at high pressure partly compensates the gain reduction with pressure. The dye laser has such a high gain and wide bandwidth that can be tunable over a frequency range of 1000–2000 cm^{-1} in the visible region [1.6]. Many review papers on tunable lasers have recently appeared [1.7].

Frequency scanning or sweep within the tuning range of the laser is often useful in spectroscopy. It is remarked, however, that a wider frequency tuning generally results in a larger instability of frequency. If a laser is tuned 100 GHz with a control voltage of 1000 V, for instance, a ripple of ± 10 mV will produce a frequency jitter of ± 1 MHz. Besides, the rate of frequency sweep must be several times lower than the value given by the resolution in frequency divided by the time constant of observation. The frequency modulation of the laser is sometimes utilized in spectroscopy, but the modulation frequency and the maximum frequency deviation must be lower than the linewidth of the sample.

The amplitude of the laser may also be modulated. The frequency of amplitude modulation by an internal modulator is limited by the bandwidth of the laser, and it is still very fast compared with the speed of electronic devices. Nanosecond pulses are produced by most gas lasers and semiconductor lasers. Picosecond and even subpicosecond pulses are produced by glass lasers and dye lasers. Time-resolved spectroscopy by using these picosecond pulses has opened a new field of so-called picosecond spectroscopy that is useful for the investigation of fast relaxation, energy transfer, chemical reaction and so on.

1.1.4 Brightness

The efficiencies of the present lasers are fairly low. The net total power emitted from a laser is not higher than that of many other light sources of comparable size. Considering the narrow beam divergence and the sharp linewidth of the laser, however, we find that the laser is incomparable in brightness. The energy density per unit frequency interval and per unit solid angle is many orders of magnitude higher than any thermal light source. For example, the effective brightness temperature of a small 1 mW gas laser with a spectral width of 1 Hz is 10^{20} K! A rather small energy of 1 mJ in a laser pulse of 1 ps duration gives a peak power of 1 GW. When this power is focused to a spot of a few micrometers in diameter, the power density will be 10^{16} W/cm^2, which is high enough to decompose any matter into ions with its optical field of 2×10^9 V/cm.

Thus it is not surprising that laser radiation exhibits a wide variety of nonlinear optical effects. Spectroscopic behaviors of nonlinear effects are in many cases utilized to obtain higher resolution. As shown in the following chapters in this book, moderate power of the laser is sufficient for the purpose. One must remember that the optical field intensity produced by any laser is just strong enough to set up the nonlinear effect that saturates the laser transition appreciably.

1.2 Historical Review of Laser Spectroscopy

Spectroscopic studies of laser transitions were the subject of research from the earliest stage, because they were essential to the understanding of laser properties. Not only identification of laser transitions and studies of relaxation and pumping processes but also coherent effects in coupled transitions that share a common level with the laser transition have been investigated.

Laser spectroscopy of materials other than the laser medium has developed gradually. Both linear and nonlinear spectroscopy were initially carried out with narrow band lasers. Recent progress in tunable lasers has definitely accelerated the development of laser spectroscopy. The history of laser spectroscopy is reviewed in the following.

1.2.1 Raman Spectroscopy

The ruby laser was first employed as a light source of Raman spectroscopy in 1962 [1.8]. The stimulated Raman effect that revealed a large Stokes gain was discovered accidentally by WOODBURY and NG, while they were studying the optical Kerr switching of the laser beam in 1962 [1.9]. Subsequent studies revealed higher-order effects and their highly directional properties [1.10]. The nonlinear nature of the process has been explained by several authors [1.11]. Transients in nonlinear propagation effects of Raman scattering is still a subject of investigation involving self phase modulation, pulse steepening, etc.

Raman spectroscopy with a cw gas laser was initiated in 1963 [1.12]. The high intensity, the narrow linewidth, and the small beam divergence of the laser have made it an unprecedented excitation source for Raman spectroscopy. The He-Ne laser at 633 nm and the argon ion laser at 514.5 nm and 488 nm are most commonly used. No tuning of the laser is required. At present many commercial laser Raman spectrometers are available for wide applications [1.13]. The cross section for electronic Raman scattering is about three orders of magnitude smaller. The electronic Raman transitions were first observed in 1963 from a single crystal of $PrCl_3$ excited by the Hg 253.7 nm radiation [1.14] and the electronic laser Raman scattering later became a useful method of studies in solid-state physics [1.15]. In addition to Raman scattering, lasers have been exclusively employed to excite Brillouin, Rayleigh and other scattering processes, which are not reviewed here.

1.2.2 Stark and Zeeman Spectroscopy

Infrared absorption spectroscopy of nonlaser materials was carried out by GERRITSEN and AHMED in 1964 with their Zeeman-tuned 3.39 μm He-Ne laser [1.16]. Many hydrocarbon molecules were found to exhibit vibrational-rotational lines of the C-H stretching mode within the tuning range of about ±3 GHz of the laser [1.17]. The technique was extended to other wavelengths by using the He-Xe laser at 3.51 μm [1.18, 19], 2.03 μm [1.20], 3.36 and 5.57 μm [1.21], and the He-Ne laser at 7.48, 7.70 and 7.77 μm [1.22].

A preliminary study of Stark effects in laser spectroscopy of molecules was reported by FELD et al. in 1965 [1.23]. The power of laser Stark spectroscopy in assigning absorption lines and in increasing the sensitivity was fully demonstrated in the study of formaldehyde with the He-Xe laser [1.19].

The technique of Stark shifting the absorption lines is particularly useful for spectroscopy with fixed frequency lasers. Molecular lasers on vibrational or rotational transitions are not tunable by the practical magnetic field. Stark-tuned spectroscopy of NH_3 with the 84.1 μm and 171.6 μm D_2O lasers was reported in 1968 [1.24]. Well-resolved Stark spectra of NH_3 with many laser lines of CO_2 were observed by SHIMIZU in 1969 [1.25]. Laser Stark spectroscopy has been further developed since then in several laboratories [1.26].

Zeeman spectroscopy of electric dipole transitions and laser magnetic resonance spectroscopy for magnetic dipole transitions of atoms and molecules with electronic magnetic moments can similarly be performed by applying a magnetic field for tuning the spectral lines. The technique was first demonstrated with a paramagnetic crystal of holmium ethyl sulphate [1.27], followed by a series of studies with stable magnetic molecules, O_2 [1.28], NO [1.29] and NO_2 [1.30]. Later, unstable free radicals, OH [1.31], CH [1.32], HO_2 and HCO [1.33] were observed.

1.2.3 Fluorescence Spectroscopy

Laser-induced fluorescence from molecules other than those of laser medium was first observed in 1966 by YARDLEY and MOORE [1.34]. They observed fluorescence decay in methane to study its vibrational relaxation and energy transfer. High resolution spectroscopy by the method of laser-induced fluorescence was initiated by EZEKIEL and WEISS in 1968 [1.35]. A molecular beam of iodine was excited by the 514.5 nm argon ion laser and its fluorescence was observed as a function of tuning, thereby demonstrating very high resolution achieved by the molecular beam technique. The development of molecular and atomic beam laser spectroscopy is given in Chapter 3.

The fluorescence spectrum from molecules excited by a laser shows a series of lines as a progression in the case of diatomic molecules. Observation of such a fluorescence spectrum by using a conventional spectrometer was shown to give the assignments of transitions and values of molecular constants from the separation of lines corresponding to the energy difference between rotational levels in a number of vibrational states [1.36–45]. Diatomic molecules K_2 [1.36, 45], Na_2 [1.37, 38, 41, 45], I_2 [1.37, 39, 41], Li_2 [1.40], Br_2 and Cl_2 [1.41], Rb_2 and Cs_2 [1.42, 45], BaO [1.43], and CuO [1.44] were observed and analysed. The method was applied to polyatomic molecules such as N_2O [1.46], H_2CO etc., but no definite analysis has yet been given because of the complex structure of their vibrational-rotational levels.

Time-resolved observation of fluorescence followed by a pulsed laser excitation yields direct determination of lifetimes of the excited states (see Sects. 7.4–6). Collisional processes of energy transfer and excitation mechanisms are studied therewith [1.34, 47]. Studies of relaxation and energy transfer in the picosecond regime have now become practicable.

1.2.4 Nonlinear Spectroscopy of High Resolution

Resolution of the above-mentioned methods of laser spectroscopy is limited by inhomogeneous linewidth as Doppler broadening of gaseous molecules. Techniques of nonlinear spectroscopy that permit resolution of line structures within the Doppler width have been developed.

Gain saturation of the gaseous laser gives rise to a sharp decrease in laser power at the middle of the tuning curve. This decrease is called the Lamb dip, since it was first theoretically found by Lamb, followed by experimental findings in 1962, and published in 1963 [1.48]. The saturated absorption of atoms in a standing-wave laser resonator was observed to show the inverted Lamb dip in 1967 by Lee and Skolnick [1.49] and shortly afterwards independently by Letokhov [1.50]. These and subsequent works of high resolution saturation spectroscopy are discussed in Chapter 4.

Level-crossing effects in the He-Ne laser were observed in the operating characteristics of the laser in an axial magnetic field [1.51]. High-resolution spectroscopy with coupled transitions in a Doppler-broadened line was theoretically analysed [1.52] and applied to study the hyperfine structure of xenon by Schlossberg and Javan in 1966 [1.53]. Two closely spaced levels make transitions to a common level, irradiated by the two-mode laser output. A sharp resonance appears when the difference frequency coincides with the separation of the two levels. It is a kind of double resonance in two coupled transitions. Subsequent development in three-level laser spectroscopy is reviewed in Chapter 6. Reviews on three-level spectroscopy and double resonance laser spectroscopy may also be found elsewhere [1.54].

Nonlinear fluorescence associated with the Lamb dip, level-crossing, and mode-crossing effects has been investigated since 1969 (see Chapts. 4 and 6). The effect of atomic coherence on fluorescence gives rise to quantum beats in the presence of unresolved sublevels [1.55]. Quantum beats in atomic fluorescence following a pulsed laser excitation were first observed from ytterbium in 1972 [1.56]. Further discussions on quantum beats and coherent transient effects are given in Chapter 7.

Two-photon absorption may be interpreted as an off-resonant behavior of double resonance in a three-level system, where only the sum frequency is resonant. Development since 1974 of two-photon spectroscopy without Doppler broadening is described in Chapter 8.

References

1.1   K. Shimoda: Sci. Papers Inst. Phys. Chem. Res. **55**, 1 (1961);
      E. I. Gordon: Bell Syst. Tech. J. **43**, 507 (1964);
      A. Blaquiere: Compt. Rend. **255**, 3141 (1962);
      H. Haken: Z. Phys. **181**, 96 (1964), and **190**, 327 (1966)

1.2   J. P. GORDON, H. J. ZEIGER, C. H. TOWNES: Phys. Rev. **99**, 1264 (1955);
        M. SARGENT, III, M. O. SCULLY, W. E. LAMB, Jr.: *Laser Physics* (Addison-Wesley
        Publ. 1974) Sect. 8–3
1.3   E. BELL, A. L. BLOOM: Appl. Optics **3**, 413 (1964);
        T. KASUYA: Appl. Phys. **2**, 339 (1973)
1.4   J. FEINLEIB, S. GROVES, W. PAUL, R. ZALLEN: Phys. Rev. **131**, 2070 (1963);
        J. M. BESSON, J. F. BUTLER, A. R. CALAWA, W. PAUL, R. H. REDIKER: Appl. Phys.
        Letters **7**, 206 (1965)
1.5   I. D. ABELLA, H. Z. CUMMINS: J. Appl. Phys. **32**, 1177 (1961);
        T. KUSHIDA: Phys. Rev. **185**, 500 (1969)
1.6   See for example: *Topics in Applied Physics*, Vol. 1: Dye Lasers, ed. by F. P. SCHÄFER
        (Springer, Berlin, Heidelberg, New York 1973)
1.7   J. KUHL, W. SCHMIDT: Appl. Phys. **3**, 251 (1974); *Laser Spectroscopy*, ed. by R. G.
        BREWER, A. MOORADIAN (Plenum Press, New York 1974) pp. 59–122, 223–272,
        513–532
1.8   S. P. S. PORTO, D. C. WOOD: J. Opt. Soc. Am. **52**, 251 (1962);
        B. P. STOICHEFF: Proc. 10-th Colloq. Spect. Internat., 399 (1962)
1.9   E. J. WOODBURY, W. K. NG: Proc. IRE **50**, 2367 (1962)
1.10   G. ECKHARDT, R. W. HELLWARTH, F. J. MCCLUNG, S. E. SCHWARTZ, D. WEINER,
        E. J. WOODBURY: Phys. Rev. Letters **9**, 455 (1962);
        B. P. STOICHEFF: Phys. Letters **7**, 186 (1963);
        B. P. STOICHEFF, in: *Quantum Electronics and Coherent Light*, ed. by P. A. MILES
        (Academic Press, New York, London 1964) pp. 306–325;
        H. J. ZEIGER, P. E. TANNENWALD, S. KERN, R. HERENDEEN: Phys. Rev. Letters **11**,
        419 (1963);
        G. ECKHARDT, D. P. BORTFELD, M. GELLER: Appl. Phys. Letters **3**, 137 (1963)
1.11   E. GARMIRE, F. PANDARESE, C. H. TOWNES: Phys. Rev. Letters **11**, 160 (1963);
        R. W. HELLWARTH: Phys. Rev. **130**, 1850 (1963);
        N. BLOEMBERGEN, Y. R. SHEN: Phys. Rev. **133**, A37 (1964)
1.12   H. KOGELNIK, S. P. S. PORTO: J. Opt. Soc. Am. **52**, 1446 (1963);
        R. C. C. LEITE, S. P. S. PORTO: J. Opt. Soc. Am. **54**, 981 (1964)
1.13   S. K. FREEMAN: *Applications of Laser Raman Spectroscopy* (John Wiley and Sons,
        Inc., New York 1974)
1.14   J. T. HOUGEN, S. SINGH: Phys. Rev. Letters **10**, 406 (1963)
1.15   J. A. KONINGSTEIN, TOA-NING NG: Solid State Commun. 7, 351 (1969)
1.16   H. J. GERRITSEN, S. A. AHMED: Phys. Letters **13**, 41 (1964)
1.17   K. SAKURAI, K. SHIMODA: Japan. J. Appl. Phys. **5**, 744 (1966);
        H. J. GERRITSEN, in: *Physics of Quantum Electronics*, ed. by P. L. KELLEY et al.
        (McGraw-Hill, New York 1966) pp. 581–590
1.18   K. SAKURAI, K. SHIMODA: Japan. J. Appl. Phys. **5**, 938 (1966)
1.19   K. SAKURAI, K. UEHARA, M. TAKAMI, K. SHIMODA: J. Phys. Soc. Japan **23**, 103 (1967)
1.20   B. F. JACOBY, R. K. LANG: Appl. Phys. Letters **8**, 202 (1966)
1.21   K. UEHARA, T. SHIMIZU, K. SHIMODA: IEEE J. Quant. Electron. QE-4, 728 (1968)
1.22   H. BRUNET: IEEE J. Quant. Electron. QE-2, 382 (1966)
1.23   M. S. FELD, K. SHIMODA, M. KOVACS, C. SHIELDS: Bull. Am. Phys. Soc. II **10**, 87 (1965)
1.24   T. SHIMIZU, K. SHIMODA, A. MINOH: J. Phys. Soc. Japan **24**, 1185 (1968)
1.25   F. SHIMIZU: J. Chem. Phys. **51**, 2754 (1969) and **52**, 3572 (1970)
1.26   R. G. BREWER, M. J. KELLY, A. JAVAN: Phys. Rev. Letters **23**, 559 (1969);
        G. DUXBURY, R. G. JONES: Chem. Phys. Letters **8**, 439 (1971);
        F. HERLEMONT, J. LEMAIRE, J. HOURIEZ, J. THIBAULT: C. R. Acad. Sci. Paris, Ser.
        B. **276**, 733 (1973);
        J. W. C. JOHNS, A. R. W. MCKELLAR: J. Mol. Spectrosc. **48**, 354 (1973);
        Y. UEDA, K. SHIMODA, in: *Laser Spectroscopy*, ed. by S. HAROCHE et al. (Springer,
        Berlin, Heidelberg, New York 1975) pp. 186–197

1.27 J. Bettcher, K. Dransfeld, K. F. Renk: Phys. Letters **26A**, 146 (1968)
1.28 K. M. Evenson, H. P. Broida, J. S. Wells, R. J. Mahler, M. Mizushima: Phys. Rev.
      Letters **21**, 1038 (1968);
      K. M. Evenson, M. Mizushima: Phys. Rev. A**6**, 2197 (1972)
1.29 M. Mizushima, K. M. Evenson, J. S. Wells: Phys. Rev. A**5**, 2276 (1972);
1.30 R. F. Curl, Jr., K. M. Evenson, J. S. Wells: J. Chem. Phys. **56**, 5143 (1972)
1.31 K. M. Evenson, J. S. Wells, H. E. Radford: Phys. Rev. Letters **25**, 199 (1970);
      T. Kasuya, K. Shimoda: Japan. J. Appl. Phys. **11**, 1571 (1972)
1.32 K. M. Evenson, H. E. Radford, M. M. Moran, Jr.: Appl. Phys. Letters **18**, 426 (1971)
1.33 K. M. Evenson, G. J. Howard, in: *Laser Spectroscopy*, ed. by R. G. Brewer, A.
      Mooradian (Plenum Press, New York 1974) pp. 535–540
1.34 Y. T. Yardley, C. B. Moore: J. Chem. Phys. **45**, 1066 (1966)
1.35 S. Ezekiel, R. Weiss: Phys. Rev. Letters **20**, 91 (1968)
1.36 W. Tango, J. K. Link, R. N. Zare: J. Chem. Phys. **49**, 4264 (1968)
1.37 K. Sakurai, H. P. Broida: J. Chem. Phys. **50**, 557 (1969)
1.38 W. Demtröder, M. McClintock, R. N. Zare: J. Chem. Phys. **51**, 5495 (1969)
1.39 J. I. Steinfeld, J. D. Campbell, N. A. Weiss: J. Mol. Spectrosc. **29**, 204 (1969)
1.40 R. Velasco, C. Ottinger, R. N. Zare: J. Chem. Phys. **51**, 5522 (1969)
1.41 W. Holzer, W. F. Murphey, H. J. Bernstein: J. Chem. Phys. **52**, 469 (1970)
1.42 M. McClintock, L. C. Balling: J. Quant. Spectrosc. Rad. Transfer **2**, 1209 (1969)
1.43 K. Sakurai, S. E. Johnson, H. P. Broida: J. Chem. Phys. **52**, 1625 (1970)
1.44 J. S. Shirk, A. M. Bass: J. Chem. Phys. **52**, 1894 (1970)
1.45 G. Baumgartner, W. Demtröder, M. Stock: Z. Physik **232**, 462 (1970)
1.46 K. Sakurai, H. P. Broida: J. Chem. Phys. **50**, 2404 (1969)
1.47 L. O. Hocker, M. A. Kovacs, C. K. Rhodes, G. W. Flynn, A. Javan: Phys. Rev.
      Letters **17**, 233 (1966);
      see a review by E. Weitz, G. Flynn: Ann. Review of Physical Chemistry **25**, 275
      (1974);
      C. B. Moore: Adv. Chem. Phys. **23**, 41 (1973)
1.48 R. A. McFarlane, W. R. Bennett, Jr., W. E. Lamb, Jr.: Appl. Phys. Letters **2**, 189
      (1963);
      A. Szöke, A. Javan: Phys. Rev. Letters **10**, 521 (1963)
1.49 P. H. Lee, M. L. Skolnick: Appl. Phys. Letters **10**, 303 (1967)
1.50 V. S. Letokhov: Zh. Eksper. I. Teor. Fiz. Pis'ma **6**, 597 (1967)
1.51 W. Culshaw, J. Kannelaud: Phys. Rev. **145**, 257 (1966)
1.52 H. R. Schlossberg, A. Javan: Phys. Rev. **150**, 267 (1966)
1.53 H. R. Schlossberg, A. Javan: Phys. Rev. Letters **17**, 1242 (1966)
1.54 I. M. Beterov, V. P. Chebotayev, in: *Progress in Quantum Electronics*, ed. by J. H.
      Sanders, S. Stenholm (Pergamon, Oxford 1974) vol. 3, part 1, pp. 1–106;
      K. Shimoda, T. Shimizu: ibid. (1972) vol. 2, part 2, pp. 45–139;
      K. Shimoda: Double-Resonance Spectroscopy of Molecules by Means of Lasers;
      in *Topics in Applied Physics*, Vol. 2, ed. by H. Walther (Springer Berlin, Heidelberg,
      New York 1976) Chapt. 3, pp. 198–252
1.55 A. Corney, G. W. Series: Proc. Roy. Soc. London, **83**, 207 (1964)
1.56 W. Gornik, D. Kaiser, W. Lange, J. Luther, H.-H. Schulz: Opt. Commun. **6**,
      327 (1972)

2. Line Broadening and Narrowing Effects

K. Shimoda

With 9 Figures

It is the purpose of this chapter to provide basic understanding of the processes involved in the methods of high resolution laser spectroscopy rather than to review diverse theories of line broadening. In particular, line broadening and narrowing effects in the presence of strong mono-chromatic light waves are considered. The saturation effect and other nonlinear optical effects provide useful methods of high resolution laser spectroscopy.

Conventional theories of lineshape and linewidth, on the other hand, are concerned mostly with small-signal behaviors of atoms and molecules. Review papers on such theories and non-laser experiments on spectral line broadening include those by CH'EN and TAKEO in 1957 [2.1] and by BREEN in 1961 and 1964 [2.2]. More recent ones are by HINDMARSH and FARR [2.3] and by BERMAN [2.4]. Neither of them covers every aspect of broadening but they are useful as complementary references. They also provide extensive bibliographies.

2.1 Homogeneous Broadening

We consider an ensemble of atoms which all behave alike. The property of any atom is assumed to be independent of the particular behavior of other atoms. The spectral line is homogeneously broadened in this case. Natural width due to the finite lifetime, saturation broadening, and the transit-time effect is described in the following. It is remarked that collisional broadening is not strictly homogeneous and will be further discussed in Subsections 2.1.3 and 2.3.1.

2.1.1 Lineshape and Width of a Classical Oscillator

a) Damped Oscillation and Lifetime Broadening

The amplitude of a classical damped oscillator is expressed by

$$E(t) = E_0 \exp(-\gamma t) \cdot \cos \omega_0 t , \qquad (2.1)$$

where γ is the damping constant and ω_0 is the proper frequency of the oscillator. The Fourier transform of (2.1) for $t>0$ is

$$f(\omega) = \frac{1}{2\pi} \int_0^\infty E(t) \exp(-i\omega t) dt$$

$$= \frac{E_0}{4\pi} \left[\frac{1}{i(\omega-\omega_0)+\gamma} + \frac{1}{i(\omega+\omega_0)+\gamma} \right].$$

When the damping is so small that $\gamma \ll \omega_0$, the second term in the brackets may be neglected for $\omega > 0$. This corresponds to the rotating-wave approximation and it is valid in almost all optical cases.

The power spectrum of (2.1) represents the line profile as given by

$$I(\omega) \propto |f(\omega)|^2 = \frac{E_0^2}{16\pi^2} \cdot \frac{1}{(\omega-\omega_0)^2+\gamma^2}. \tag{2.2}$$

It shows a Lorentzian lineshape with the *half width at half maximum* (hereafter denoted by HWHM) γ, that is equal to the reciprocal decay time.

b) Transit-Time Broadening

When an undamped oscillator $E_0 \cos(\omega_0 t + \phi)$ is observed during a finite time interval T, the observed lineshape will be

$$I(\omega) \propto E_0^2 \frac{\sin^2(\omega-\omega_0)T/2}{(\omega-\omega_0)^2}. \tag{2.3}$$

The magnitude of HWHM is calculated from (2.3) to be $2.79/T$. The lineshape (2.3) applies to the case when a transition of atoms or molecules in a univelocity beam is observed within an interaction length. The transit time of an atom with velocity v through the interaction length L is $T = L/v$, and the HWHM becomes $\Delta\omega = 2.79v/L$, or the full width at half maximum (FWHM) is $2\Delta v = 0.89v/L$ in linear frequency.

In any experimental conditions, atomic or molecular velocities in the beam are not uniform but are more or less distributed. The lineshape and the linewidth are obtained by integration over the velocity distribution which may not be Maxwellian. It should be noted that this integration will be complicated, when collision, saturation, and other nonlinear effects are involved, because molecules of different velocities suffer different amounts of collision, saturation, and other effects.

c) Random Perturbation—Autocorrelation Function

The lineshape may also be expressed by the power spectrum of the autocorrelation function which is defined by

$$R(T) = \langle E(t+T) \cdot E(t) \rangle \, ,$$

where $\langle \; \rangle$ means the ensemble average.

As a simplest model, assume that oscillation of frequency ω_0 are randomly perturbed. Then the duration T of free oscillation is not constant, but randomly distributed as given by the distribution function

$$f(T) = \tau^{-1} \exp(-T/\tau) \, .$$

The autocorrelation function in this case becomes

$$R(T) = \tfrac{1}{2} E_0^2 \exp(-T/\tau) \cdot \cos \omega_0 T$$

and the power spectrum is obtained to become Lorentzian as

$$G(\omega) = \frac{2}{\pi} \int_0^\infty R(T) \cdot \cos \omega_0 T \cdot dT$$

$$= \frac{E_0^2}{2\pi} \cdot \frac{\gamma}{(\omega - \omega_0)^2 + \gamma^2}$$

where $\gamma = \tau^{-1}$ is the HWHM.

d) Transit-Time Broadening with a Gaussian Beam

Consider a Gaussian beam of the laser which is propagating along the z-direction. The wave surface is planar at its beam waist. The field distribution there is written as

$$E(x, y, t) = E_0 \exp(-r^2/w^2) \cos \omega_0 t \, ,$$

where $r^2 = x^2 + y^2$, and w is the beam width. Then the autocorrelation function of the field as seen by an atom traversing the beam with the velocity v is calculated by using $r^2 = a^2 + v^2 t^2$ to be

$$R(T) \propto \exp(-v^2 T^2/2w^2) \cdot \cos \omega_0 T \, , \qquad (2.4)$$

where a is the minimum distance of the atom from the beam axis. The Fourier transform of (2.4) gives a Gaussian lineshape

$$G(\omega) \propto \exp[-(\omega-\omega_0)^2 w^2/2v^2].$$

The spectral lineshape of moving atoms observed by the Gaussian beam in the perpendicular direction to the atomic velocity is evidently the same as above. The HWHM is

$$\Delta\omega = \sqrt{2\ln 2}\,(v/w). \tag{2.5}$$

In terms of the beam diameter $D = 2w$, the HWHM is written as $\Delta\omega = 2.35v/D$ or being nearly equal to that of the homogeneous distribution for $L = D$ in case b).

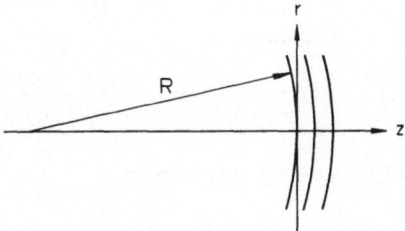

Fig. 2.1. A molecule traveling across a spherical wave surface

e) Broadening Effect by Spherical Wave Surface

The wave surface of a Gaussian beam is curved except at its beam waist. The curvature of the wave surface restricts the effective transit time and broadens the line. As can be seen in Fig. 2.1, the phase shift at radius r is $kr^2/2R$, where R is the radius of curvature of the wave surface and $k = \omega/c$ is the magnitude of the wave vector.

The Gaussian beam propagating along the z-direction is then expressed by

$$E(r, z, t) = E_0 \exp\left[-\frac{r^2}{w^2} + i\left(\omega_0 t - kz - \frac{kr^2}{2R}\right)\right],$$

where w is the laser beam radius and R is the radius of curvature of the wave surface at the position of the atom. The Fourier transform of the above field as seen by an atom traversing the beam[1] is given by substitut-

---

[1] Even if the atomic velocity is not perpendicular to the z-axis, the parallel beam of atoms exhibits the same lineshape and only the line center is shifted.

ing $r^2 = a^2 + v^2 t^2$ as

$$F(\omega) = \frac{1}{2\pi} \int_{-\infty}^{\infty} E_0 \exp\left[-i(\omega - \omega_0)t - (a^2 + v^2 t^2)\left(\frac{1}{w^2} + \frac{ik}{2R}\right)\right] dt$$

$$= A \exp\left[-\frac{(\omega - \omega_0)^2 w^2 R\lambda}{4v^2(R\lambda + i\pi w^2)}\right]$$

$$A = \frac{E_0 w}{2v}\left(\frac{R\lambda}{\pi R\lambda + i\pi^2 w^2}\right)^{1/2} \exp\left(-\frac{a^2}{w^2} - \frac{i\pi a^2}{R\lambda}\right),$$

where v is the atomic velocity. The lineshape in linear spectroscopy is therefore obtained to become Gaussian as

$$I(\omega) = |F(\omega)|^2 \propto \exp\left[-\frac{(\omega - \omega_0)^2}{2v^2} \cdot \frac{w^2 R^2 \lambda^2}{R^2 \lambda^2 + \pi^2 w^4}\right].$$

The HWHM becomes

$$\Delta\omega = \sqrt{2\ln 2}\, v(w^{-2} + w_R^{-2})^{1/2}, \tag{2.6}$$

where

$$w_R = R\lambda/\pi w.$$

In the case when the radius of curvature is so large that $R \gg \pi w^2/\lambda$, we have $w_R \gg w$, and (2.6) becomes equal to (2.5).

It should be remarked, however, that the HWHM is quite often limited by the curvature rather than the beam radius. Take $R = 10$ cm, $w = 1$ mm, and $\lambda = 1$ µm for example, then we obtain $w_R = 3 \times 10^{-2}$ cm. Since w_R is much smaller than w in such a case, the HWHM is approximately given by

$$\Delta\omega = \sqrt{2\ln 2} \cdot v/w_R = \pi\sqrt{2\ln 2} \cdot vw/R\lambda. \tag{2.7}$$

In order that the curvature of the wave surface does not much broaden the line, the radius of curvature must be

$$R \gg \pi w^2/\lambda. \tag{2.8}$$

A numerical example for $w = 1$ cm, $\lambda = 3$ µm requires that the radius of curvature must be much larger than 100 m. A very flat wave surface must be used for high resolution spectroscopy of atomic beams and saturated absorption.

2.1.2 Natural Width and Saturation Broadening

A quantum mechanical description of the lineshape of the transition between levels 1 and 2 may be given by the ensemble averaged density matrix. The equation of motion of the 2×2 density matrix ϱ is

$$d\varrho/dt = i\hbar^{-1}[H, \varrho] - (\Gamma \varrho + \varrho \Gamma)/2. \tag{2.9}$$

Here H is the Hamiltonian and Γ is a diagonal matrix, where the atoms in levels 1 and 2 decay, respectively, at constant rates γ_1 and γ_2 to the rest of the levels, as expressed by

$$\Gamma = \begin{bmatrix} \gamma_1 & 0 \\ 0 & \gamma_2 \end{bmatrix}. \tag{2.10}$$

The perturbed Hamiltonian in the presence of a nearly resonant optical field

$$E(t) = E_0 \cos \omega t = (E_0/2) \exp(i\omega t) + c.c.$$

is

$$H = \begin{bmatrix} W_1 & -\mu \cdot E \\ -\mu \cdot E & W_2 \end{bmatrix}, \tag{2.11}$$

where W_1 and W_2 are eigen energies, and μ is the electric dipole operator. For $W_1 < W_2$ as shown in Fig. 2.2, the center frequency of the transition is $\omega_0 = (W_2 - W_1)/\hbar$. From (2.9), (2.10), and (2.11) we find the differential equations for the density matrix elements:

$$d\varrho_{12}/dt = (i\omega_0 - \gamma)\varrho_{12} - i\hbar^{-1}\mu \cdot E \cdot (\varrho_{11} - \varrho_{22}), \tag{2.12a}$$

$$d\varrho_{11}/dt = -\gamma_1(\varrho_{11} - \varrho_{11}^0) - i\hbar^{-1}\mu \cdot E \cdot (\varrho_{12} - \varrho_{21}), \tag{2.12b}$$

$$d\varrho_{22}/dt = -\gamma_2(\varrho_{22} - \varrho_{22}^0) + i\hbar^{-1}\mu \cdot E \cdot (\varrho_{12} - \varrho_{21}), \tag{2.12c}$$

where

$$\gamma = (\gamma_1 + \gamma_2)/2 \tag{2.12d}$$

is *the transverse relaxation rate*, and ϱ_{11}^0 and ϱ_{22}^0 are, respectively, the values of ϱ_{11} and ϱ_{22} in the absence of the optical field; they represent atomic populations in levels 1 and 2 caused by pumping or excitation.

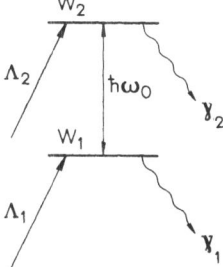

Fig. 2.2. Two-level system

The excitation rates to levels 1 and 2 are $\Lambda_1 = \gamma_1 \varrho_{11}^0$ and $\Lambda_2 = \gamma_2 \varrho_{22}^0$, respectively.

The steady-state solution of (2.12a–c) in the rotating-wave approximation gives

$$\varrho_{12} = -\frac{ix \exp(i\omega t)}{i(\omega - \omega_0) + \gamma} (\varrho_{11} - \varrho_{22})/2 \tag{2.13a}$$

$$\varrho_{11} = \varrho_{11}^0 - \frac{\gamma}{2\gamma_1} \Delta\varrho^0 \frac{|x|^2}{(\omega - \omega_0)^2 + \gamma^2 + \gamma\tau|x|^2}, \tag{2.13b}$$

$$\varrho_{22} = \varrho_{22}^0 + \frac{\gamma}{2\gamma_2} \Delta\varrho^0 \frac{|x|^2}{(\omega - \omega_0)^2 + \gamma^2 + \gamma\tau|x|^2}, \tag{2.13c}$$

where $x = \mu_{12} E_0/\hbar$ is sometimes called the characteristic Rabi frequency; for a linearly polarized light, μ_{12} is the component of μ along the optical field, where levels 1 and 2 are degenerate in general; $\Delta\varrho^0 = \varrho_{11}^0 - \varrho_{22}^0$ is the population difference under any pumping effect as well as thermal excitation, and

$$\tau = (\gamma_1^{-1} + \gamma_2^{-1})/2 \tag{2.14}$$

is the *longitudinal relaxation time*[2].

The atomic dipole moment induced by the optical field is expressed by

$$p = \mathrm{Tr}(\varrho\mu) = \varrho_{12}\mu_{21} + \text{c.c.} \tag{2.15}$$

---

[2] The transverse relaxation time γ^{-1} is not equal to the longitudinal relaxation time τ in this simple model when $\gamma_1 \neq \gamma_2$.

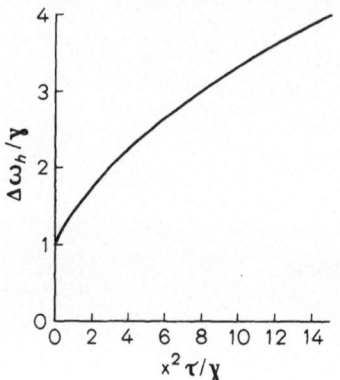

Fig. 2.3. Homogeneous linewidth $\Delta\omega_h$ as a function of the power. The abscissa is $x^2\tau/\gamma = P/P_s$

Substitution of (2.13) and (2.14) into (2.15) gives

$$p = -\frac{|\mu_{12}|^2}{2\hbar}\Delta\varrho^0\frac{\omega-\omega_0+i\gamma}{(\omega-\omega_0)^2+\gamma^2+\gamma\tau|x|^2}E_0\exp(i\omega t) + \text{c.c.}\qquad(2.16)$$

The gas of N such two-level atoms in a unit volume shows the macroscopic polarization Np, and the corresponding (electric) susceptibility $\chi = \chi' - i\chi''$ in the MKSA system of units becomes

$$\chi(\omega) = -\frac{N\Delta\varrho^0|\mu_{12}|^2}{\varepsilon_0\hbar}\cdot\frac{\omega-\omega_0+i\gamma}{(\omega-\omega_0)^2+\gamma^2+\gamma\tau|x|^2}.$$

Thus the power absorption constant is obtained from its imaginary part as

$$\alpha(\omega) = k\chi''(\omega) = \frac{N\Delta\varrho^0|\mu_{12}|^2k}{\varepsilon_0\hbar}\cdot\frac{\gamma}{(\omega-\omega_0)^2+\gamma^2+\gamma\tau|x|^2}.\qquad(2.17)$$

It has a Lorentzian lineshape with the HWHM of

$$\Delta\omega_h = (\gamma^2+\gamma\tau|x|^2)^{1/2}.\qquad(2.18)$$

We find, therefore, that the linewidth increases with the optical field intensity as shown in Fig. 2.3. This is normally called *saturation broadening*, but it is sometimes called *power broadening* since $|x|^2$ is proportional to the incident power P. Equation (2.18) is rewritten as

$$\Delta\omega_h = \gamma\sqrt{1+P/P_s}\qquad(2.19)$$

where

$$P_s = \frac{\varepsilon_0 c \hbar^2 \gamma A}{2|\mu_{12}|^2 \tau} \tag{2.20}$$

is the *saturation power*[3] through the cross section A, and P/P_s is the parameter that indicates the *degree of saturation*. The degree of saturation is denoted by G in Chapter 4, S in Chapter 5, and κ in Chapter 6,

$$G \quad \text{or} \quad \dot{S} \quad \text{or} \quad \kappa = P/P_s = |x|^2 \tau / \gamma ,$$

while it is denoted in Chapter 3 for $\gamma = \tau^{-1}$ by $\chi = |x|^2/\gamma^2$.

The HWHM of the small-signal lineshape is evidently (2.12d), that is equal to the average of reciprocal lifetimes of the states 1 and 2. It may be interpreted that the two levels are broadened by $\pm \gamma_1/2$ and $\pm \gamma_2/2$ so that the transition between these two levels shows the width of their sum.

This type of width as determined by the decay times of the involved levels is the *natural width*. For a transition at frequency ω_0 from an excited state to the ground state, the average lifetime of the excited atom before it spontaneously emits a photon is

$$\tau_s = \frac{\pi \varepsilon_0 c^3 \hbar}{|\mu_{12}|^2 \omega_0^3}$$

in the MKSA system of units[4]. Then the natural width (HWHM) of the line is given by the probability of spontaneous emission as

$$\Delta \omega_s = 1/2\tau_s ,$$

if transition to other levels are negligible.

Since the natural width is proportional to the cubic frequency, it may be neglected in the far-infrared, becomes appreciable in the infrared, and can be fairly large in the visible region.

2.1.3 Phenomenological Treatment of Collision Broadening

We assume that the two-level atoms (or molecules) are at rest and that they are perturbed by collisions with other atoms (or molecules). Although

---

[3] Some authors denote P_s the saturation parameter.

[4] It is useful to remember that 1 debye ($= 10^{-18}$ cgs esu) corresponds to 3.335×10^{-30} C·m in the MKSA or SI unit.

such a condition cannot be fully realized in any experiment, it simplifies the theory and provides useful approximations and interpretations of collision-broadening effects. Any resonant exchange of excitation between the colliding and collided atoms is not considered.

Neglect of the excitation exchange as well as the Doppler effect greatly simplifies the theory. Nevertheless, this theory is applicable to most practical cases with a fairly good approximation. Two types of collisions are discussed as follows.

a) Collision-Induced Transitions

If the energy difference between atomic levels is not very large compared with the kinetic energy of the colliding atom, the collision induces a transition between the atomic levels. The collision-induced transitions between levels 1 and 2 are relatively rare and negligible in the case of optical transitions in atoms. In gases of diatomic and polyatomic molecules, on the other hand, the molecular collision induces transitions between rotational levels and less frequently between vibrational levels. Since the collision frequency is proportional to the gas pressure p, the effect of population relaxation can be treated by modifying the relaxation rates in (2.10) in the form

$$\gamma_1 = \gamma_1^0 + C_1 p$$
$$\gamma_2 = \gamma_2^0 + C_2 p$$

where γ_1^0 and γ_2^0 are the unperturbed relaxation rates, and C_1 and C_2 represent collision-induced transition probabilities from the levels 1 and 2, respectively, at unit gas pressure. The HWHM of the line is immediately obtained from (2.12d) to be

$$\gamma = (\gamma_1^0 + \gamma_2^0)/2 + (C_1 + C_2)p/2, \qquad (2.21)$$

which represents *pressure broadening*.

b) Phase-Changing Collisions

The phase of the atomic wavefunction is more frequently perturbed than is the population of the atomic level by collisions with other atoms. The phase of the state 1 suffers a phase shift θ_1 by collision. The simultaneous phase shift θ_2 of the state 2 generally differs from θ_1 so that the change in ϱ_{12} for such collisions is

$$(\Delta\varrho_{12})_{col} = \langle \exp(i\Delta\theta) - 1 \rangle \varrho_{12}, \qquad (2.22)$$

where the phase shift $\Delta\theta = \theta_1 - \theta_2$ is a statistical quantity which depends on the impact parameter and the velocity of the perturbing atom.

The phase-changing collisions are classified into three cases: one is the *weak collision* which induces a small phase change $\Delta\theta \ll 1$; the second is the *strong collision* which induces a large phase changes $\Delta\theta \gg 1$; and the third is the general case including $\Delta\theta \approx 1$.

The weak collision occurs in a distant collision with a large impact parameter. The statistical average of (2.22) for weak collisions with $\Delta\theta \ll 1$ shows

$$(\Delta\varrho_{12})_{col} \simeq i\langle\Delta\theta\rangle\varrho_{12} .$$

This results in a frequency shift as can be seen by substituting $(\delta\varrho_{12}/\delta t)_{col} = if\langle\Delta\theta\rangle\varrho_{12}$ into (2.12a), where f is the collision frequency. The spectral line is shifted to $\omega_0 + f\langle\Delta\theta\rangle$, but remains Lorentzian of the same width.

In the general case when $\Delta\theta$ is not very small, both the phase and the amplitude of ϱ_{12} change by collisions:

$$\begin{aligned}(\delta\varrho_{12}/\delta t)_{col} &= f\langle\exp(i\Delta\theta) - 1\rangle\varrho_{12} \\ &= -(\gamma'_{col} - i\gamma''_{col})\varrho_{12} .\end{aligned}$$

Here γ'_{col} and γ''_{col} are, respectively, the real and imaginary parts of the collisional relaxation rate.

For very strong collisions when the dispersion of the value of $\Delta\theta$ is much larger than 2π, the phase of the oscillating dipole is essentially interrupted. For such phase-interrupting collisions, we find $\gamma''_{col} = 0$ and $\gamma'_{col} = f$, because $\langle\exp(i\Delta\theta)\rangle = 0$.

In the general case, both γ'_{col} and γ''_{col} are proportional to pressure since only binary collisions are considered at low pressure: $f \propto p$. Thus we have $\gamma'_{col} = C'_{12}p$, and $\gamma''_{col} = C''_{12}p$. The lineshape is obtained by substituting

$$\omega_0 = (W_2 - W_1)/\hbar + C''_{12}p \qquad (2.23)$$

and

$$\gamma = (\gamma_1 + \gamma_2)/2 + C'_{12}p \qquad (2.24)$$

into (2.17). The longitudinal relaxation time τ remains unchanged by the phase-changing collisions.

c) Effect of Phase-Changing and Population-Changing Collisions

On the plausible assumption that the collision-induced transitions and the phase shifts of wavefunctions occur independently, we may superpose those effects. The line center is the same as (2.23), since the decay in population does not shift the line, but the line is broadened. The HWHM is found from (2.21) and (2.24) to be

$$\gamma = (\gamma_1^0 + \gamma_2^0)/2 + (2C_{12}' + C_1 + C_2)p/2 . \tag{2.25}$$

Thus the homogeneous width of the line increases linearly with pressure. The longitudinal relaxation time in this general case is written as

$$\tau = [(\gamma_1^0 + C_1 p)^{-1} + (\gamma_2^0 + C_2 p)^{-1}]/2 . \tag{2.26}$$

It should be remembered that phase-interrupting collisions are often dominant in broadening and $C_{12}' p$ is the largest except at very low pressure.

Experimental values of these pressure-broadening constants, C_1, C_2, and C_{12}' are found in the range from 1 to 30 MHz/Torr depending on the kind of molecules and the temperature. We should be reminded in concluding this section that they are not constants in a strict sense, but they are phenomenological constants which differ more or less in different experimental situations. A notable example is the variation of pressure-broadening constant of the Lamb dip as a function of the beam width. In a wider beam of light, molecules take a longer transit time and the transit-time broadening is smaller. In such a case, a smaller change in the molecular velocity by collision with a larger impact parameter contributes to broadening, and this results in a larger value of the pressure-broadening parameter.

2.2 Inhomogeneous Broadening

When the transition frequencies of different atoms in the ensemble are different, the spectral line is apparently broadened. Stark and Zeeman effects of atoms in an inhomogeneous field are obvious examples of inhomogeneous broadening. In the absence of any inhomogeneous external field, the spectral line of a low pressure gas is inhomogeneously broadened by the Doppler effect due to molecular velocities.

2.2.1 Doppler Broadening

The proper frequency ω_0 of a moving molecule (atom) at velocity v is observed to be shifted by the Doppler effect to $\omega_0 + k \cdot v$, where k is the wave vector of the light. As a simple case, we assume that the collision effect is negligible and that a plane wave interacts with the molecules. Thus only the component v of the velocity v along the direction of propagation of the plane wave has to be considered.

Components of molecular velocities along a fixed direction in the gas at thermal equilibrium at temperature T obey the Maxwell distribution

$$w(v) = (1/\sqrt{\pi} u) \exp(-v^2/u^2), \tag{2.27}$$

$$u^2 = 2\kappa T/m = 2RT/M,$$

where m is the molecular mass, M the molecular weight (in a.m.u.), κ the Boltzmann constant, and R the gas constant.

The induced dipole moment of a moving molecule in a strong plane-wave field can be written from (2.16) to become

$$p(v) = -\frac{|\mu_{12}|^2}{2\hbar} \Delta\varrho^0 \frac{\omega - \omega_0 - kv + i\gamma}{(\omega - \omega_0 - kv)^2 + \gamma^2 + \gamma\tau|x|^2} E_0 \exp(i\omega t) + \text{c.c.}$$

The susceptibility of the gas of molecules with a velocity distribution is thus expressed by

$$\chi(\omega) = -\frac{N\Delta\varrho^0 |\mu_{12}|^2}{\varepsilon_0 \hbar} \int_{-\infty}^{\infty} w(v) \frac{\omega - \omega_0 - kv + i\gamma}{(\omega - \omega_0 - kv)^2 + \gamma^2 + \gamma\tau|x|^2} dv. \tag{2.28}$$

The power absorption constant is obtained by substituting (2.27) into (2.28) in the form of the Voigt integral

$$\alpha(\omega) = \frac{(N_1 - N_2)|\mu_{12}|^2 k\gamma}{\sqrt{\pi}\varepsilon_0 \hbar u} \int_{-\infty}^{\infty} \frac{\exp(-v^2/u^2)}{(\omega - \omega_0 - kv)^2 + \gamma^2 + \gamma\tau|x|^2} dv, \tag{2.29}$$

where $N_1 - N_2 = N\Delta\varrho^0$ is the (uninverted) population difference between the two levels.

The result obtained above includes Doppler broadening, transverse and longitudinal relaxation effects, and the saturation effect. It may reasonably be extended to include collisional effects by modifying γ and τ as shown in (2.25) and (2.26), respectively. Such modification is only approximately valid, however, because of the effect of velocity-changing collisions which is discussed in Section 2.3.

Although the optical field is often strong in laser spectroscopy, it is worth knowing the small-signal behavior of the gas. In the absence of saturation, (2.28) is reduced to the small-signal susceptibility, that is, the linear susceptibility.

$$\chi_L(\omega) = \frac{(N_2 - N_1)|\mu_{12}|^2}{\sqrt{\pi}\,\varepsilon_0 \hbar u} \int_{-\infty}^{\infty} \frac{\exp(-v^2/u^2)}{\omega - \omega_0 - kv - i\gamma}\, dv \,. \tag{2.30}$$

The correlation function for the small-signal response is written as

$$R(t) = \pi(i\hbar)^{-1}(N_2 - N_1)|\mu_{12}|^2 \exp(i\omega_0 t - \gamma t - k^2 u^2 t^2/4)$$

and the Fourier transform of $R(t)$ gives (2.30). It can be expressed by using the plasma dispersion function [2.5] to be

$$\chi_L(\omega) = (N_2 - N_1)\frac{|\mu_{12}|^2}{\varepsilon_0 \hbar k u} Z(\zeta) \tag{2.31}$$

$$\zeta = (\omega_0 - \omega + i\gamma)/ku \,.$$

Here the plasma dispersion function of a complex variable is defined by

$$Z(\zeta) = 2i \exp(-\zeta^2)\int_{-\infty}^{i\zeta} \exp(-z^2)dz \,,$$

which may be rewritten as integrals of real variables in the form

$$Z(\zeta) = \frac{1}{\sqrt{\pi}} \int_{-\infty}^{\infty} \frac{\exp(-x^2)}{x - \zeta}\, dx$$

$$= i \int_0^{\infty} \exp(i\zeta y - y^2/4)dy \,.$$

The HWHM of the Voigt integral of (2.29) for the large signal is known numerically, as shown in Fig. 2.4, as a function of the homogeneous width relative to the Doppler width. The line width $\Delta\omega$ for the convolution of (2.29) is neither $\Delta\omega_D + \Delta\omega_h$ nor $(\Delta\omega_D^2 + \Delta\omega_h^2)^{1/2}$, but it is found between these two simple expressions as illustrated in Fig. 2.4.

In the limiting case when the homogeneous broadening is much smaller than the Doppler broadening so that $\Delta\omega_h \ll ku$, (2.29) is reduces to

$$\alpha(\omega) = \alpha_0 \exp\left[-(\omega - \omega_0)^2/k^2 u^2\right], \tag{2.32}$$

where $\Delta\omega_h$ is the homogeneous width given by (2.18) or (2.19) including the saturation broadening. The lineshape (2.32) is Gaussian and the

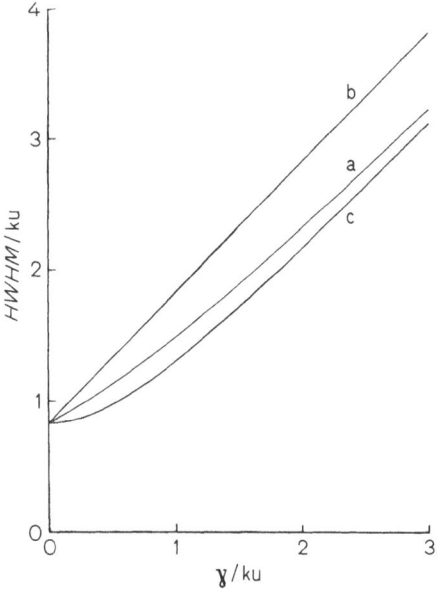

Fig. 2.4. Doppler-broadened HWHM as functions of the homogeneous width: a) Voigt integral, b) $\Delta\omega = \Delta\omega_D + \Delta\omega_h$, and c) $\Delta\omega = (\Delta\omega_D^2 + \Delta\omega_h^2)^{1/2}$

Doppler width as given by the HWHM is

$$\Delta\omega_D = ku\sqrt{\ln 2} = 0.833\omega_0 u/c \ . \tag{2.33}$$

The most probable velocity of molecules is normally between 100 and 1000 m/s. The Doppler width is thus about $10^{-6}\omega_0$, that is ~ 30 MHz at 10 µm, ~ 100 MHz at 3 µm, and ~ 600 MHz at 0.5 µm.

The peak absorption constant α_0 in (2.32) includes the saturation effect as given by

$$\alpha_0 = \alpha_L \gamma/\Delta\omega_h = \alpha_L(1 + P/P_s)^{-1/2} \ , \tag{2.34}$$

where the linear absorption constant α_L is

$$\alpha_L = \sqrt{\pi}(N_1 - N_2)|\mu_{12}|^2/\varepsilon_0 \hbar u \ , \tag{2.35}$$

and the saturation power P_s is the same as (2.20).

2.2.2 Hole Burning

The phenomenon of *hole burning* is well known in magnetic resonance since it was initially observed and interpreted by Bloembergen et al. [2.6] in 1948. The effect of hole burning in the standing-wave field of the laser was discussed in 1962 by Bennett [2.7].

The molecular populations in levels 1 and 2 as functions of the velocity component v are obtained from (2.13b) and (2.13c) by substituting $\omega_0 + kv$ into ω_0:

$$N_1(v) = N_1^0(v) - \frac{\gamma}{2\gamma_1} \Delta N^0(v) \frac{|x|^2}{(\omega - \omega_0 - kv)^2 + \Delta\omega_h^2}, \tag{2.36a}$$

$$N_2(v) = N_2^0(v) + \frac{\gamma}{2\gamma_2} \Delta N^0(v) \frac{|x|^2}{(\omega - \omega_0 - kv)^2 + \Delta\omega_h^2}, \tag{2.36b}$$

where $N_1^0(v)$ and $N_2^0(v)$ are, respectively, the velocity distribution of molecules in the levels 1 and 2 in the absence of the signal light. We may assume the Maxwell distribution in most practical cases so that

$$N_1^0(v) = N\varrho_{11}^0 w(v), \qquad N_2^0(v) = N\varrho_{22}^0 w(v). \tag{2.37}$$

When a strong monochromatic light at frequency ω is incident, the population of the lower state (2.36a) is reduced at the velocity component of $v = (\omega - \omega_0)/k$, whereas that of the upper state (2.36b) is raised. As illustrated in Fig. 2.5, the incident light burns a hole in the velocity distribution of the lower state and piles up a peak in that of the upper state. The HWHM of the hole is $\Delta\omega_h/k$. The depth of the hole increases with the light intensity as $|x|^2/(\gamma^2 + \gamma\tau|x|^2)$ and the width increases as $(\gamma^2 + \gamma\tau|x|^2)^{1/2}$. The peak height in level 2 and the hole depth in level 1 are not equal but different in proportion to their relaxation times, γ_2^{-1} and γ_1^{-1}, respectively.

It is noted that exactly the same results as those of the density matrix calculation mentioned above can be obtained from the following rate equations:

$$dN_1(v)/dt = -\gamma_1[N_1(v) - N_1^0(v)] - S(v)\Delta N(v), \tag{2.38a}$$

$$dN_2(v)/dt = -\gamma_2[N_2(v) - N_2^0(v)] + S(v)\Delta N(v), \tag{2.38b}$$

where

$$S(v) = \gamma|x|^2/2[(\omega - \omega_0 - kv)^2 + \gamma^2] \tag{2.39}$$

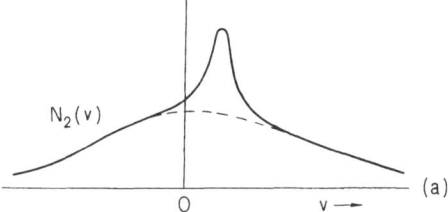

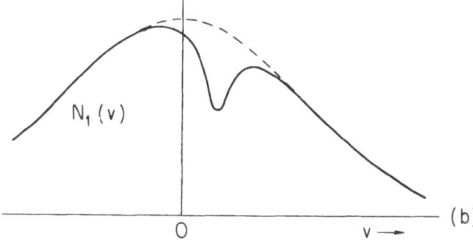

Fig. 2.5a and b. Molecular velocity distributions in the presence of strong monochromatic light: (a) in the upper state, and (b) in the lower state

is the unsaturated transition probability. Then the populations in the steady state are found from (2.38a and b) to be

$$N_1(v) = N_1^0(v) - \Delta N(v)S(v)/\gamma_1 \tag{2.40a}$$

$$N_2(v) = N_2^0(v) + \Delta N(v)S(v)/\gamma_2 . \tag{2.40b}$$

The difference between these two is

$$\Delta N(v) = \Delta N^0(v)/[1 + 2\tau S(v)] \tag{2.41}$$

where $\tau = (\gamma_1^{-1} + \gamma_2^{-1})/2$ and $\Delta N^0(v) = N_1^0(v) - N_2^0(v) = N\Delta\varrho^0 w(v)$. Substitution of (2.41) into (2.40a and b) gives identical results with (2.36a and b) which include saturation effects.

The rate of absorption in a unit volume is expressed by

$$\Delta P = \int_{-\infty}^{\infty} \hbar\omega S(v)\Delta N(v)dv . \tag{2.42}$$

Substituting (2.39) and (2.41) into (2.42), we find

$$\Delta P = \frac{\omega|\mu_{12}|^2\gamma N\Delta\varrho^0 P}{\varepsilon_0\hbar c}\int_{-\infty}^{\infty}\frac{w(v)dv}{(\omega - \omega_0 - kv)^2 + \gamma^2 + \gamma\tau|x|^2} , \tag{2.43}$$

where $P = c\varepsilon_0 E_0^2/2$ is the incident power density. Equation (2.43) results in the absorption constant as given in the previous Subsection by (2.29).

Therefore saturated absorption of a single traveling wave by a two-level system can be fully interpreted by the hole-burning effect using the unsaturated transition probability. It is not exactly valid, however, when two waves are incident. A typical example is the *Lamb dip* of the two-level atoms in a standing-wave field which is composed of two waves traveling in directions opposite to each other. For rigorous discussions of such cases, higher-order effects including the multiphoton processes must be taken into consideration. In the case when one of the two incident waves is weak, the hole-burning effect is sufficient to explain the nonlinear response of the two-level system. This is discussed in detail in Chapter 4.

In the case when the second radiation at a different frequency ω' is weak, the effect is satisfactorily explained as the hole-burning effect by using the rate equations. The absorbed power of this second probing light by the two-level system saturated with the first radiation is

$$\Delta P' = \int_{-\infty}^{\infty} \hbar\omega' S'(v)\Delta N(v)\mathrm{d}v\,,$$

where

$$S'(v) = \gamma|x'|^2/2[(\omega' - \omega_0 - k'v)^2 + \gamma^2]\,.$$

As the probing frequency ω' is swept, the velocity distribution $\Delta N(v)$ is traced by observing the absorption at the frequency ω'. When the weak radiation and the strong radiation are traveling in the same direction, the absorption of the weak radiation $\Delta P'$ shows a sharp dip at $\omega' = \omega$ with the HWHM of 2γ.

The very important case where the weak probing radiation is in the opposite direction to the strong radiation will be described in detail in Chapter 4.

2.3 Velocity-Changing Collision

The effect of molecular collisions has been considered in the previous section only to quench the states and shorten their lifetimes. Such collisional effect is independent of the inhomogeneous broadening. Molecules in the gas make elastic collisions as well as inelastic collisions.

The elastic collision does not change the quantum state, but it changes the molecular velocity and it may shift the phase of the coherent interaction. The effect of collisional phase shift was described in Subsection 2.1.3. The velocity change of molecules interacting with the radiation influences the lineshape in two ways. At first it tends to reduce the Doppler broadening and secondly it diffuses the burnt hole, resulting in cross relaxation.

2.3.1 Dicke Narrowing

Since the elastic collision does not perturb the optical polarization of the molecule, the molecular dipole moment oscillates coherently throughout the collision. But the Doppler shift $k \cdot v$ is altered at the instant of collision which changes the molecular velocity v. When the elastic collisions are frequent, the Doppler shift is altered many times during the decay time of the oscillating dipole moment and it is effectively averaged. The result is the reduction of Doppler broadening.

Such *collisional narrowing* was first pointed out by DICKE [2.8] and subsequently observed in magnetic dipole transitions in the microwave region [2.9]. The physics of Dicke narrowing is the same as motional narrowing which is often applied in high resolution magnetic resonance spectroscopy. In either case, when the oscillating dipole is phase modulated, in which the modulation frequency is higher than the relaxation frequency (rate) of the dipole, the variation of its instantaneous frequency is not observable but it is more or less averaged. In order that collisional narrowing be effective, the Doppler shift kv must be smaller than the uncertainty of frequency that corresponds to the observation during the interval between elastic collisions τ_{el}. In other words, Dicke narrowing becomes significant at pressures where the mean free path $(\bar{l} = u\tau_{el})$ is less than the wavelength $(\lambda = 2\pi/k)$. Thus Dicke narrowing is remarkable when

$$\gamma < kv < 1/\tau_{el},$$

where γ represents the homogeneous width including the collision broadening as described in Subsection 2.1.3.

Dicke narrowing in optical transitions was observed in vibrational Raman spectroscopy of H_2 [2.10] and in high-J rotational transitions of H_2O [2.11]. In almost all other cases of optical transitions, however, Dicke narrowing is overwhelmed by collision broadening. In particular, Dicke narrowing is not considered to occur in electronic transitions. Thus it is not attractive as a general method of high resolution spectroscopy.

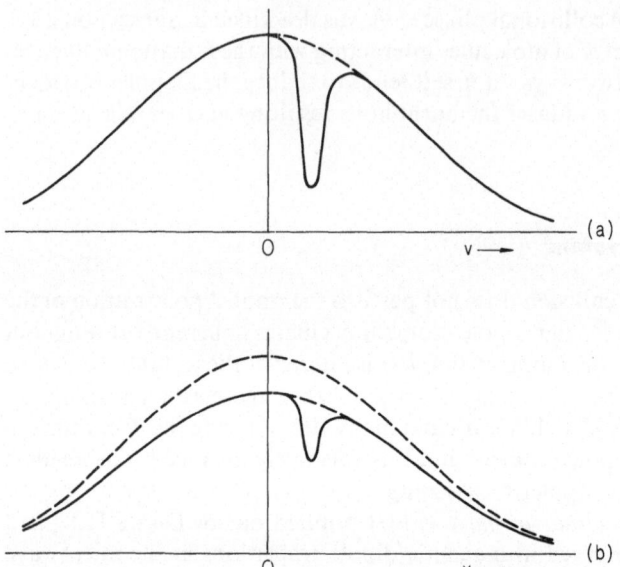

Fig. 2.6a and b. Hole burning in the molecular velocity distribution: (a) in the absence of cross relaxation, and (b) in the presence of cross relaxation

2.3.2 Cross Relaxation—Diffusion of the Burnt Hole

In the absence of velocity-changing collisions, a hole is burnt as shown in Fig. 2.6(a), when the molecules are in a strong monochromatic traveling wave of light. As the molecular velocity is varied by collisions without inducing any transitions, the burnt hole is spread over the whole velocity distribution as illustrated in Fig. 2.6(b). This is a process of *cross relaxation* between molecules of different velocity groups. It is known to strongly affect the gain of a laser amplifier and the output characteristics of a laser oscillator [2.12, 13].

The rate of cross relaxation is faster at higher gas pressure. The laser radiation first saturates the molecules in the narrow homogeneous width and then it will saturate those of other velocities by the process of cross relaxation. Thus the hole depth decreases rapidly with gas pressure and a broad overall depletion of the population in the lower state is produced. The effect of cross relaxation results in the increase in saturation power and the diminution of nonlinear absorption with pressure. It is important, therefore, to reduce the sample gas pressure in the experiment of saturation spectroscopy to a lower value than that estimated from a single constant of pressure broadening ignoring cross relaxation.

2.4 Sensitivity of Laser Spectrometers

Sensitivity is an essential factor for high resolution spectroscopy, because all the broadening effects besides the natural width can be reduced only at the sacrifice of sensitivity. An ample reserve in sensitivity of laser spectroscopy has made it possible to develop such methods as saturated absorption, atomic or molecular beam spectroscopy, two-photon spectroscopy, and other methods of nonlinear spectroscopy.

At first, the laser power must be lowered in order to reduce saturation broadening; secondly, the gas pressure of the sample must be reduced so as to reduce collision broadening; and thirdly, in order to reduce transit time broadening, the diameter of the laser beam must be expanded and both the laser power and the gas pressure must further be reduced. It is evident, therefore, that higher resolution can be attained as the sensitivity is increased.

The intrinsic sensitivity of a laser spectrometer is very high. It was pointed out by the author in 1964 that some hundreds of atoms could be detected with an idealized laser spectrometer [2.14]. The theoretical considerations of sensitivity were later extended to include practical conditions and methods of laser spectroscopy [2.15]. These results are briefly described in Subsections 2.4.1 and 2.4.2 in order to discuss factors that determine the sensitivity and to reveal conditions for its optimization. Experimental results are described in Subsection 2.4.3.

2.4.1 Noise in Laser Spectrometers

Any laser spectrometer consists of a laser as the light source, an absorption cell or a sample, a detector, and some other associated components. Sensitivity of the laser spectrometer is limited by the noises in these components: in particular the laser noise and the detector noise are dominant. The laser noise is a limiting factor, when a noisy laser such as the dye laser is used. Solid-state lasers and the semiconductor diode and spin-flip lasers are often quite noisy. The laser noises are ascribed to the instabilities in gain and refractive index of the active medium and to the mechanical disturbances so that they can be reduced to a negligible level in principle. Gas lasers are less noisy in general. Particularly, a molecular laser of rigid construction with a stable power supply and good acoustic/mechanical isolation is very stable both in its amplitude and frequency. Although the noise level of such a stable gas laser is still several orders of magnitude higher than the theoretical quantum noise of the laser oscillator, it is so low that it can be indifferent in a laser spectrometer.

Ultimate sensitivity of laser spectrometers, therefore, is not limited by the laser noise but determined mainly by noise in the detector. The detector noise or the noise equivalent power P_n of an infrared detector is conventionally expressed by the detectivity D in the form

$$P_n = 1/D = \sqrt{A\Delta f}/D^*,$$

where A is the effective area of the detector and Δf is the bandwidth of the post-detection amplifier.

a) Noise Figure

The noise equivalent power of a detector in the radiofrequency and microwave regions is usually expressed by using the noise figure F as

$$P_n = F\kappa TB, \tag{2.44}$$

where B is the effective bandwidth. It is not true that the absorbed power larger than the noise equivalent power P_n is detectable, but the change in amplitude of the optical field E for detectable absorption must be larger than the amplitude flucturation e_n corresponding to the noise equivalent power. When this criterion for the threshold of detection is adopted, the minimum detectable change ΔP_{min} in the optical power $P = GE^2$ is expressed by

$$\Delta P_{min} = GE^2 - G(E - e_n)^2 \approx 2GEe_n,$$

where G is a proportionality constant. Since the noise equivalent power is $P_n = Ge_n^2$, we obtain

$$\Delta P_{min} = \sqrt{4PP_n} \tag{2.45}$$

which may be rewritten as

$$\Delta P_{min} = \sqrt{4PF\kappa TB}$$

by substituting (2.44).

b) Quantum Efficiency of a Photoelectric Detector

The sensitivity of a photoelectric detector is determined by fluctuations in photoelectron counts, if dark current may be ignored. The number of photoelectrons in a time interval Δt is $C = \eta P\Delta t/\hbar\omega$, where η is the quantum efficiency, P is the total power entering the detector, and $\hbar\omega$ is the photon energy. Thus the standard deviation of counts becomes

$$\Delta C = \sqrt{\eta P \Delta t / \hbar\omega} \,.$$

The minimum detectable change in power is therefore given by

$$\Delta P_{min} = \Delta C \hbar\omega / \eta \Delta t = \sqrt{4 P \hbar\omega B / \eta} \,, \tag{2.46}$$

where $B = \int_0^\infty [1 + (2\pi f \Delta t)^2]^{-1} df = (4\Delta t)^{-1}$ is the bandwidth. The above expression (2.46) is rewritten as $\Delta P_{min} = \sqrt{4 P P_n}$, where

$$P_n = \hbar\omega B / \eta \tag{2.47}$$

is the noise power of the photoelectric detector.

Fluctuations in the number of photogenerated carriers in a semiconductor may be expressed in the same way, although noise in the photoconductive detector includes statistical fluctuation of the current pulse due to random recombination of photogenerated carriers.

c) Quantum Noise in Ideal Detection

The phenomenological treatments of noise mentioned above are adequate in practice for consideration of spectrometer sensitivity. But let us here consider the background noise and the detector noise in the ideal case.

The average thermal energy in a single optical mode is well known from the Planck's law of radiation to be

$$W_n = \hbar\omega / [\exp(\hbar\omega / \kappa T) - 1] \,.$$

Now the thermal noise of the background, $W_n B$, and the shot noise of an ideal detector, $\hbar\omega B$, are assumed to be statistically independent. The noise equivalent power of an ideal detector ($\eta = 1$) with the background at the temperature T is thus writtes as

$$P_n = \hbar\omega B + W_n B$$
$$= \hbar\omega / [1 - \exp(-\hbar\omega / \kappa T)] \,. \tag{2.48}$$

It may be noted here that this noise power is just equal to the noise power of a high-gain maser amplifier in which the negative temperature is $T_m = -T$. The equivalent noise power of a maser with gain G is given by [2.16, 17]

$$P_{maser} = [(G-1)/G][N_2 / (N_2 - N_1)] \hbar\omega B \,. \tag{2.49}$$

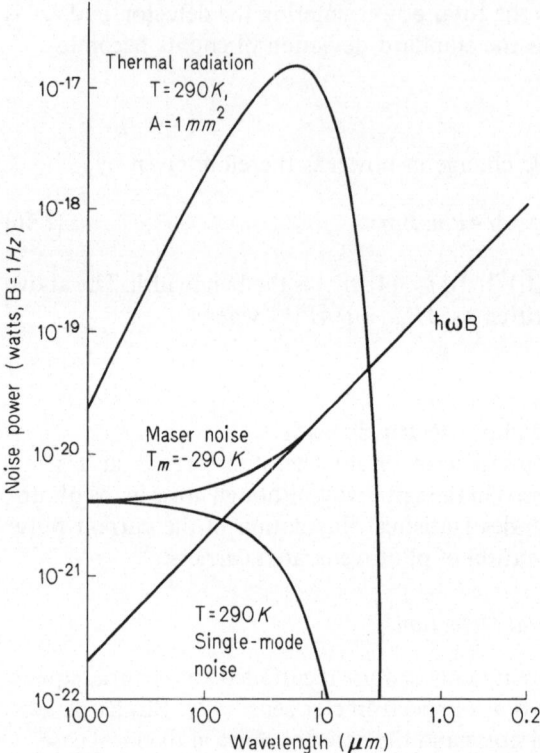

Fig. 2.7. Noise powers in a bandwidth of $B=1$ Hz. Thermal noise at $T=290$ K in a single mode, photon shot noise $\hbar\omega B$, the total noise or the maser noise for $T_m = -290$ K, and the multimode thermal radiation on an area of 1 mm² at $T = 290$ K are shown

For $G \gg 1$ and $N_2/N_1 = \exp(-\hbar\omega/\kappa T_m)$ the above equation (2.49) becomes equivalent to (2.48).

In Fig. 2.7 are shown the thermal noise in a single mode, the photon shot noise, and the total noise of (2.48) or the maser noise of (2.49), as functions of the wavelength for $B=1$ Hz and $T=290$ K. The power of multimode thermal radiation falling on an area of 1 mm² is also shown for comparison.

2.4.2 Theoretical Sensitivity of Laser Spectrometers

Theoretical sensitivity is considered below for two typical types of spectrometers; one is the simple spectrometer for *linear absorption* and the other is the hole-burning spectrometer for *saturated absorption*.

a) Sensitivity for Linear Absorption

The power absorbed in the gas of length L in the absence of inhomogeneous broadening is obtained from (2.17) to be

$$\Delta P_{abs} = \alpha_L L P / (1 + P/P_s),$$

where α_L is the small-signal absorption constant, and P/P_s is the saturation parameter. Note that saturation is considered here in order to find the optimum condition for so-called linear absorption. From (2.45) the signal-to-noise ratio of the absorption line is expressed by

$$\frac{S}{N} = \frac{\Delta P_{abs}}{\Delta P_{min}} = \frac{\alpha_L L \sqrt{P}}{2\sqrt{P_n}(1 + P/P_s)}. \tag{2.50}$$

This takes a maximum value at $P = P_s$.

If the absorption line is predominantly broadened by the Doppler effect, on the other hand, the absorbed power is expressed by using (2.34) in the form

$$\Delta P_{abs} \approx \alpha_L L P / \sqrt{1 + P/P_s}.$$

In this case the signal-to-noise ratio would increase monotonously with the power, if the noise equivalent power P_n were constant. The actual noise of a detector increases with the incident power; and the line is much broadened when the power P is much larger than the saturation power P_s. Therefore the optimum power for the highest sensitivity in the case of an inhomogeneously broadened line may also be found somewhere around $P = P_s$. Thus the minimum detectable absorption by using the incident power $P \approx P_s$ is approximately given from (2.50) by

$$\alpha_{min} \approx 2\Delta P_{min} / L P = (4/L)\sqrt{P_n/P}. \tag{2.51}$$

The absorption constant may be written as

$$\alpha = (N_1 - N_2)\sigma$$

where σ is the absorption cross section. At a wavelength shorter than about 30 μm, the number of molecules in the upper level is negligible so that $N_2 \ll N_1$ and the minimum detectable number of molecules in a unit volume is given by

$$N_{min} = (4/\sigma L)\sqrt{P_n/P}. \tag{2.52}$$

As a numerical example, take $P_n = 10^{-18}$ W at $B = 1$ Hz (see Fig. 2.7), and $P = P_s = 1$ W. Then we find $\alpha_{min} = 10^{-10}$ cm^{-1} for a cell length of $L = 40$ cm from (2.51). This shows that we can detect as small as 10^4 molecules/cm^3 if the molecular absorption cross section is $\sigma = 10^{-14}$ cm^2.

When the Doppler broadening is much larger than the homogeneous broadening, the absorption cross section is obtained from (2.35) to be

$$\sigma = \sqrt{\pi} |\mu_{12}|^2 / \varepsilon_0 \hbar u \tag{2.53}$$

and the saturation power density becomes

$$P_s/A = (\sqrt{\pi}/2)\hbar c \gamma / \sigma u \tau, \tag{2.54}$$

where A is the area of the laser beam cross section. The minimum detectable number of molecules in the laser beam, $n_{min} = N_{min} AL$, is thus obtained from (2.52), (2.45), and (2.54) for $P = P_s$ to be

$$n_{min} = (4u\tau/\sqrt{\pi}\hbar c\gamma)\Delta P_{min}. \tag{2.55}$$

For values of $\mu = 0.1$ debye and $u = 210$ m/s, corresponding to the molecular velocity for $M = 100$ and $T = 273$ K, the absorption cross section is calculated from (2.53) to be $\sigma = 1.0 \times 10^{-14}$ cm^2. Hence the saturation power through an area of $A = 0.1$ cm^2 will be 1 W, when $\tau/\gamma = 1.33 \times 10^{-15}$ s^2. This corresponds to a pressure of a few tenths of Torr for which $\tau \approx \gamma^{-1} \approx 3.7 \times 10^{-8}$ s. Using a pressure-broadening constant of 10 MHz/Torr, the gas pressure for the saturation power density of 10 W/cm^2 is calculated to be $p = 0.4$ Torr for the transition having $\sigma = 1 \times 10^{-14}$ cm^2.

In the case of a homogeneously broadened line, on the other hand, the saturation power density is

$$P_s/A = \hbar\omega/2\sigma\tau$$

and the minimum detectable number of molecules is

$$n_{min} = (4\tau/\hbar\omega)\Delta P_{min}.$$

b) Sensitivity for Saturated Absorption

Saturated absorption is a powerful method of Doppler-free spectroscopy of high resolution; its detailed description appears in Chapter 4. The inverted Lamb dip of saturated absorption can be explained by the hole-burning effect.

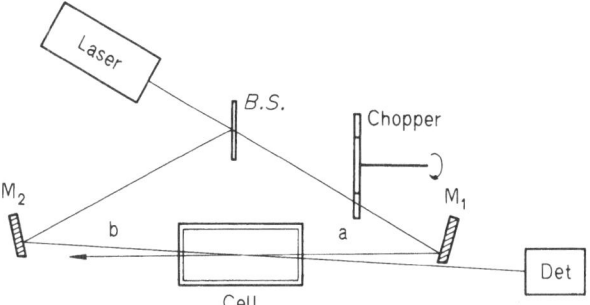

Fig. 2.8. Schematic diagram of the experimental setup for hole-burning spectroscopy

If the cross-relaxation and higher-order coherence effects are ignored, the rate-equation approach of Subsection 2.2.2 is applicable. In the case of saturation spectroscopy, however, two waves a and b of the same frequency ω and of opposite direction of propagation are incident as shown typically in Fig. 2.8. Thus the transition probability appearing in rate equations (2.38a and b) is now

$$S(v) = S_a(v) + S_b(v), \tag{2.56}$$

where

$$S_a(v) = \gamma |x_a|^2 / 2[(\omega - \omega_0 - kv)^2 + \gamma^2] \tag{2.57a}$$

and

$$S_b(v) = \gamma |x_b|^2 / 2[(\omega - \omega_0 + kv^2) + \gamma^2]. \tag{2.57b}$$

The population difference under saturation is given by (2.41), and the absorbed power of the probe wave b in the volume V is likewise given by

$$\Delta P_b(\omega) = V \int_{-\infty}^{\infty} \hbar \omega S_b(v) \Delta N(v) dv$$
$$= V \int_{-\infty}^{\infty} \frac{\hbar \omega S_b(v) \Delta N^0(v) dv}{1 + 2\tau S_a(v) + 2\tau S_b(v)}.$$

Since the linear absorption of the probe wave in the absence of the saturating wave a is

$$\Delta P_b^L(\omega) = V \int_{-\infty}^{\infty} \frac{\hbar \omega S_b(v) \Delta N^0(v) dv}{1 + 2\tau S_b(v)},$$

the nonlinear part of saturated absorption is given by

$$\Delta P_b^{NL}(\omega) = \Delta P_b^L(\omega) - \Delta P_b(\omega)$$
$$= V \int_{-\infty}^{\infty} \frac{2\hbar\omega\tau S_a(v) S_b(v) \Delta N^0(v) dv}{[1 + 2\tau S_a(v) + 2\tau S_b(v)][1 + 2\tau S_b(v)]}. \qquad (2.58)$$

Integration of (2.58) over the Maxwell distribution, $\Delta N^0(v) = (\Delta N^0 / \sqrt{\pi}u) \exp(-v^2/u^2)$, at the Doppler limit (for $\gamma \ll ku$) becomes

$$\Delta P_b^{NL}(\omega) = \frac{\sqrt{\pi} V \hbar\omega\gamma^2 \tau \Delta N^0 |x_a|^2 |x_b|^2 (\Delta\omega_a + \Delta\omega_b)}{2ku\Delta\omega_a\Delta\omega_b[4(\omega - \omega_0)^2 + (\Delta\omega_a + \Delta\omega_b)^2]}. \qquad (2.59)$$

where

$$(\Delta\omega_a)^2 = \gamma^2 + \gamma\tau|x_a|^2, \quad (\Delta\omega_b)^2 = \gamma^2 + \gamma\tau|x_b|^2.$$

The HWHM of saturated absorption from (2.59) is

$$\Delta\omega = (\Delta\omega_a + \Delta\omega_b)/2.$$

The optimum sensitivity can thus be found from a similar consideration as before at $\gamma\tau|x_a|^2 \gtrsim \gamma\tau|x_b|^2 \approx \gamma^2$. A reasonable estimate of the optimum sensitivity may thus be calculated by taking two powers to be equal to the saturation power so that

$$|x_a|^2 = |x_b|^2 = \gamma/\tau.$$

The maximum signal of saturated absorption at $\omega = \omega_0$ is therefore calculated from (2.59) to be

$$\Delta P_b^{NL}(\omega_0) = \sqrt{\pi} V \hbar\omega\gamma\Delta N^0 / 8\sqrt{2}ku\tau.$$

The signal larger than ΔP_{min} is detectable so that the minimum detectable number of molecules in the volume V is

$$n_{min} = (8\sqrt{2}u\tau/\sqrt{\pi}\hbar c\gamma)\Delta P_{min}, \qquad (2.60)$$

when the upper state population is negligible. This result is equivalent to (2.55) except for a difference in factor by $2\sqrt{2}$.

It is concluded therefore that saturation spectroscopy is only about 3 times less sensitive than linear spectroscopy, when the equal gas pressure and the equal power are used. While a few tenths of Torr is the typical pressure in linear spectroscopy of Doppler-broadened linewidth, a few mTorr or even lower pressure must be used in saturation spectroscopy to ensure a narrow homogeneous width ($\lesssim 1$ MHz) for high resolution. This means that both τ^{-1} and γ in (2.60) become small in proportion to the reduced pressure, if the natural width is neglected. The minimum detectable power under the incident power equal to the saturation power, $\Delta P_{min} = \sqrt{4 P_s P_n}$, is inversely proportional to pressure, because P_s is proportional to the square of pressure as seen from (2.54). Thus the theoretical sensitivity of saturation spectroscopy at low pressure is proportional to pressure.

c) Sensitivity of Fluorescence Spectroscopy

Laser induced fluorescence has been employed in a variety of spectroscopic studies. Here we restrict ourselves to the observation of narrow saturated absorption by molecules in a standing wave field [see c) in Subsect. 4.3.1].

The rate of excitation by the standing wave field composed of two opposing waves a and b is calculated from (2.41) and (2.56) to be

$$\frac{dn}{dt} = V \int_{-\infty}^{\infty} \frac{[S_a(v) + S_b(v)] N^0(v)}{1 + 2\tau[S_a(v) + S_b(v)]} \, dv .$$

The observable power of fluorescence from the excited state is expressed by

$$P_F = \frac{\Omega_F}{4\pi} \eta_F \hbar \omega_F \frac{dn}{dt}$$

where Ω_F is the solid angle of observation, η_F the quantum efficiency and $\hbar \omega_F$ the photon energy of fluorescence. The fluorescence light has been assumed to be isotropic.

In order to pick out sensitively the nonlinear component of fluorescence, the laser beam a is chopped at a frequency f_a and the beam b is chopped at another frequency f_b. Then the intensity of fluorescence at the frequency $f_a + f_b$ is observed by a method [2.18] as shown in Fig. 2.9. The Fourier amplitude of fluorescence intensity at the frequency $f_a + f_b$ is approximately calculated to be [2.15]

$$P_F = \frac{\Omega_F \eta_F \hbar \omega_F \gamma^2 \tau |x_a|^2 |x_b|^2 (\Delta \omega_a + \Delta \omega_b) V \Delta N^0}{8 \sqrt{\pi} k u \Delta \omega_a \Delta \omega_b [4(\omega - \omega_0)^2 + (\Delta \omega_a + \Delta \omega_b)^2]} . \tag{2.61}$$

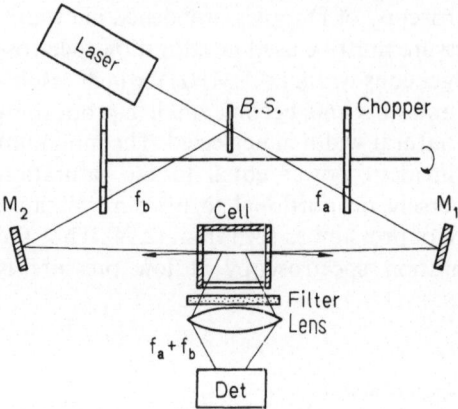

Fig. 2.9. Schematic diagram of the setup for nonlinear fluorescence spectroscopy. One of the laser beams from the beam splitter is modulated at frequency f_a, while the other is modulated at frequency f_b. Fluorescence is observed at $f_a + f_b$

Because of the absence of background radiation excepting the thermal radiation at the frequency of fluorescence ω_F, the fluorescence power of higher than $P_n = \hbar\omega_F B/\eta$ of (2.47) rather than ΔP_{min} of (2.45) is detectable. Therefore, the minimum number of detectable molecules by fluorescence in the effective volume V is given by

$$n_{min} = (32\sqrt{2\pi}/\Omega_F\eta_F\eta)(ku/\gamma)B\tau \tag{2.62}$$

with the saturation powers for $|x_a|^2 = |x_b|^2 = \gamma/\tau$.

The sensitivity in detecting saturated absorption by fluorescence at a gas pressure of about 0.01 Torr is evaluated. Using $\tau = 10^{-6}$ s, $ku/\gamma = 10^{-2}$ and assuming $\Omega_F = 0.1$ sterad, $\eta_F\eta = 10^{-3}$, and $B = 1$ Hz, we find $n_{min} = 80$ from (2.62). This sensitivity is several orders of magnitude higher than that of either linear absorption or saturated absorption, but the effective volume V in observing fluorescence is smaller. It is noted that the sensitivity of fluorescence spectroscopy is proportional to the square of the gas pressure as can be seen in (2.62).

It should be remarked in conclusion that higher sensitivity and higher resolution cannot be achieved, unless the quantum efficiency of fluorescence from the laser-excited state is very high. In the case of two-photon spectroscopy, on the other hand, detection by fluorescence is the most common technique. This is because the final state of the two-photon absorption cannot usually make radiative decay to the initial state (see Chapt. 8). Fluorescence is often useful also in atomic beam spectroscopy (see Chapt. 3).

2.4.3 Experimental Results and Discussions

Most laser spectrometers have not yet been fully engineered to permit quantitative evaluation of sensitivity, although they all have demonstrated much higher sensitivity than that of the conventional spectrometers. The sensitivity of laser spectrometers up to the present has been limited by the instability of the laser and the optical components. Only a few papers have so far reported experimental values of sensitivity. They seem to indicate that the sensitivity of laser spectrometers can reach the above-mentioned theoretical limit at least within the order of magnitude.

a) Stark Modulation Laser Spectrometer

One of the laser spectrometers approaching the theoretical limit is the Stark-modulation spectrometer. A laser spectrometer with Stark modulation was first reported in 1966 by SAKURAI and SHIMODA [2.19]. The minimum detectable absorption constant of their spectrometer was later reported to be $\alpha_{min} = 1 \times 10^{-6}$ cm^{-1} [2.20]. By using a stable molecular laser as the light source of a Stark-modulation spectrometer, FREUND et al. [2.21] obtained a sensitivity that permits detection of 10^9 molecules in the 10 μm region, while JOHNS and MCKELLAR [2.22] obtained a value of less than 10^7 molecules in the 5 μm region with an integration time of 1 s.

They also showed that the sensitivity for Lamb-dip spectroscopy of saturated absorption was just as high as that of linear spectroscopy at the same gas pressure. This is obviously consistent with the theoretical expectation mentioned in the preceding subsection.

A Stark-modulation spectrometer in the author's laboratory at the University of Tokyo was carefully constructed to reduce excessive noises of all sorts. The minimum detectable absorption has been measured to be $\alpha_{min} = 3 \times 10^{-11}$ cm^{-1} with an integration time of 100 s [2.23]. Since the effective volume in the Stark cell is 0.3 cm^3, this value permits detection of 10^3 molecules having $\sigma = 10^{-14}$ cm^2. An outline of the spectrometer is given below.

Feedback stabilized CO_2 and N_2O lasers can generate $0.1 - 1$ W single-mode single-line output at 280 different frequencies between 900 and 1100 cm^{-1}. The laser is 3 m long with an external grating. The focused laser beam transmitted through 30-cm-long Stark electrodes of 1 mm separation is incident on a copper-doped germanium detector at liquid helium temperature. A Stark modulation voltage is provided by a 124 kHz quartz oscillator. A crystal filter is employed so as not to saturate the preamplifier with the detector noise under the incident laser power of 100–500 mW.

With an integration time of 100 s at the lock-in amplifier, the noise equivalent voltage on the infrared detector is 10 nV. Since a laser power of 100 mW is measured to give a detector output of ~ 10 V, the minimum detectable absorption with the 30-cm cell is

$\alpha_{min} = 3 \times 10^{-11}$ cm^{-1}. The sensitivity is somewhat higher when a higher laser power can be used.

This high sensitivity has been verified by the observed signal-to-noise ratio of the "forbidden" $\Delta k = 3$ transitions of $^{14}NH_3$ at a pressure of about 4 mTorr [2.23, 24]. The observed sensitivity agrees with the theoretical value in the order of magnitude.

b) Detection by Fluorescence

The laser noise is not a limiting factor of sensitivity in the detection of laser-induced fluorescence. Thus it is easier to achieve an experimental sensitivity approaching the theoretical limit. Essentials to the high sensitivity of this method are the high quantum efficiency of fluorescence and efficient collection of fluorescent light into the detector.

An experimental value of sensitivity in detecting sodium atoms by laser-induced fluorescence was given by Jennings and Keller [2.25]. Fluorescence from sodium vapor excited by a dye laser at 589.6 nm was visually observed at the lowest concentration of 4×10^8 atoms/cm^3. The minimum detectable number of sodium atoms in the working volume of 1 mm^3 was thus 4×10^5. The quantum efficiency of fluorescence and the sensitivity of detection by a human eye were rather high, but the solid angle of observation was as low as perhaps 10^{-4} sterad.

The method of Freed and Javan [2.26] for observation of nonlinear fluorescence in CO_2 has demonstrated its high sensitivity. An absorption cell filled with CO_2 at a low pressure is placed inside the laser resonator. The CO_2 molecules in the lower (100) vibrational state absorb the laser radiation at 10.6 μm and populate the upper (001) vibrational state. Fluorescence from the (001) state at a wavelength of 4.3 μm is detected to observe absorption of 10.6 μm. They reported the minimum detectable absorption constant of the 10.6 μm CO_2 transition to be 1.5×10^{-6} cm^{-1} with an absorption path length of 3 cm.

Observation of nonlinear fluorescence at the sum (or difference) of two frequencies, with which two incident laser beams were chopped, was successfully employed by Sorem and Schawlow [2.18] to observe and resolve hyperfine components of I_2 lines within the tuning range of the 514.5 nm argon ion laser.

The high sensitivity of fluorescence detection has enabled observation of sharper lines by using a more collimated molecular beam in a high vacuum. The 514.5 nm lines of I_2 in a molecular beam have continuously been investigated by Ezekiel and his collaborators. The observed FWHM in their early work in 1967 [2.27] was 25 MHz; it was reduced to 3 MHz in 1972 [2.28]; and recently a value of 150 kHz which is close to the natural width was obtained by further reducing the beam divergence to 10^{-4} rad [2.29]. No experimental values of the number of molecules are reported, however.

c) Supplementary Methods

In the case when Stark modulation is not applicable, other methods of modulation such as Zeeman modulation may be employed. Because of the large inductance of the modulation coil, the frequency of Zeeman modulation cannot be high. When the modulation frequency is lower, however, the spectrometer is less sensitive, because the detector is more noisy at a lower frequency with the so-called flicker noise.

Unfortunately at present, neither good photoconductive nor photo-emissive detectors are available in some frequency range of the electro-magnetic spectrum. Thermal detectors are very slow and less sensitive in general. Optoacoustic detectors are somewhat faster and fairly sensitive.

The optoacoustic effect was discovered by TYNDALL in 1881 [2.30]. When the microphone and the amplifier became available in the 1930's, the spectrophone that was a microphone attached to an absorption cell was used to detect weak absorption of a chopped incident radiation. It was combined with the laser by KREUZER in 1970, who showed that the limit in sensitivity is primarily determined by the Brownian motion [2.31]. His experiment with the 3.39 μm He-Ne laser of 15 mW output gave a minimum detectable concentration of 10^{-8} of methane in nitrogen. The optoacoustic detector is not appropriate for a low pressure gas of narrow linewidth, but it is favorable and convenient for observation of absorption by molecules of low concentrations in gas of a moderate pressure. Thus they have been used in the observation of infrared spectra of pollutants [2.32, 33]. The optoacoustic detector will be useful in the far-infrared and submillimeter wave regions, although it is not a very attractive detector for high resolution spectroscopy.

2.5 Ultimate Resolution of Laser Spectroscopy

One of the landmarks of high resolution spectroscopy will be to reach the limit of natural width determined by the spontaneous decay rates of the involved levels. In principle, a linewidth narrower than the natural width may be observed in an atomic beam experiment, if only long-lived atoms after a pulse of coherent excitation are observed. This method corresponds to an optical analogue of Ramsey resonance in two oscillating fields at a distance between them. This experiment requires a means of discriminating the coherent component from the incoherent component of the signals induced by the two interaction fields. Extremely high sensitivity is also necessary because only a very small number of atoms is observed. Further details are described in Section 7.6.

2.5.1 Practical Limit of Resolution of Saturation Spectroscopy

A narrower linewidth is observed with a lower gas pressure and a lower laser power with a larger beam diameter. Since the intensity of the observed signal decreases when the pressure and the power are lowered, there is a practical limit of resolution in observing the spectral line.

The saturation power is expressed by (2.20), where both γ and τ^{-1} vary with pressure as shown by (2.25) and (2.26). The minimum detectable number of molecules is now rewritten from (2.20), (2.45), and (2.60) to be

$$n_{min} = 16u|\mu_{12}|^{-1}(\varepsilon_0 P_n \tau A/\pi c\gamma)^{1/2} . \tag{2.63}$$

The population of molecules N_1 on level 1 is a small fraction f_1 of the total number of molecules. Thus the number of molecules on level 1 in the volume V is

$$n_1 = V f_1 N_0 p , \tag{2.64}$$

where p is the pressure in Torr, N_0 is the total number of molecules in a unit volume at 1 Torr pressure, and f_1 may be written in the form

$$f_1 = (g_1/Z) \exp(-E_1/\kappa T) . \tag{2.65}$$

Here g_1 is the degeneracy, E_1 is the energy of the lower level, and Z is the partition function.

If natural width is neglected, a theoretical expression for the signal-to-noise ratio is given by the ratio of (2.64) to (2.63) to become

$$\frac{S}{N} = \frac{L f_1 N_0 |\mu_{12}| p}{16u} \left(\frac{\pi A c\gamma}{\varepsilon_0 P_n \tau}\right)^{1/2} . \tag{2.66}$$

When the radiative decay is negligible ($\gamma_1^0 = \gamma_2^0 = 0$), $\sqrt{\gamma/\tau}$ will simply be proportional to pressure. For high resolution spectroscopy under a small collision broadening at low pressure, however, the effect of transit-time broadening γ_t must be taken into account. Thus the signal-to-noise ratio of (2.66) must be modified in the form

$$\frac{S}{N} = \frac{\pi w L f_1 N_0 |\mu_{12}|}{16u} \left(\frac{c}{\varepsilon_0 P_n \gamma \tau}\right)^{1/2} \frac{\gamma^2 p}{\gamma + \gamma_t} . \tag{2.67}$$

Here $\gamma\tau$, that is, the ratio of the longitudinal relaxation time to the transverse relaxation time, is of the order of unity in gas molecules.

Let us consider, for example, the 3.39 μm line of the $F_2^{(2)}$ component[5] of the P(7), v_3 line of $^{12}CH_4$. Since the radiative decay is slow in methane, we may assume

$$\gamma = 1/\tau = Cp \,, \tag{2.68}$$

where $C = (C_1 + C_2)/2$ from (2.25). Then the signal-to-noise ratio at a lower pressure $(p < \gamma_t/C)$ is nearly proportional to p^3, while at a higher pressure it is proportional to p^2.

The transit-time broadening is calculated from (2.5)[6] to be

$$\gamma_t = 1.18 \langle v \rangle / w \,, \tag{2.69}$$

where $\langle v \rangle$ is the effective average velocity of molecules. The magnitude of $\langle v \rangle$ is neither equal to u nor the simple average velocity $(\sqrt{\pi}/2)u$, because molecules of different velocities contribute to the signal with different weights, when saturation is appreciable. Faster molecules cannot spend enough time for saturation with a low laser power, while the slower molecules are sufficiently saturated. This results in a smaller value of $\langle v \rangle$ than u. As the laser power is increased, however, slower molecules are oversaturated, and faster molecules become saturated so that the effective value of the average velocity $\langle v \rangle$ increases.

The above-mentioned decrease of transit-time broadening, as the laser power is reduced, was recently observed in saturated absorption of methane in an expanded laser beam by HALL et al. and by CHEBOTAYEV et al. This effect had been observed in molecular beam experiments of radiofrequency and microwave spectroscopy.

The numerical values for the 3.39 μm line of $^{12}CH_4$ are: $g_1 = 3 \times 15$, $Z = 606$ and $\exp(-E_1/\kappa T) = 0.24$ at $T = 300$ K, so that $f_1 = 1.78 \times 10^{-2}$; $N_0 = 3.22 \times 10^{22}$ m^{-3} Torr^{-1}, $|\mu_{12}| = 0.025$ debye $= 8.3 \times 10^{-32}$ C·m, $u = 550$ m/s, and $C/2\pi = 3.7$ MHz/Torr [2.34]. Recent experimental studies with a large absorption cell are described in Section 5.4, in which we have $L = 13$ m and $w = 10$ cm. Since the effective average velocity is a function of the power density of the laser beam, we take $\langle v \rangle = 300$ m/s as an example, which gives $\gamma_t/2\pi = 0.56$ kHz from (2.69). Substituting these values into (2.67), and assuming $P_n = 10^{-18}$ W (see Subsect. 2.4.2),

---

[5] This is the $F_1^{(2)}(v_3 = 1) \leftarrow F_2^{(2)}(v = 0)$ component which has been denoted by $F_1^{(2)}$ in most previous references from Hecht and Jahn, while the notation $F_2^{(2)}$ adopted in this book is from Herzberg and Hougen. Unfortunately other notations, $F_2^{(1)}$ and F_2', are used in some literature for the same component.

[6] Expressions of other authors are found in b) in Subsection 4.4.1.

we obtain

$$S/N = 5.5 p_\mu^3/(3.7 p_\mu + 560),\tag{2.70}$$

where p_μ is the pressure in μTorr so that $p_\mu = 10^6 p$.

This means that the signal is not observable at a pressure below 4.7 μTorr. In order to find a reasonable value of $S/N = 100$, the gas pressure must be $p = 22.7 \mu$Torr, at which the pressure broadening is 84 Hz and the HWHM of unsaturated absorption becomes 644 Hz. The resultant HWHM of saturated absorption under the optimum power density is thus calculated to be 911 Hz, which explains the experimental results of Section 5.4 reasonably well.

Further increase of the beam size for higher resolution imposes a stringent problem on the beam geometry. The radius of curvature of the wave surface must be larger than 10 km according to (2.8). In view of the nonlinear dispersion effect near the molecular resonance frequency, it is extremely difficult to establish very flat surfaces of the propagating waves in the nonlinear medium.

2.5.2 Spectroscopy of Trapped Particles

The linewidth of atomic hydrogen at 1420 MHz can be reduced to a fraction of hertz by storing the atoms in a bulb coated with teflon or drifilm (dimethyldichlorosilane), etc. [2.35]. A technique of buffer gas for line narrowing of hydrogen and alkali atoms is well known to microwave spectroscopists [2.9].

Dicke narrowing is effective when the mean free path l is shorter than the wavelength λ: the expression $kv < 1/\tau_{el}$ in Subsection 2.3.1 is rewritten as $l < \lambda/2\pi$. A similar condition applies to the line-narrowing effect of stored atoms or molecules. The atoms or molecules must be trapped within a dimension shorter than the wavelength of observation. The condition is severer, therefore, at shorter wavelengths.

Atomic and molecular ions can be trapped in an electromagnetic field by a method known as ion storage for high resolution microwave spectroscopy [2.36, 37]. In order to confine ions in a small volume, a very strong field is necessary. Otherwise, only a very small fraction of ions having small kinetic energies is trapped.

Trapping of neutral atoms and molecules in a strong standing-wave field was proposed by Letokhov [2.38]. The neutral particles are pulled toward the loops of the standing-wave field because of their normally positive polarizabilities. Even with the large gradient of squared field intensity produced by a cw laser, the number of stored particles will be quite small, because the gas pressure must be very low in order to prevent collisions from kicking out the trapped particles.

Although the first-order Doppler broadening will be reduced by these storage techniques, the frequency shift due to the inhomogeneous electromagnetic field for trapping particles will broaden the line. Reduction of the field intensity to narrow the line will further lose a considerable number of trapped particles. Because of such difficulties, high resolution spectroscopy with stored particles has not yet materialized in the optical region. It would be a promising technique, however, in the far-infrared region where the wavelength is fairly long and storage in a cavity is still difficult.

2.5.3 Resolution of Doppler-Free Two-Photon Spectroscopy

Two-photon spectroscopy discussed in Chapter 8 is another technique of eliminating Doppler broadening in the presence of velocity distribution. Because all atoms of different velocities make the two-photon transition without Doppler shift, high sensitivity of Doppler-free two-photon spectroscopy is assured. Besides, the transit-time broadening is much smaller than that in saturated absorption. While only atoms perpendicularly traversing the laser beam are observed in saturation spectroscopy, atoms moving along the direction of two opposing laser beams for a longer time are observed in two-photon spectroscopy.

It is noted that there still remains the second-order Doppler effect which broadens the line by

$$\Delta\omega_{2D} \approx (u/c)^2 \omega_0/2 . \tag{2.71}$$

This only amounts to less than 10^{-12} of the frequency, and hence it is negligible for the present.

A larger broadening arises from the level shift by the optical Stark effect (see Sect. 8.4). As will be discussed in Chapter 8, the coefficient for the optical Stark shift and that for the probability of two-photon transition are nearly the same. Here again we must sacrifice much of the signal intensity by reducing the laser power in order to reduce the optical Stark broadening.

It is true that curvature of the surface of the standing wave does not give rise to any shift nor broadening of the Doppler-free two-photon transition, because two photons traveling in opposite directions are simultaneously absorbed. If there exists any discrepancy between the surfaces of two waves of opposite directions of propagation, however, the frequency shift and broadening of the two-photon transition will result. Discrepancy with an angle $\Delta\phi$ gives a frequency shift $kv\Delta\phi$, where v is the component of the molecular velocity perpendicular to the laser beam axis. For the beam radius w with the radius of curvature R, discrepancy of wave surfaces with respect to the plane wave is $\Delta\phi \approx w/R$. Thus the broadening due to the inhomogeneous frequency shift will be approximately given by

$$\Delta\omega \approx \langle kv\Delta\phi \rangle \approx k\bar{v}w/R . \tag{2.72}$$

It is found therefore that the broadening in this case is almost equal to that in the Lamb dip given by (2.7) in Subsection 2.1.1. It should be remembered, however, that those molecules traveling along the direction of the laser beam produce little broadening due to the wave-surface mismatch. A numerical example for a beam radius of 3 mm with a discrepancy in curvatures of two components of $(10 \text{ m})^{-1}$ for a wavelength of 3 μm gives a linewidth of 10 kHz for $\bar{v} = 100$ m/s from (2.72). It will not be very difficult to achieve higher resolutions by further improvements in the methods of laser spectroscopy.

References

2.1 S. Y. Ch'en, M. Takeo: Rev. Mod. Phys. **29**, 20 (1957)
2.2 R. G. Breen, Jr.: *The Shift and Shape of Spectral Lines* (Pergamon Press, Oxford 1961); in *Handbuch der Physik*, XXVII, ed. by S. Flügge (Springer, Berlin, Heidelberg, New York 1964) pp. 1–79
2.3 W. R. Hindmarsh, J. M. Farr: Collision Broadening of Spectral Lines by Neutral Atoms, in *Progress in Quantum Electronics*, vol. 2, part 4, ed. by J. H. Sanders, S. Stenholm (Pergamon Press, Oxford 1973)
2.4 P. R. Berman: Appl. Phys. **6**, 283 (1975)
2.5 B. D. Fried, S. D. Conte: *The Plasma Dispersion Function* (Academic Press, New York 1961)
2.6 N. Bloembergen, E. M. Purcell, R. V. Pound: Phys. Rev. **73**, 679 (1948)
2.7 W. R. Bennett, Jr.: Phys. Rev. **126**, 580 (1962); Appl. Optics, Suppl. **1**, 24 (1962)
2.8 R. H. Dicke: Phys. Rev. **89**, 472 (1953)
2.9 J. P. Wittke, R. H. Dicke: Phys. Rev. **103**, 620 (1956)
2.10 V. G. Cooper, A. D. May, E. H. Hara, K. F. P. Knapp: Canad. J. Phys. **46**, 2019 (1968);
     J. R. Murray, A. Javan: J. Mol. Spectrosc. **42**, 1 (1972);
     F. De Martini, F. Simoni, E. Santamato: Opt. Commun. **9**, 176 (1973)
2.11 R. S. Eng, A. R. Calawa, T. C. Harman, P. L. Kelley: Appl. Phys. Letters **21**, 303 (1972)
2.12 H. Granek, C. Freed, H. A. Haus: IEEE J. QE-**8**, 404 (1972)
2.13 P. W. Smith: IEEE J. QE-**8**, 704 (1972)
2.14 K. Shimoda: Kagaku-To-Kogyo (Chemistry and Chemical Industry, in Japanese), **17**, 114 (1964)

2.15 K. SHIMODA: Appl. Phys. **1**, 77 (1973)
2.16 K. SHIMODA, H. TAKAHASI, C. H. TOWNES: J. Phys. Soc. Japan **12**, 686 (1957);
see Eq. (21) by substituting $a = N_2$, $b = N_1$, and $k = G$
2.17 E. I. GORDON: Bell Syst. Tech. J. **43**, 507 (1964)
2.18 M. S. SOREM, A. L. SCHAWLOW: Opt. Commun. **5**, 148 (1972)
2.19 K. SAKURAI, K. SHIMODA: Japan. J. Appl. Phys. **5**, 938 (1966)
2.20 K. UEHARA, T. SHIMIZU, K. SHIMODA: IEEE J. QE-**4**, 728 (1968)
2.21 S. M. FREUND, G. DUXBURY, M. RÖMHELD, J. T. TIEDJE, T. OKA: J. Mol. Spectrosc. **52**, 38 (1974)
2.22 J. W. C. JOHNS, A. R. W. MCKELLAR: J. Chem. Phys. **63**, 1682 (1975)
2.23 Y. UEDA, K. SHIMODA: in *Laser Spectroscopy*, Lecture Notes in Physics, Vol. 43, ed. by S. HAROCHE et al. (Springer, Berlin, Heidelberg, New York 1975) pp. 186–197
2.24 Y. UEDA, J. IWAHORI, K. SHIMODA: (to be published)
2.25 D. A. JENNINGS, R. A. KELLER: J. Am. Chem. Soc. **94**, 9249 (1972)
2.26 C. FREED, A. JAVAN: Appl. Phys. Letters **17**, 53 (1970)
2.27 S. EZEKIEL, R. WEISS: Phys. Rev. Letters **20**, 91 (1968)
2.28 D. G. YOUMANS, L. A. HACKEL, S. EZEKIEL: J. Appl. Phys. **44**, 2319 (1973)
2.29 L. A. HACKEL, K. H. CASLETON, S. G. KUKOLICH, S. EZEKIEL: Phys. Rev. Letters **35**, 568 (1975)
2.30 J. TYNDALL: Proc. Roy. Soc. London **31**, 307 (1881)
2.31 L. B. KREUZER: J. Appl. Phys. **42**, 2934 (1971)
2.32 E. D. HINKLEY, P. L. KELLEY: Science **171**, 635 (1971)
2.33 L. B. KREUZER, C. K. N. PATEL: Science **173**, 45 (1971)
2.34 H. J. GERRITSEN, S. A. AHMED: Phys. Letters **13**, 41 (1964)
2.35 D. KLEPPNER, H. C. BERG, S. B. CRAMPTON, N. F. RAMSEY: Phys. Rev. **138**, A972 (1965)
2.36 R. NOVICK, E. D. COMMINS: Phys. Rev. **111**, 822 (1958);
H. A. SCHUESSLER, E. N. FORTSON, H. G. DEHMELT: Phys. Rev. **187**, 5 (1969);
M. H. PRIOR, E. C. WANG: Phys. Rev. Letters **35**, 29 (1975)
2.37 H. G. DEHMELT: Advan. Atomic Molec. Phys. **3**, 53–72 (1967), and **5**, 109–154 (1969)
2.38 V. S. LETOKHOV: Zh. Eksper. I. Teor. Fiz. Pis'ma **7**, 348 (1968);
V. S. LETOKHOV, B. D. PAVLIK: Appl. Phys. **9**, 229 (1976)

3. Atomic Beam Spectroscopy

P. JACQUINOT

With 20 Figures

Long before the advent of lasers, atomic (or molecular) beams[1] had already been used in high-resolution spectroscopy. In the pioneer work of JACKSON and KUHN (see, for instance, [3.1]) the absorption by atomic beams was analyzed by means of a high-resolution Fabry-Perot interferometer. Zeeman effect of hyperfine structure in low magnetic fields, for instance, was analyzed by this method long before the use of radiofrequency spectroscopy. Even recently, measurements on isotope shifts and hyperfine structures in noble gases have been made by putting the atomic beam between the plates of a spherical Fabry-Perot [3.2]. Emission spectra have also been obtained by bombarding the atomic beam with electrons; see, for instance, MEISSNER [3.3], STANLEY [3.4], ODINTSOV [3.5].

In these experiments the Doppler broadening is greatly reduced, but the resolving power is still limited by the interference spectrometer and by the lack of energy. Nevertheless, the atomic beam was not unknown as a very useful tool in optical spectroscopy, and when suitable lasers became available it was obvious that fruitful results could be obtained by combining the remarkable properties of both tools. The first application of an atomic (molecular) beam in laser spectroscopy was described by EZEKIEL and WEISS [3.6] in the case of the I_2 molecule.

The complete history of the use of beams in atomic or molecular physics will not be repeated here. As most of the experiments were more or less derived from the famous experiment of Stern and Gerlach in 1924, with introduction of the magnetic resonance by Rabi in 1938, it is not surprising that the most comprehensive book on the subject has been written by a specialist in this field. Almost every-thing on atomic (molecular) beams can be found in Ramsey's *Molecular Beams* [3.7] except for the optical spectroscopic applications. Applications of atomic beams in conventional spectroscopy have been dealt with by KOPFERMAN in *Nuclear Moments* [3.8].

---

[1] In the following the expression "atomic beam" will be used throughout but it also includes molecular beams. Indeed it would be better to use the latter expression since atoms are a particular case of molecules, and not the reverse.

In some cases comparable results can be obtained without using an atomic beam by making use of such methods as saturated absorption or two-photon transitions. In the next section we shall endeavor to show the specific features and advantages of the atomic beam method.

3.1 Reasons for Using Atomic Beams in High-Resolution Laser Spectroscopy

In most laser spectroscopy methods one observes the response of the atoms to the tunable monochromatic excitation provided by a laser, instead of analyzing the complex light emitted or absorbed by the atoms. Since the excitation can be really considered as monochromatic and the observation of the response does not usually include any high-resolution spectrometer, the resolving power of the method is limited only by homogeneous or inhomogeneous line broadening due to the atoms themselves or their environment (see Chapts. 1 and 2). In an atomic beam some of these causes of broadening can be greatly reduced, and this is the main reason for using it.

3.1.1 Reduction of the Doppler Width

Since all atoms of an atomic beam travel in the same direction, there should be no Doppler effect provided the laser beam is exactly perpendicular to the atomic beam. In fact the atomic beam has an angular aperture α so that there remains a residual Doppler broadening of the order of

$$\delta v = v\alpha \bar{v}/c$$

where $\bar{v}$ is the average velocity of the beam. In some cases the angular aperture of the laser beam must also be considered, and there is always a velocity distribution of the atoms, so that in general the contribution of the Doppler effect to the line shape is more complicated.

The point of interest here is to compare the atomic beam method to other methods which give the same suppression of Doppler broadening in a vapor. Whereas, with an atomic beam there really exists only one class of atoms, all traveling in one direction, so that the Doppler broadening is reduced for any kind of interaction—including purely linear interaction—between light and atoms, all other methods make use of nonlinear interaction. These methods are studied in the following

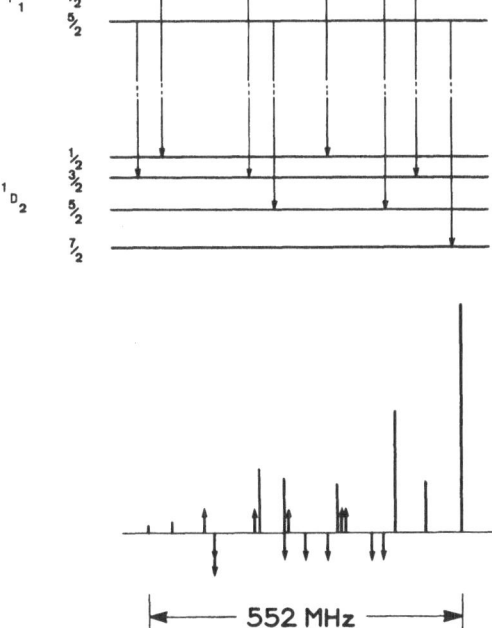

Fig. 3.1. Calculated pattern for the 1.50 μm line of Ba, as expected in saturated absorption. Arrows indicate the expected positions of the spurious signals [3.9]

chapters of this book and will not be described here. We only mention some shortcomings that atomic beams do not present. These methods can be divided into two groups according to whether the light interacts only with one particular class of velocity among all classes existing in a vapor or, on the contrary, interacts with all classes of atoms.

A) In the first group we find essentially saturated absorption spectroscopy, fluorescence line narrowing, and the different methods involving three-level systems.

a) If two components within the Doppler width share a common level, spurious resonances (or cross resonances) appear. This is due to the fact that the same frequency interacts with atoms making transitions to different levels (within the Doppler width) and belonging to different classes of velocity. This effect is particularly well known in saturated absorption spectroscopy but occurs in all the methods involving a selective interaction with a particular class of velocity. In some cases the cross resonances are well separated from the others; they can then be easily identified and may even be helpful. But in very narrow structures, the spurious signals may be unresolved and in these cases the interpretation of the whole pattern may be impossible. An example of such a case is given by Fig. 3.1 which shows the theoretical saturated absorption pattern of the $6s6p\,^1P_1$—$6s5d\,^1D_2$ (1.50 μm) line hyperfine structure of Ba [3.9].

b) In contrast to what happens in a purely linear process, the intensities of the signals observed are not simply related to the intensities that would be observed in an ordinary emission or absorption spectrum. For instance in saturated absorption, relative intensities are proportional to the square of the oscillator strengths. However for a mixture of different species, different isotopes for example, intensities are really proportional to the abundances (provided the different lines are well within a Doppler width). With other methods involving three-level systems there may be a strong distortion of intensities making the observed signal very difficult to interpret if the components are not completely resolved.

B) Doppler-free two-photon spectroscopy has the unique feature that the Doppler effect is exactly cancelled regardless of the direction and velocity of the atoms, if the two photons are of equal frequency and propagate in opposite directions; it thus works in a vapor without ignoring any atom. In this respect it is a very efficient method. However, the cross section for such a process is usually very small compared to that of a direct transition, unless a real level is happily located about midway between the two levels of the transition. Beautiful results have been obtained in favorable cases but this type of spectroscopy is not of general use. In addition, it is limited to transitions between levels of the same parity, and the selection rules for the two-photon transitions are rather restrictive (see *Measurements on other resonance lines* of Subsect. 3.4.1).

3.1.2 Reduction of Collisions

This effect may be, in some cases, at least as important as the reduction of the Doppler width, and it can be obtained only with atomic beams. The density in a beam is usually very small, and it is not easy to have more than 10^8 atoms/cm^3. At the same density the pressure $p=nKT$ in a gas would be $4\cdot10^{-8}$ Torr at $T=300$ K. It seems that no attempt has been made to calculate the number of collisions in a beam by taking into account its angular aperture and velocity distribution.

But since all atoms travel in about the same direction there are much fewer collisions than in a gas at the equivalent pressure, so that one usually says that the atom can be considered as isolated. In many of the experiments described in Section 3.4 the atomic beam has been chosen essentially for this reason. This is particularly true when highly excited states are studied, since these states have very large radii (increasing as n^2) and are very sensitive to collisions.

3.1.3 Possibility of Using Deflection of Trajectories

Two of the methods of detection of interaction between atoms and light in atomic beams (see Subsects. 3.3.3 and 3.3.4) rest upon deflection of trajectories of the atoms which have suffered a transition. This, of course, is possible only with atomic beams, and can be of considerable interest since atoms instead of photons bring the information to a suitable detector. In this type of detection each atom affected by a transition reaches the detector, whereas fluorescence photons radiate in 4π steradians and very few of them are received by the photodetector. More sensitive detection is then possible, allowing the high-resolution spectroscopy of rare atoms (see *Experiments on the sodium resonance lines* of Subsect. 3.4.1).

3.1.4 Possibility of Space Resolution Along the Beam

Since the position z of an atom along the beam at a time t is given by $z = vt$, time evolution of the radiation emitted by an atom can be observed as space evolution. If the atoms cross a cw laser beam at $z = 0$ the observation of the emitted light as a function of z is equivalent to the observation as a function of t after a pulse at $t = 0$. This would allow a space resolution of phenomena such as quantum beats or radiative decay. But this time→space conversion is useful only if the order of magnitude of v is correct for the observation of these atomic phenomena. With thermal atomic beams, v does not exceed 10^5 cm/s so that a decay of $\tau = 10^{-8}$ s would occupy only 0.01 mm, and quantum beats between two levels distant from 1000 MHz would occupy only one micrometer. Thermal beams are thus inadequate for this type of experiment; only very long-lived states, and quantum beats between such states very closely spaced, could be observed in this manner. However, qualitative observation of slower phenomena, such as optical pumping taking place during the traversal of a wide laser beam by the atoms, has been made [3.10].

With very fast beams, and essentially with ion beams accelerated by a high voltage, on the other hand, this space resolution may be fruitfully used. Radiative lifetimes and quantum beats in ions and neutral atoms have been made by ANDRÄ and coworkers [3.11] on ion beams struck by a laser beam (neutral atoms were obtained by the traversal of a thin carbon foil by the ion beam). In this technique, closely related to the *Beam Foil Spectroscopy* [3.12], the velocity spread is much smaller and Doppler tuning of the laser to the atoms by oblique incidence has been used. But no more will be said about these experiments since ion beams, and also quantum beats, are not actually in the scope of this chapter.

3.1.5 Purely Technological Reasons

In some cases it may be easier to handle certain substances in the form of a beam than in the vapor state in a cell; the atomic beam is then preferred even if it does not bring any of the above-mentioned advantages. A good example is given by the lifetime measurements in iron spectrum by Figger et al. [3.13]. They use a pulsed excitation and measure the time decay. An atomic beam is used although a resolving power beyond the Doppler limit is not needed. Very refractory metals may, for instance, be heated by electron bombardment of a moving piece according to the technique used by Büttgenbach et al. [3.14]. In that case it is as easy—or easier—to make an atomic beam apparatus than a vapor cell.

3.2 Relevant Characteristics of Atomic Beams

The design, construction and theory of atomic beams will not be dealt with here. In this section we suppose that the general arrangement of an atomic beam is known and we only recall very briefly some elementary properties which are useful in the spectroscopic use of atomic beams with lasers.

3.2.1 Some Useful Quantities

Let us call n_0 the number of atoms in the "oven" per cm^3
$\qquad s_1 l_1$ the width and length of the oven slit
$\qquad s_2 l_2$ the width and length of the collimator slit
$\qquad d$ their distance
$\qquad \lambda$ the mean free path in the oven
$\qquad \bar{v}$ the mean velocity of the atoms in the oven (it is not useful here to distinguish between the different velocities)
$\qquad \delta$ the effective atomic diameter.
The following characteristics of the beam are useful. Hereafter the beam is supposed to have a single oven slit.

A) Beam Shape, and Angular Profile

At a given distance from the collimating slit the number of atoms arriving at a given point has a trapezoidal shape the dimensions of which are determined by elementary geometry; this shape changes with the distance (Fig. 3.2). It is also interesting to consider an "angular profile" which represents the number of atoms whose trajectory makes an angle

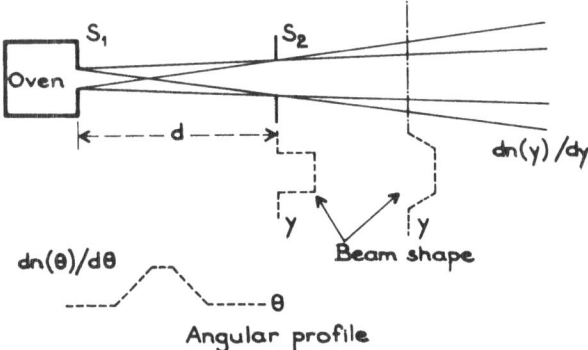

Fig. 3.2. Atomic beam geometry

α with the axis. Clearly this shape does not change with the distance and is also a trapezoid. The half-sum of the bases is s/d, s being the width of the wider slit. The trapezoid reduces to a triangle of half-base s/d if the two slit are equally wide $s_1 = s_2 = s$, a case which is often realized in practice.

B) Collimation Ratio

We define it as $C = d/s$, so that high values of C correspond to "highly collimated" beams. In practice C may vary between very low values 5 or 10 to very large values like 10^4. This quantity is very important since the residual Doppler broadening depends on it.

C) Flux

The number of atoms carried by the beam per second is

$$\Phi = n_0 \bar{v} l_1 s_1 l_2 s_2 / 4\pi d^2 . \tag{3.1}$$

If $s_1 = s_2 = s$, $l_1 = l_2 = l$,

$$\Phi = n_0 \bar{v} l^2 / 4\pi C^2 , \tag{3.1a}$$

Φ can be also expressed as a function of the pressure in the oven by

$$\Phi \simeq 10^{22} p l^2 / C^2 \sqrt{MT} \tag{3.2}$$

p being expressed in Torr. Φ cannot be increased indefinitely by increasing n_0 or s_1 since s_1 cannot be larger than the mean free path λ of the atoms in the oven, in order to have a well-defined beam.

Since $\lambda=(\sqrt{2}\pi n_0\delta^2)^{-1}$, the product $n_0 s_1$ has a maximum value $a=(\sqrt{2}\pi d^2)^{-1}$, depending on the gas, so that the maximum flux is

$$\Phi_M = a\bar{v}l_1 l_2 s_2/4\pi d^2 . \tag{3.3}$$

This shows that if all dimensions of an apparatus are multiplied by a number k, C remains unchanged and Φ_M is multiplied by k; but the pressure in the oven and also the residual pressure in the apparatus have to be divided by k.

In some cases, for instance in the case of very rare elements, n_0 is much smaller than its maximum permitted value, and s_1 can be increased almost at will; in such cases ΦC^2 depends only of the lengths of the slits. Values of Φ of the order of $10^{10}-10^{13}$ atoms/s may be considered as high. In some cases one has to use a flux as weak as 10^3 atoms/s.

D) Density

The number of atoms per unit volume (cm^3) in the beam is

$$n = \Phi(l_2 s_2 \bar{v})^{-1} = n_0 l_1 s_1/4\pi d^2 . \tag{3.4}$$

Here also there is a maximum value

$$n_M = a l_1/4\pi d^2 . \tag{3.5}$$

If all dimensions of the apparatus are multiplied by k, n_M is divided by k. A value of 10^8 atoms/cm^3 is already high and one has sometimes to work with densities as low as 1 atom/cm^3.

E) Intensity

I is defined as the flux per unit area of the beam

$$I = \Phi/l_2 s_2 = n\bar{v} . \tag{3.6}$$

Its maximum value is $I_M = a l_1 \bar{v}/4\pi d^2$. The same considerations as above are valid. Intensities of the order of 10^{13} atoms. cm^{-2}s^{-1} are considered as high and it is difficult to have larger values without the use of multiple-channel oven slits.

F) Velocity Distribution

In a gas the velocity distribution is given by

$$N(v)dv \propto v^2 \exp(-v^2/u^2)dv \tag{3.7}$$

u being the most probable velocity $u = \sqrt{2RT/M}$. In a thermal beam the distribution is the same. This means that in any direction the number of atoms having the velocity v is given by the same expression. A different result is obtained if one considers the *intensity* of the beam due to atoms of velocity v. This intensity is proportional to $vN(v)dv$ so that

$$I(v)dv \propto v^3 \exp(-v^2/u^2)dv . \tag{3.8}$$

The difference between these two "distributions" must be well understood; in the calculation of Doppler shifts $N(v)$ must be used, not $I(v)$. Both $N(v)$ and $I(v)$ are broad distributions and a thermal beam is very far from being monokinetic. Practically the width (half-maximum full-width) δv of both curves is equal to u. This makes it rather difficult to use beam properties depending on v, like Doppler tuning or lifetime measurements, unless a velocity selection is made by some mechanical means [3.15] or by the methods indicated in Subsection 3.5.1.

This velocity spread must not be forgotten when the interaction time T of the atoms with the laser beams plays a role, for instance, when there is an optical pumping changing the relative population of different states or substates.

The situation is completely different if non-thermal beams are used. With supersonic jets (see, for instance, [3.16]) the velocity distribution is much narrower. With electrically accelerated beams it is easier to achieve monoenergetic conditions, at least for energies higher than 100 eV. Non-thermal beams have so far been very little used in high-resolution laser spectroscopy and no more will be said here about them.

3.2.2 Multiple-Channel Slits

If high density beams are desired, it is possible to escape the condition $(n_0 s_1)_{\text{Max}} = (\sqrt{2\pi\delta^2})^{-1} = a$ by dividing the slit into a number of elementary slits each of which obeys the above condition. With a slit composed of $\underline{m}$ such elements the maximum values of the flux, density and intensity are multiplied by $\underline{m}$. The width s_1' of the entire slit is now $s_1' = ms_1/t$, where t is usually of the order of 0.5 because there must be walls between the different elementary slits. The collimation ratio then becomes $C' = dt/s_1$, assuming that $s_1 \geq s_2$. This shows that, *for a given collimation ratio*, the maximum values of Φ, I and n are only multiplied by mt. However, the net gain can be considerable if very large values of m, which means very small values of $s_1 = s_1'/m$, can be realized. In fact one does not use multiple slits but rather multiple-channel slits. Each elementary slit is actually made of a number of channels so that the total slit is composed of a two-dimensional array of channels. This is because it is almost the only practical way of making a multiple-element slit, but there is an additional advantage in using channels instead of holes or slits of negligible thickness. When atoms effuse from the oven through a small aperture in a thin wall, their angular distribution follows a cosine law, but in the effusion through a "long" channel the angular distribution is much narrower, the intensity in the normal direction being unaffected. This reduces the consumption of material by a factor of the order of L/w, L being the length of the channel and w its smaller dimension. Substantial economy may thus be obtained in the case of rare materials. In practice multi-channel slits are made of stacked crimped metal foils. It is possible to make slits composed of hundreds of elements a few micrometers wide. Practical details on the construction of such slits can be found in [3.4, 17, 18].

3.2.3 Residual Doppler Shift and Broadening

a) If the angular apertures of both the atomic beam and the laser beam were exactly zero, and the two beams exactly perpendicular, there would be no Doppler shift and no Doppler broadening.

b) With two beams of zero apertures but making an angle $\alpha + \pi/2$ one has to account for the velocity distribution given in F of the preceding subsection. The Doppler shifts are given by $x = v - v_0 = v_0 v \sin \alpha / c$ and the number of atoms tuned to $v_0 + x$ is

$$N(v)dv \propto v^2 \exp(-v^2/u^2)dv \tag{3.9}$$

where v has to be replaced by $xc/v_0 \sin \alpha$. The response curve of the beam to the laser excitation is then of the type

$$\mathcal{R}(x) \propto x^2 \exp(-x^2/\delta^2) \tag{3.10}$$

where $\delta = v_0 u \sin \alpha / c$. (It is the component of the most probable velocity in the beam on the line of sight that counts.)

There is thus a *frequency shift* δ proportional to $\sin \alpha$ and a *frequency spread* almost equal to δ.

c) Let us now consider a laser beam of zero aperture, exactly perpendicular to the mean direction of the atomic beam, and an atomic beam of finite aperture but with a rectangular angular shape; this is the case where one of the two slits is of negligible width, compared to the other. Of course, this case is never realized but it helps to understand the general case. Here all directions are equivalent in the angular range $1/C$. The calculation of the Doppler broadening is thus exactly the same as in a vapor, except that each speed v has to be replaced by $v \sin \alpha$ as in the former case. The response $\mathcal{R}(x)$ has thus a Gaussian shape but the width is reduced in the ratio $1/\sin \alpha \simeq 1/\alpha$ compared to the vapor in the oven. The residual Doppler width is

$$\delta v(\text{beam}) = \delta v(\text{vapor})/C . \tag{3.11}$$

d) It is necessary now to introduce two additional causes of broadening: the width of the other slit, which gives the angular shape of the beam a trapezoidal or triangular $(s_1 = s_2)$ shape, and also the angular aperture of the laser beam. If one of these two additional causes was introduced separately it would produce, for an angular apertures, the same effect as the cause considered in c) since this effect depends only on angles between light rays and atomic rays. Unfortunately these angles introduce themselves in the equations in such a way that the combination of the different effects is not given by a convolution which would allow a intuitive and correct understanding. A calculation of the "line shape" has been made by MINKOWSKI and BRUCK [3.19] for the emission lines of an excited atomic beam. Unfortunately this calculation is not easy to interpret and moreover it does not consider the effect of the finite angular aperture of the light beam. One result of this calculation is that the response curve of the atomic beam to a perpendicular laser beam of zero aperture is not Gaussian—sharper near the center and wider in the wings —and that its full width at half maximum is about the same as if it were Gaussian. In cases where the laser beam is focused onto the atomic beam there will be a slight additional broadening as long as its angular aperture is smaller than the atomic beam aperture. If it were larger the dominant effect would be given by $\delta v = \delta v(\text{vapor})/C'$, C' being the collimation ratio of the laser beam. But this would be far from the optimum conditions. Since the three causes of broadening play exactly the same role, the best compromise between resolution and intensity of the response is obtained when the three separate broadening effects are the same.

In the above considerations we have neglected the natural width of the frequency response of an atom to monochromatic excitation. Obviously the natural shape has to be convoluted with the residual Doppler shape. When the highest resolution is desired, the latter must be a few times sharper than the natural shape which may impose rather severe conditions on the collimation of the atomic beam and on the eventual focusing of the laser beam. For instance in the case of sodium D-lines the natural width is 10 MHz. In order not to broaden the lines too much, the residual Doppler width must be reduced to, say, 3 MHz assuming that the frequency jitter of the laser is less than 1 MHz. This imposes a collimation ratio on the atomic beam of at least 300/1 and preferably a non-focused laser beam. For molecular transitions with natural widths of the order of 0.1 MHz the atomic beam collimation ratio must be much higher, not less than 10^4.

3.2.4 Production of Metastable Atoms

The relative populations of the different states in the beam are given by the Boltzmann law at the temperature of the oven. Since 1000 K is approximately equivalent to $1000 \, \text{cm}^{-1}$, only the low-lying levels are thermally populated and different means have to be used to increase the population of the others.

This can been done by electron bombardment of the beam after it has been formed; an electron gun produces an electron beam which crosses the atomic beam after the second slit. This is the case in the experiments of Champeau and Keller [3.20] (see also [3.4]) on neon, where transitions are observed from the metastable $2p^5 3s^3 P_2$ situated $134\,000 \, \text{cm}^{-1}$ above the ground state. It is also the case in the experiments presently carried on by Barger et al. [3.21] on the Balmer line $H\alpha$.

In other devices, a dc discharge is produced in a region separating the oven from the first slit. An efficient source of metastable Ca atoms in the $4s4p^3 P_{0,1,2}$ and $4s3d^1 D_2$ states has been described by Brinkmann et al. [3.22]. Ba metastables have also been produced by Ishii and Ohlendorf very efficiently with a similar device [3.23].

Other devices make use of high frequency discharge along the beam.

Direct optical excitation from the ground state towards the metastable is, of course, impossible. But there are cases where it is possible to use a cascade like in Ba with the transitions $6^1 S_0 \rightarrow 6^1 P_1 \rightarrow 6^1 D_2$ or $6^1 S_0 \rightarrow 6^3 P_1 \rightarrow 6^3 D_{1,2}$. In such cases it would be possible to combine the production of metastables with a velocity selection for these states (see Subsect. 3.5.1).

3.2.5 Laser Requirements—Saturation—Power Broadening

The resolution is ultimately limited by the natural width of the transitions if the laser line occupies a spectral range much smaller than this natural width. The excitation will be called "monochromatic" and have the corresponding properties only if this condition is fulfilled. The maximum frequency jitter of the laser corresponding to this condition depends on the type of transition under study. In atomic spectra the natural linewidths are usually a few MHz and it is not too difficult to have an effective laser linewidth of 1 MHz or less. But with molecules the lifetimes are much longer and natural widths of 0.1 MHz are normal, so that the short-term stability requirement for the laser is much more severe, but not unrealistic.

In the opposite case where the frequency jitter is larger than the natural width, not only is the resolution limited by the laser, but also the response of the atoms or molecules is weaker since it has to be averaged between useful and useless frequencies.

The power requirement is much easier to meet in most spectroscopic experiments since rather low laser intensities (power flow per unit area) are necessary to saturate atomic or molecular transitions. Also, since the size of an atomic beam is usually small (a few millimeters at maximum in its larger transverse dimension) the total power of the laser need not be very high; a few milliwatts is usually sufficient.

In a two-level system, the number N_2 of atoms transferred from the ground state 1 onto the upper level 2 under the action of a *monochromatic* excitation is

$$N_2 = (N/2)[\chi/(1+\chi)] \tag{3.12}$$

with $\chi = 8\pi\mu^2 P/c\hbar^2\gamma^2$ in cgs system of units where μ is the matrix element of the electric dipole moment, P the power flow per unit area, and γ the natural width of the transition. In the case of a laser line much broader than γ the expression for χ becomes

$$\chi = 8\pi\mu^2 P/c\hbar^2\gamma\Gamma \quad (\Gamma \text{ laser linewidth}). \tag{3.13}$$

It is not useful to have $\chi > 1$ since at this value the system is already half-saturated and a much lower power is more than sufficient. Moreover, at this power there is already a significant broadening of the transition, since the power broadening is given by $\gamma'^2 = \gamma^2(1+\chi)$. The broadening is then closely related to the saturation (see Subsect. 2.1.2). One must be very careful about this effect since it is very easy to obtain a recorded width twice as large as the natural width at power densities as low as 1 mW/mm^2. This has been observed by LANGE et al. [3.24] and also by PICQUÉ [3.25] who has measured, on the hyperfine structure of the Na D line, broadenings by a factor of 10, using a Spectra Physics dye laser, model 370.

Moderately powerful lasers are thus usually sufficient for spectroscopy. However if the effects of interaction of atoms with intense light fields are studied for themselves (see Subsect. 3.4.3) much more power is needed but is easily delivered by some available single mode lasers.

3.3 Detection of the Interaction of an Atomic Beam with a Laser Beam

Studying the spectrum of the atoms in the beam consists in observing how their interaction with the photons in the laser beam varies when the frequency of the laser is changed. The sensitivity of the whole process thus depends on the sensitivity of the detection of this interaction.

This detection can be either a direct *optical detection* (by absorption or fluorescence) or a *triggered detection* whereby atoms themselves which have suffered the transition are detected by means of some change in their behaviour.

3.3.1 Optical Detection by Absorption

The absorption of the incident light by the atomic beam is usually very weak because the atomic beam is of low density and usually rather thin; the signal/noise ratio may thus be very poor. If the fraction a $(a \ll 1)$ of the N incident photons (per second) is absorbed, the absorption signal aN is superimposed on a background N so that the signal/noise ratio for an integration time of 1 second is

$$s/n = a\eta N/(\eta N)^{1/2} = a(\eta N)^{1/2} , \qquad (3.14)$$

where η is the quantum efficiency of the photodetector.

However, if multiple-channel atomic beams (see Subsect. 3.2.2) are used, the thickness of the beam can be large enough to give a reasonable value of a so that a good signal-to-noise ratio is obtained [3.26].

3.3.2 Optical Detection by Fluorescence

This is the most usual way of detection. The photons emitted by fluorescence from the upper state of the transition are collected by an optical system, the aperture Ω of which must be as large as possible since these photons are emitted isotropically. If sufficient precautions are taken to avoid parasitic light, there is no background as in the detection by absorption and the signal/noise ratio may be much higher. Assuming that this condition is fulfilled the number of photoelectrons given by the photodetector is (with the same notation as is the preceding section)

$$N' = a\eta N\Omega/4\pi . \qquad (3.15)$$

For the same experimental conditions $\underline{a}$ has the same value as in the preceding section since, regardless of saturation, in an equilibrium state, the number of spontaneously emitted photons must be equal to the number of absorbed photons. But when one approaches saturation, aN tends towards a finite limit $A\mathcal{N}/2$ (A is the Einstein coefficient, and $\mathcal{N}$ is the total number of atoms in the interaction volume) whereas $\underline{a}$ tends to zero.

The signal to noise ratio is

$$(s/n)_{\mathrm{fl}} = a^{\frac{1}{2}}(\eta N)^{\frac{1}{2}}(\Omega/4\pi)^{\frac{1}{2}} . \tag{3.16}$$

Compared to $(s/n)_{\mathrm{abs}} = a(\eta N)^{\frac{1}{2}}$ the gain is

$$(s/n)_{\mathrm{fl}}/(s/n)_{\mathrm{abs}} = a^{-\frac{1}{2}}(\Omega/4\pi)^{\frac{1}{2}} . \tag{3.17}$$

Even with $\Omega/4\pi$ as small as $1/400$ this gain may be considerable due to the very small value of $\underline{a}$, usually of the order of 10^{-6} or 10^{-8}, even far from saturation. Of course, this gain is still higher if one approaches saturation since then $\underline{a}$ tends to zero.

Considerable care must be taken to avoid stray light coming from the laser. It is best to observe the fluorescence on a line different from the excitating one whenever this is possible, because even a coarse spectral filtering may be very efficient.

This method may be much more sensitive than appears at first sight, because a given atom can emit several fluorescent photons during its interaction time T with a sufficiently intense laser beam.

In the hypothesis of a pure two-level system and in the limiting case where the upper level is saturated, this number is $T/2\tau$, τ being the lifetime of the state. Since T is of the order of a few microseconds and τ a few nanoseconds the same atom can give about one thousand photons. But this is no longer true if the upper level can also combine with another level, or sublevel, of the ground state. In that case high power gives rise to an optical pumping. The lower level can be emptied after a few pumping cycles and the number of photons emitted by each atom is only a few units. In a given structure some transitions may then give rise to weak signals whereas some others give intense signals, and very strong distortion of relative intensities can be produced.

It is thus preferable to use lower power at the expense of sensitivity, except in some particular cases. One must also remember that saturation is always accompanied by a broadening of the transition, as mentioned in Subsection 3.2.5.

3.3.3 Detection by Resonant Recoil

In the process of absorption of a photon by a resonant atom, not only an energy hv, but also a linear momentum $p = hv/c$ is transferred to the atom. If this momentum is perpendicular to the momentum mv of the atom this produces a change in the direction of the atom by an angle $\alpha = p/mv$. Although this angle is very small, about 10^{-5} rad for thermal velocities, deflection of the beam can be observed if the beam is highly collimated, and used for monitoring the absorption. Of course, one must examine what happens when the same atom returns to its initial state by re-emitting a photon:

a) In the process of spontaneous emission the photon is emitted isotropically; the recoil suffered by the atom is then isotropically distributed so that a spread of the beam arises, without changing its mean direction;

b) any process of induced emission gives a recoil in the direction opposite to the incident photons so that it exactly cancels the recoil produced by absorption.

Both processes exist during the interaction time of a given atom with the laser beam. Let us first consider a pure two-level system submitted to a monochromatic resonant light beam. The effective number of recoils per second is $\dot{N}_{eff} = \dot{N}_r - \dot{N}_{ar}$, where the indices r and ar mean recoil and anti-recoil. In a steady state this number is equal to the number $\dot{N}_{sp}$ of spontaneous emissions per second since

$$\dot{N}_{sp} + \dot{N}_{ar} = \dot{N}_r \tag{3.18}$$

and $\dot{N}_{sp} = N_2 A$, A being the usual Einstein coefficient. As we have seen in Subsection 3.2.5

$$N_2 = (N/2)[\chi/(1 + \chi)] \tag{3.19}$$

where $\chi = 8\pi\mu^2 P/c\hbar^2 A^2$. As χ varies from 0 to a value much larger than 1, $\dot{N}_{eff}$ varies according to the saturation curve of Fig. 3.3. The total number of effective recoils during the interaction time T of atoms with the laser beam is

$$N_{eff} = AN\chi T/2(1 + \chi).$$

The mean number of effective recoils suffered by each atom is $n = N_{eff}/N$, which tends to $AT/2 = T/2\tau$ at saturation, and is equal to 1 for

$$\chi = \frac{2}{AT - 2} \simeq \frac{2\tau}{T}. \tag{3.20}$$

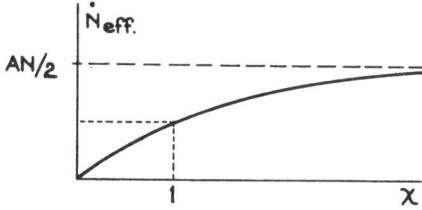

Fig. 3.3. Number of effective recoils per second, as a function of the saturation parameter

For typical values of T ($\simeq 1$ μs) and τ (a few ns) n can be 100–1000 at saturation whereas very low powers can produce $n \simeq 1$.

It is thus possible to obtain large deflections with high power lasers. In that case a very high collimation is no longer necessary so that more intense beams can be used. But, as has been explained in the preceding subsection, this is true only in the hypothesis of a pure two-level system, and the same drawback of working near saturation, will be encountered: strong intensity distortions and power broadening. For spectroscopic measurements it is thus preferable to use low power and very high collimation of the atomic beam, at the expense of sensitivity.

The deflection of an atomic beam was first demonstrated experimentally by FRISCH in 1933 [3.27]. ASHKIN [3.28] proposed to use the resonant beam deflection for isotope separation with an arrangement using several laser beams oriented in such a way that the deflected atomic beam is always perpendicular to the laser beam it sees. This effect was later studied by PICQUÉ and VIALLE [3.29] under weak illumination by a discharge tube emitting the resonance line where each atom gives approximately one recoil, and by SCHIEDER et al. [3.30], under illumination by a 10 mW laser in which each atom undergoes approximately 60 recoils. Deflection was clearly seen and measured in both experiments, and spread was also demonstrated in the first one.

NEBENZAHL and SZÖKE [3.31] have suggested a method which is capable of high energy-efficiency. In this method pulses properly shaped and frequency swept are reflected back and forth on the beam by two mirrors. An adiabatic passage is obtained and this is equivalent to using the same photon several times. This system might be applicable to isotope separation by beam deflection, but probably not to spectroscopy.

The first recording of an hyperfine structure by beam deflection was made by JACQUINOT et al. [3.32] on sodium D lines. The collimation ratio was 10^4 and the atom detector was a 10 μm hot wire. The very simple arrangement and a result are shown on Fig. 3.4.

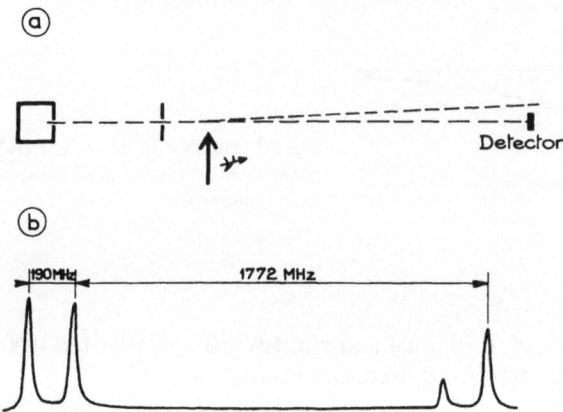

Fig. 3.4. (a) Deflection of an atomic beam by resonant recoil (the deflection angle is greatly exaggerated). (b) Single recording of the Na D_1 line

3.3.4 Detection by Magnetic Deflection

It is well known since the famous experiments of RABI that any change in the magnetic substate $|m_J\rangle$ of an atom in an atomic beam can be detected by a change in the deflection of the beam in a inhomogeneous magnetic field; this is the detection of magnetic resonance transitions by magnetic deflection. RABI himself suggested in 1952 [3.33] that this could be extended to the study of excited states by magnetic resonances if atoms were illuminated by a resonance line in the interaction region. He suggested also that the same device could be used to study isotope shifts.

In 1965 MARRUS [3.34] was able, with exactly the same arrangement, to detect and measure optical transitions by making use of the change in the magnetic substate of the ground level that arises when the atom returns to its ground level. The aim of these experiments was not to increase the resolution but to make possible measurements on atoms in very small quantities since the detection of the atoms was very sensitive, especially in the case of radioactive isotopes. Since at that time single-mode tunable lasers were not yet available, a "monochromatic" radiation from a discharge tube at a fixed frequency was used to produce the optical transition, and the frequency of this transition was varied by Stark scanning. The resolution was limited not by the residual Doppler broadening in the atomic beam, but by the width of the radiation produced by the discharge lamp. However, it was necessary to use a very well collimated atomic beam, as in any Rabi-type experiment. Some measurements were made with success by MARRUS and coworkers [3.34].

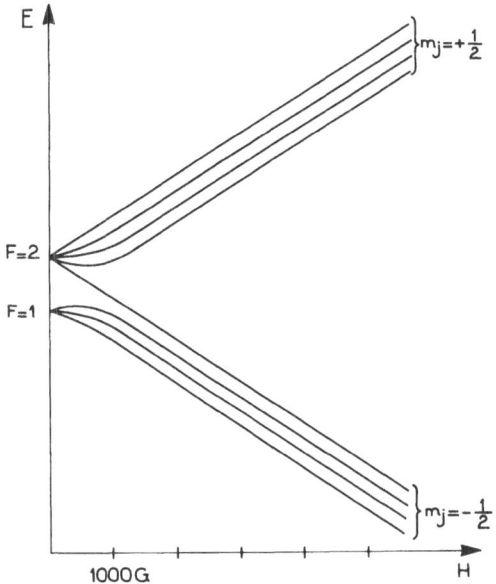

Fig. 3.5. Breit-Rabi diagram of a $^2S_{1/2}$ state with $I = 3/2$

The situation dramatically changed with the advent of single-mode tunable lasers; it then became possible to use a very sharp tunable optical excitation and to increase the resolution while retaining the other advantages of the method. A demonstration of the feasibility of such experiments was given in 1973 by DUONG et al. [3.35] (see Fig. 3.6).

For atoms without hyperfine structure the properties of these methods are obvious. In the case of hyperfine structure the main properties can be understood by considering the Breit-Rabi diagram showing the different substates of the ground state as a function of the magnetic field. This diagram is given in Fig. 3.5 for the $^2S_{1/2}$ state of ^{23}Na ($I = 3/2$); for different values of I or L_J, the main features of the diagrams would be the same. In the transition from "weak" field to "strong" field, four of the five Zeeman sublevels of $|F = 2\rangle$ go towards $|m_J = +\frac{1}{2}\rangle$ while one of them joints the group $|m_J = -\frac{1}{2}\rangle$ which comes mainly from $|F = 1\rangle$. Let us call $|\pm\rangle$ the four states which end in a $|m_J = \pm\frac{1}{2}\rangle$ state. If a transition takes place in the weak field region, an atom started from a $|\pm\rangle$ state has a certain probability of falling back on a state of opposite sign $|\mp\rangle$. When the atom in the beam enters the strong field it remains adiabatically in the same $|m_F\rangle$ state and its trajectory in the strong inhomogeneous field depends on its m_J which is now a good quantum number.

Two methods may be used to profit from this fact:

a) The first is the one used by MARRUS et al. [3.34] and by DUONG et al. [3.35]. The arrangement is exactly the same as in the original Rabi

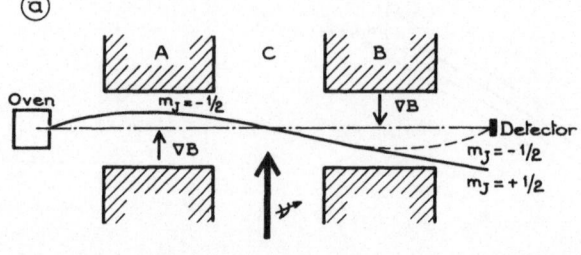

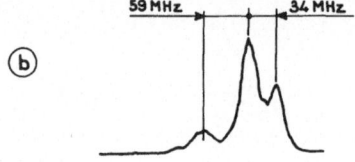

Fig. 3.6. (a) Rabi-type experiment for the detection of optical resonances. (b) Single recording of the $F = 3, 2, 1 \leftrightarrow F = 2$ components of the Na D_2 line

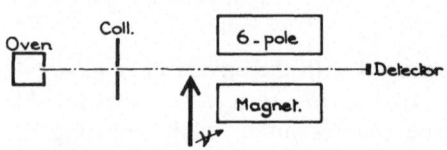

Fig. 3.7. Another method for magnetic detection of optical resonances

experiment except that there is no magnetic field in the C-field region and that the radiofrequency field is replaced by the laser beam (Fig. 3.6). The arrangement can be used in the flop-out or in the flop-in variant. It should be noted that a $|+\rangle \rightarrow |-\rangle$ or a $|-\rangle \rightarrow |+\rangle$ change have the same effect and that the signal is always negative (flop-out) or positive (flop in), regardless of the hyperfine transition observed.

b) The second method uses a different arrangement which is simpler and allows use of an atomic beam with a larger flux giving a larger sensitivity, at the expense, of course of the resolution.

There is only one inhomogeneous magnetic field and the interaction of the atoms with the laser beam takes place before they enter this field (Fig. 3.7). A similar configuration has been used by BUCKA [3.36] for measuring magnetic resonances in an excited state. Here the inhomogeneous field is produced by a six-pole magnet of the type used by RAMSEY [3.37] as a filter of magnetic substates in his hydrogen maser. In the six-pole magnet the magnetic field is zero on the axis and increases with the distance from the axis. As a consequence the magnet has focusing

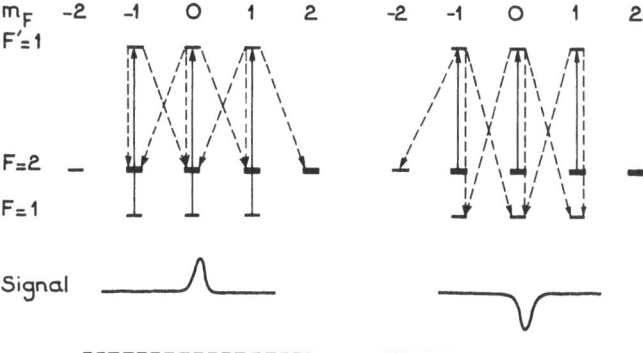

Fig. 3.8. Transitions changing the sign of m_J for components $\underline{a}$ and $\underline{c}$ (see Fig. 3.12) of the D_1 line of Na, in π illumination, and signals obtained

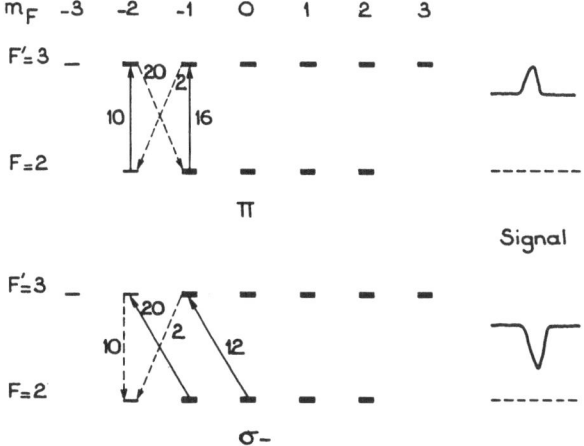

Fig. 3.9. The same as in Fig. 3.8 for the $F' = 3 \leftrightarrow F = 2$ component of the D_2 line in π and σ_- illumination

properties for the atoms with m_J positive and defocusing properties for the others. An atom detector of suitable size is placed at the focusing point. If no optical transition arises, there are as many atoms in the $|+\rangle$ states as in the $|-\rangle$ states, and the detector receives one-half of the atom flux. If, following an optical transition, atoms are transferred from $|-\rangle$ to $|+\rangle$ states, the number of detected atoms increases (positive signal); in the opposite case, a negative signal is recorded. The signs and amplitudes of obtained signals corresponding to the different hyperfine transitions have to be calculated in each particular case. Transitions corresponding

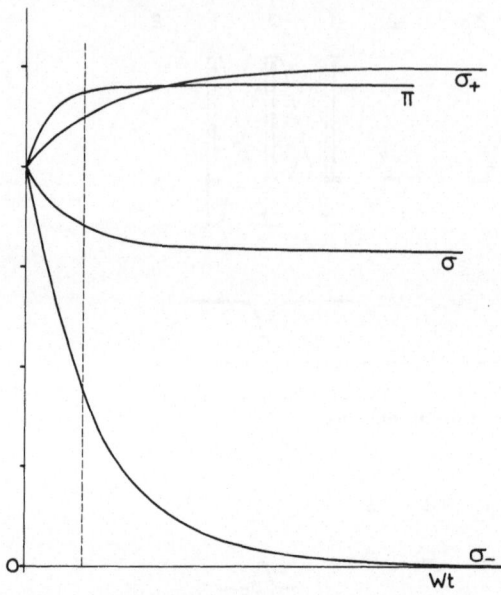

Fig. 3.10. Evolution of the signal, due to optical pumping for the $F'=3 \leftrightarrow F=2$ component of D_2. The curves represent the number of atoms arriving on the detector as a function of the product Wt of the laser power density W by the transit time t. σ means linear polarization perpendicular to the magnetic field. The dotted line corresponds to the value of Wt in the experiment described in Subsect. 3.4.1 [3.38]

to some of the hyperfine components of the D_1 and D_2 lines are shown in Fig. 3.8 in the case of exciting light polarized parallel (π) to the field for components $\underline{a}$ and $\underline{c}$ of D_1 and in Fig. 3.9 for component $2 \leftrightarrow 3$ of D_2 in π light and in σ_- light. In order to have a well-defined direction of the magnetic field in the interaction region, a field of a few Gauss is created by a coil in this region. Among all the transitions from the upper state to the lower state, only those changing the sign of m_J have been represented, for sake of clarity of the figure. One sees that in the case of component $\underline{a}$ eight transitions are of the type $|-\rangle \rightarrow |+\rangle$ so that more atoms are focused by the six-pole magnet if the laser is tuned to the component $\underline{a}$. A positive signal is thus produced; one would find the same result for component $\underline{b}$. The situation is exactly reversed for component $\underline{c}$ (and $\underline{d}$): eight transitions are of the type $|+\rangle \rightarrow |-\rangle$ so that a negative signal is produced. The case of component $2 \leftrightarrow 3$ of D_2 is different. With π light there is only one transition $|+\rangle \rightarrow |-\rangle$ and one transition $|-\rangle \rightarrow |+\rangle$, whereas eleven other transitions do not change the sign of m_J. Because of different probabilities of the two sign-changing transitions, a weak positive signal is to be expected. In order to make this component more

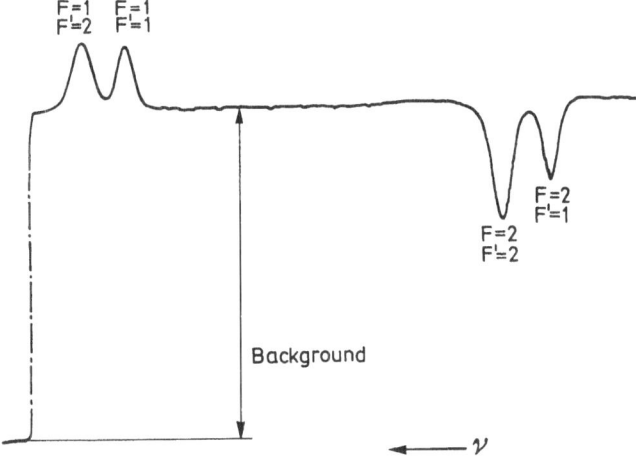

Fig. 3.11. Recording of the D_1 line of natural Na [3.39]

easily detectable σ light must be used. The transitions in σ_- illumination for this component are shown on Fig. 3.9, where there are two $|+\rangle \rightarrow |-\rangle$ transitions (for ten transitions without change of sign) so that a negative signal can be obtained; in σ_+ illumination one would obtain a positive signal.

A complete calculation of these different signals must take into account not only the different transition probabilities, but also the effect of optical pumping which creates unequal populations between the different $|m_F\rangle$ substates ("orientation" for $\sigma_\pm$ light, "alignment" for π light). The evolution of the signal for the different hyperfine components as a function of the product Wt (W: laser power, t: interaction time) has been calculated [3.38]. An example is given for the $2 \leftrightarrow 3$ component in Fig. 3.10. In this calculation it has been assumed that the power density corresponds to a situation very far from saturation of the transitions, as is actually the case in these experiments.

This method was proposed by DUONG and VIALLE and was first tested in the case of the D_1 line of natural Na [3.39]. Good signals were obtained (Fig. 3.11) and the feasibility of the experiment was thus demonstrated. Afterwards the method was applied with success to the study of short-lived isotopes of sodium (see *Experiments on the sodium resonance lines* of Subsection 3.4.1).

A common feature of these two methods of detection by magnetic deflection is that they are applicable only to paramagnetic atoms. In addition, it is impossible to observe those transitions for which atoms can fall back only on the same $|m_F\rangle$ state as the initial one, for instance the transition $F = 1 \rightarrow F = 0$ of ^{23}Na. However, it has been proved that the second method has a good sensitivity permitting one to measure the hyperfine structure with atoms in small quantities.

3.3.5 Detection by Photoionization (or by Field Ionization) from the Upper State

In this method, the atomic beam is submitted to a second light beam, the photons of which have just enough energy to photoionize the excited atoms from the upper level of the transition under study, and the ions so produced are collected. The number of ions is then proportional to the number of transitions induced by the tunable laser. Since ion counting is a very efficient process one can expect that the method is usable. However photoionization cross sections from excited levels are usually very small, of the order of 10^{-18} cm^2 and the probability of ionization from the excited level is usually much weaker than the probability of excitation. The overall efficiency of this type of detection is then entirely determined by the efficiency of the ionization. An example of the use of this method for high-resolution spectroscopic studies has been given by DUONG et al. [3.40] in the case of sodium D lines. The source used for photoionization was a high-pressure mercury lamp giving its maximum of energy around $32\,000$ cm^{-1} (with filters opaque to the UV), the ionization limit being at $25\,500$ cm^{-1}. Although very good recordings were obtained with natural sodium, it is by no means certain that, with the presently available sources for photoionization, the method could work with atoms in small quantities. Advances with this method will strongly depend on the extension of the spectral range of powerful tunable lasers. The same process has been proposed and demonstrated with success [3.41, 42] for isotope separation, a problem very close to the preceding one. Here also the success of the process depends on the availability of suitable ionizing sources. Decisive progress can be made if there exist autoionizing states since the photoionization cross section can be increased by several orders of magnitude by tuning the ionizing source to one of these states. Such states have been studied in particular by STEBBINGS and DUNNING [3.43] in the case of rare gases.

An interesting observation has also been made by BRINKMANN et al. [3.42] in the case of calcium. They used a Ca atomic beam source in which a high density of metastable $4s4p\,^3P_2$ (5×10^9 atoms/cm^3) is produced by a dc discharge following the oven [3.36]. A dye laser ($\lambda = 6162$ Å) excites the atoms from the metastable to the short-lived $4s5s\,^3S_1$ state and an Ar$^+$ laser is used to produce the photoionization.

When the $4880\,\text{Å}$ line of Ar^+ is used, the broad autoionizing level $3d5p\,^3P_1$ is approximately reached (within $118\,\text{cm}^{-1}$) and the photo-ionization is greatly enhanced. The authors reported that in this experiment 10^{-2} of the atoms in the metastable state are ionized. Autoionizing states have also been found by SOLARZ et al. [3.43] in the case of uranium.

In the special case where highly excited states ("Rydberg states") very close to the limit of ionization are studied, the photoionization cross section is very weak being roughly proportional to n^{-3}. DUCAS et al. [3.44] have shown, however, that in this case field (or Stark) ionization provides close to 100% detection efficiency and very low background; for instance, an electric field of $300\,\text{V/cm}$ is sufficient to ionize a Na atom in the 35 s state.

3.3.6 Combination of Saturated Absorption and the Atomic Beam

Although each of these two methods permits suppression of the first-order (or longitudinal) Doppler effect it can be advantageous to use them simultaneously in some very special cases where the shift due to the slightest lack of orthogonality of the two beams must absolutely be avoided. One example will be given in Subsect. 3.4.1, another in Subsect. 3.4.5.

3.4 Examples of High-Resolution Experiments

The availability of single-mode tunable lasers is still quite recent. Moreover the wavelengths coverage of easy-to-operate lasers is still rather limited so that high-resolution experiments have so far been restricted to a few particular cases. Many of these experiments have thus a methodological character, and are made on transitions easily accessible to the dye laser working with rhodamine 6G as in sodium atoms, or to the Ar^+ laser as in iodine molecules I_2. However some experiments have permitted the study of interesting phenomena. The evolution may be quite rapid and a rich harvest of new results will probably be obtained in the next few years.

Without pretending to be exhaustive, we now examine different high-resolution spectroscopy experiments which use both lasers and atomic beams. Some of these experiments have already been mentioned in the preceding sections while reviewing the methods of detection.

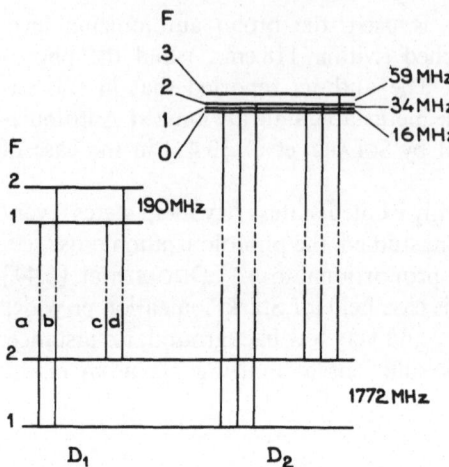

Fig. 3.12. Hyperfine structure of the D_1 and D_2 lines of natural sodium

3.4.1 Atomic Fine or Hyperfine Structure and Isotope Shift

Experiments on the Sodium Resonance Lines

The lines represent an ideal case for testing the different methods of Doppler-free spectroscopy; their hyperfine structure is recalled on Fig. 3.12. It can be seen that the coarser features, depending on the hyperfine splitting (1772 MHz) of the ground level $^2S_{1/2}$, are just within a normal Doppler width at 600 K whereas the smaller separation, 15 MHz, is a little larger than the natural linewidth, 10 MHz. In addition, the wavelength lies in the best working region of dye lasers. The first experiments at high resolution on these lines were made by HÄNSCH et al. [3.45] by saturated absorption in sodium vapor and the smallest linewidth obtained was about 40 MHz.

Observations with atomic beams and detection by fluorescence were made by SCHUDA et al. [3.46], HARTIG and WALTHER [3.47] and LANGE et al. [3.24] with about the same arrangement. The highest resolution was achieved by LANGE et al. with a laser giving a linewidth of 7 MHz and a beam having a collimating ratio of 400/1, corresponding to a residual Doppler width of 5 MHz. Their result is reproduced in Fig. 3.13 where it can be seen that the two components separated by 15 MHz are clearly resolved. It should be noted that in these last experiments the exciting light was attenuated to less than 1 mW/mm^2 in order to obtain this resolution; at higher intensities the lines were power broadened, and intensity distortions produced by optical pumping appeared.

The same lines have been studied with non-optical detection of the transition, as mentioned in Subsections 3.3.3, 3.3.4, and 3.3.5, by DUONG

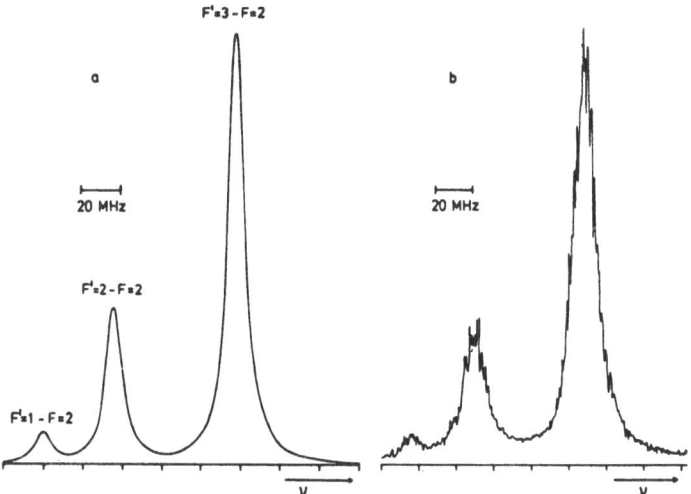

Fig. 3.13. Portion of the hyperfine structure of the Na D_2 line recorded by fluorescent detection. a) Calculated signals with natural linewidth. b) Recorded signal, 3 sweeps averaged, sweep time 8 s. From [3.24]

Table 3.1

A	20	21	22	23	24	25
$T_{1/2}$	0.4s	23s	26y		15h	60s
I	2	3/2	3	3/2	4	5/2
R	10^{-3}	0.1	1		0.88	0.18

et al. [3.35] (detection by magnetic deflection in a Rabi-type arrangement), by JACQUINOT et al. [3.32] (beam deflection by recoil) and by DUONG et al. [3.40] (photoionization). In all these experiments the linewidth obtained was of the order of 20–50 MHz, but the signal-to-noise ratio was very high.

Experiments under more severe conditions have been recently made on some unstable, short-lived isotopes of sodium. A long chain of these isotopes is known, from $A = 19$ to $A = 33$, and their masses have been measured with great accuracy [3.48]. Only those ranging from $A = 20$ to $A = 25$ have their nuclear spin known. The spins and lifetime are given in the table below. The table gives also the relative production rate R in the spallation reaction ^{27}Al, $(p, 3pxn)$ Na with 150 MeV protons.

The hyperfine structure of the ground level of all these isotopes had already been measured by magnetic resonance (quite recently $|\mu_I|$ has been determined for ^{25}Na [3.49]) but no spectroscopic study had ever been

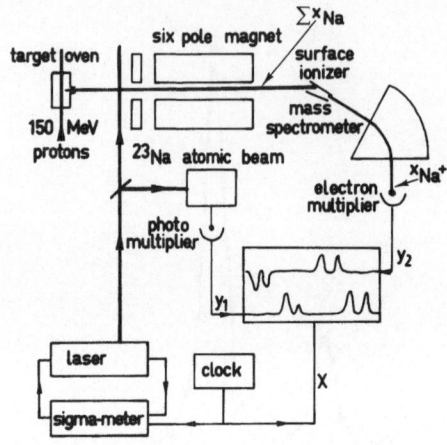

Fig. 3.14. General diagram of the experiment on unstable isotopes of sodium. From [3.50]

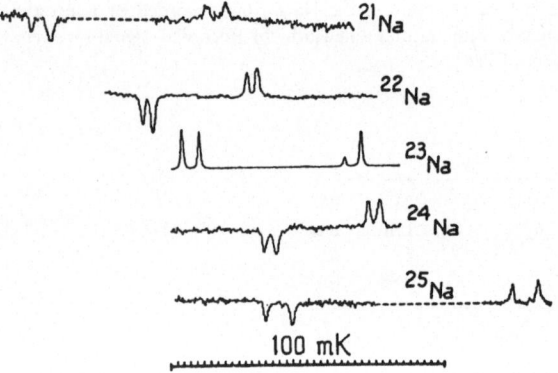

Fig. 3.15. Hyperfine structure and isotope shifts of $^{21-25}$Na isotopes. For each isotope a single scan has been made together with the natural ^{23}Na line (detected by fluorescence) and the four recordings have been put together on a common wavenumber scale. Scanning time was about 10 mn. From [3.50]

done on them. The interest of such optical studies is to give the nuclear electric quadrupole moment, reflected in the structure of the $^2P_{3/2}$ level, and also the isotope shift. Preliminary experiments have been made by HUBER et al. [3.50] on the isotopes 21, 22, 23 and 25 produced with the synchrocyclotron at Orsay. Due to the low production rate of these isotopes the method of detection by deflection of trajectories in a six-pole magnet (cf. Subsect. 3.3.4) was chosen because of its good sensitivity. The collimation ratio was chosen as low as 30/1 so that the limit of resolution was larger than 40 MHz, but this was judged to be sufficient for the desired results. The general scheme of the experiment is given in Fig. 3.14. In this first series of measurements only the D_1 line was studied. Under the conditions of production of the isotopes, the total flux leaving the oven carries about 10^8 atoms/s of ^{25}Na which corresponds to a very low density of the order of 1 atoms/cm^3 in the beam. Because of different losses in the whole apparatus (mass spectrometer and so on), the corresponding counting rate at the detector was of the order of 1000 per second. For each isotope the hyperfine structure of its D_1 line was recorded in a single scan, of about 10 mm, together with the corresponding line of ^{23}Na observed with a separated atomic beam by fluorescent detection. The results for all the isotopes are shown together in Fig. 3.15. A new series of experiments has been made quite recently by the same authors, but not yet published; here the D_2 line was studied using σ_- light for all isotopes. This line has been resolved for ^{21}Na and ^{25}Na, the magnetic moments of which are larger than for ^{23}Na.

Measurements on Other Resonance Lines

The $6s^2\,^1S_0 - 6s6p\,^3P_1$ ($\lambda = 555.6$ nm) transition of natural ytterbium has recently been studied by BROADHURST et al. [3.51]. The residual Doppler width due to the collimation was 5 MHz, and the frequency jitter of the cw dye laser was 17 MHz. Isotope shifts and hyperfine structure of all the natural isotopes have been measured with an accuracy of ± 0.6 MHz which is about an order of magnitude better than with conventional spectroscopy. The hyperfine anomaly between ^{171}Yb and ^{173}Yb was found to be different ($-0.47 \pm 0.02\%$ instead of $-0.367 \pm 0.009\%$) from that determined by a level crossing technique. This discrepancy has not been explained.

Measurements on Non-Resonant Levels

Those levels which are not accessible by a direct optical transition from the ground state can be reached either by a two-photon transition or by a stepwise excitation. The first possibility is treated in a special chapter of this book and has many attractive features. In the case of stepwise excitation the first step can be either a resonant absorption or the excitation of a metastable state by electron bombardment.

A) *Neon*. Electron bombardment has been used by CHAMPEAU and KELLER [3.20] to study some levels of the $2p^5 3p$ configuration of neon. The levels involved in their experiment are represented in Fig. 3.16. The aim of the experiment was the study of the hyperfine structure of ^{21}Ne which is of great theoretical interest because of the influence of relativistic corrections and of configuration mixing. A gaseous atomic beam of the

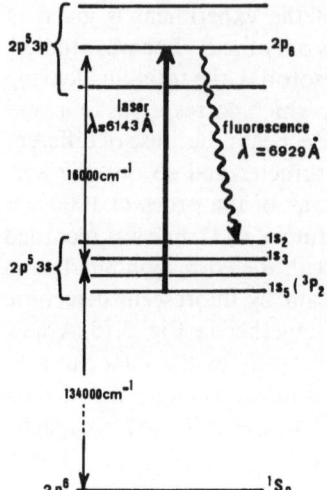

Fig. 3.16. Levels involved in the experiment on neon. The transition $134000\,\mathrm{cm}^{-1}$ is made by electron bombardment. From [3.20]

type described by STANLEY [3.4] was used with enriched ^{21}Ne. The collimation ratio was 25/1 and the density of the order of 10^{14} atoms/cm^3. The 30 mA electron beam was accelerated by 500 volts. The electron beam, the laser beam, and the atomic beam are perpendicular to each other but do not cross at the same point. After leaving the collimation slit, the atoms first cross the electron beam and, a few millimeters down-stream, the laser beam. The observation of the fluorescence is made only at this second crossing point. The emission of the atomic beam at the point where it crosses the electron beam is thus avoided, without losing many of the metastable atoms.

In such an experiment where the excitation to the intermediate level is not isotropic, the different Zeeman states $|m_F\rangle$ are not populated so that the population of the upper level itself has an isotropic distribution, depending on the polarization of the laser light. The total fluorescence light emitted is related only to the total population of each hyperfine upper level, but this is no longer true if the light is collected in a limited solid angle. As a consequence the relative intensities observed are not the same as in an experiment made on a resonance line. This had to be taken into consideration by CHAMPEAU and KELLER in the fitting procedure to reconstitute the recorded profile by a number of hyperfine components. The intensities were considered as free parameters instead of giving them the values calculated using isotropic population and relative transition probabilities. An example of the results obtained is shown in Fig. 3.17. From these measurements, values of the hyperfine constants A and B

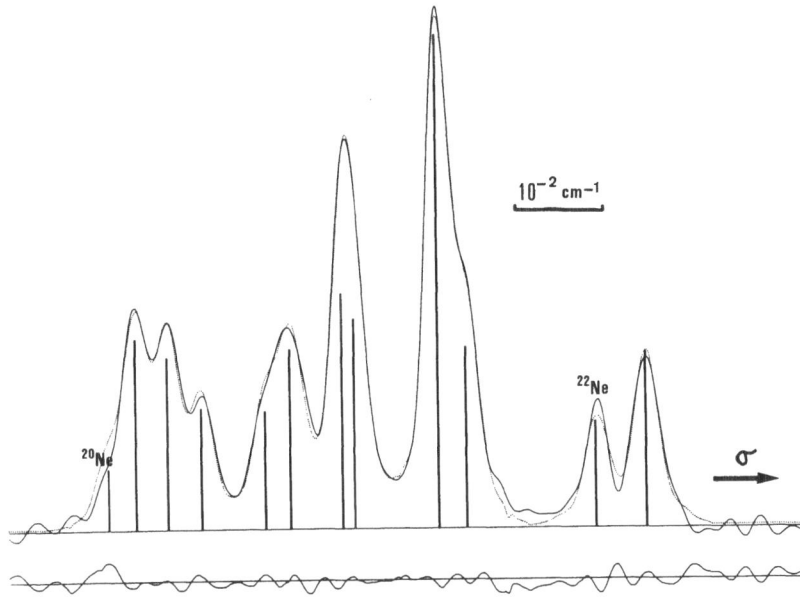

Fig. 3.17. Hyperfine structure of the line $\lambda = 614.3$ nm (solid line: direct recording, filtered); (dotted line: pattern fitted by a calculated profile); (the lower trace shows the difference). From [3.20]

for the three $2p^5 3p$ levels indicated in Fig. 3.16 were deduced with an precision of ± 0.1 cm^{-1}.

B) *Hydrogen*. A sophisticated experiment was undertaken by BAR-GER et al. [3.21] in order to measure the Rydberg constant by means of the Hα line, with an accuracy of a few parts in 10^{10}. An atomic beam is used to eliminate systematic errors due to collisions or fields encountered in discharges. Saturated absorption is used to ensure that there is no Doppler shift due to a lack of orthogonality. The H atoms are produced with an rf discharge in a cell having a multi-channel slit (see Subsect. 3.2.2). Then the atoms pass through an electron excitation region where a few percent are excited onto the $2^2 S_{1/2}$ state. The natural linewidth of the hyperfine component of Hα, due to the rather short-lived $3^2 P$ level, is 30 MHz, a value too high for the accuracy desired. But a transition is produced between the $3p$ and the $3s$ levels by a resonant rf field. In this way, as has been shown by ROBERTS and FORTSON [3.52], the linewidth of the double quantum transition $2S \rightarrow 3S$ depends only on the lifetime of the $2S$ and $3S$ levels which is much longer than for $3P$ so that the linewidth is reduced to 1 MHz.

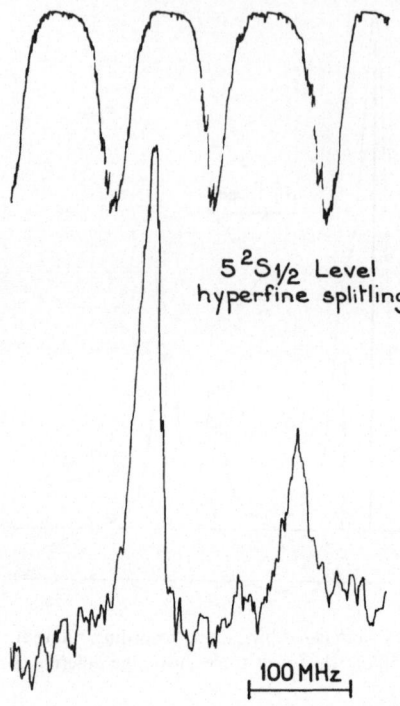

5 $^2S_{1/2}$ Level
hyperfine splitting

100 MHz

Fig. 3.18. Hyperfine structure of the $5\,^2S_{1/2}$ level of Na obtained by two-step excitation. The upper trace gives the calibration from a spherical Fabry-Perot. From [3.53]

C) *Measurement of the Hyperfine Structure of the $5\,^2S_{1/2}$ Level of Na.* This level is connected to $3\,^2P_{1/2}$ which is not metastable. An experiment has been done by Duong et al. [3.49] on this transition at 615.4 nm. The $3\,^2P_{1/2}$ state was populated by a first dye laser of fixed frequency adjusted to the hyperfine transition $F=1\leftrightarrow F=2$ or $F=2\leftrightarrow F=2$ of the D_1 line at 589.6 nm. A second dye laser was scanned (frequency: v_2) continuously over the entire structure of the $3\,^2P_{1/2}\leftrightarrow 5\,^2S_{1/2}$ line at 615.4 nm. The transitions were detected by the fluorescence from $5\,^2S_{1/2}$ on 614.5 nm. (It would have been much preferable to use only the decay to lower levels, so as to avoid stray light from the second laser and obtain a much better signal/noise ratio; but this was only a preliminary experiment.)

In the recordings obtained, the signal/noise ratio is not very good, (Fig. 3.18) for the reason explained above, and the linewidth of 24 MHz was consistent with the collimation ratio (300/1) and the combined linewidths of the two lasers. From such recordings, the hyperfine structure of the $5\,^2S_{1/2}$ level can be directly deduced (159 ± 6 MHz with 10 recordings).

D) *Fine Structure of the $3\,^2D$ Level of Li.* An experiment of the same kind has been made by HARTIG et al. [3.54] to determine this fine structure which is of the order of 1000 MHz and thus narrower than the Doppler width. In this experiment three dye lasers were used. The first one populated the $2\,{}^2P_{3/2}$ level; the second one was locked to the transition $2\,{}^2P_{3/2} - 3\,{}^2D_{5/2}$ by the servo-system in use in the Walther's group [3.47]; the third laser was scanned through the transition $2\,{}^2P_{3/2} - 3\,{}^2D_{3/2}$ and the beat note between the lasers 2 and 3 was measured. In this way a direct measurement of the distance between the two levels of the fine structure was obtained with a precision of about ± 3 MHz.

It should be noted that an atomic beam is not strictly necessary for such experiments involving three levels, since cancellation of the Doppler broadening is obtained even in a vapor if the two laser beams are colinear. In this case, in effect, the first beam populates the intermediate level with atoms of only one class of velocity and only these atoms are available for the second transition. But spurious resonances occur if the intermediate level has several components within the Doppler width since in that case the different sublevels are populated by atoms belonging to different classes of velocity. It is easy to show that, if there are two intermediate levels separated by δv ($\delta v \ll \delta v_{\text{Dopp}}$), there appears a spurious resonance at a frequency

$$v_2' = v_2 + \delta v(v_2/v_1 - 1) \tag{3.21}$$

when the two laser beams propagate in opposite directions. This spurious signal would disappear if $v_2 = v_1$. This is precisely the case of a two-photon transition except that in this latter case the intermediate level, which is virtual, can be considered as a continuum of components spread over more than the Doppler width, so that all atoms contribute to the signal without adding unwanted components. Coming back to the two-step process we see that it is preferable to use an atomic beam since v_2 is never equal to v_1.

We can now compare the use of a "two-step" process in an atomic beam to the use of a "two-photon" process in a vapor since both permit one to reach levels such as nS or nD which are not directly connected to the ground state by allowed transitions. It is well known that the two-photon method has been applied with considerable success (see Chapter 8). However it relies on the existence of a "relay" level not too far from the middle of the distance between the two levels to be connected, and usually the probability of the two-step transition will be much higher than for two-photon transition. Moreover the selection rules in two-photon transitions are more restrictive. For instance, the hyperfine structure of $5\,{}^2S_{1/2}$ cannot be measured directly but has to be obtained as the difference between a measured quantity and the known value of the hyperfine structure of $2\,{}^2S_{1/2}$.

This type of experiment lends itself very well to the study of such phenomena as the Autler-Townes effect; this will be explained in Subsection 3.4.4.

3.4.2 Molecular Spectra

This is the field where the largest number of very high resolution studies have been made by nonlinear methods, particularly saturated absorption. This can be seen in Chapter 4 and here we restrict ourselves to examining the experiments using molecular beams, which are much less numerous.

A) Absorption Spectroscopy on Polyatomic Molecules with Multiple Molecular Beams. CHU and OKA [3.26] have recently reported experiments on the Q (8, 7) transitions of NH_3 and R (4, 3) of $^{13}CH_3F$ in the medium infrared. The lasers were molecular lasers (N_2O and CO_2) swept over 75 MHz by displacement of one of the mirrors. The main feature of the experiments is that the detection is made by absorption instead of fluorescence. This is made possible by using several multiple channels so as to obtain an absorption equivalent to a 10 cm path length in the vapor at 10^{-4} Torr pressure. This is obtained by traversing five beams, each of them (2 cm × 0.5 cm) being produced by a bundle of many tubes 0.18 cm long and 5 μm radius. The divergence of these beams was less than 0.1 rad. The observed absorption linewidth was 2.5 MHz with a signal/noise ratio of 30 at 300 ms time constant.

B) Experiments on I_2. This molecule has been extensively studied by saturated absorption. In the experiments of EZEKIEL and his group [3.6], YOUMANS et al. [3.55], RUBEN et al. [3.56], and WU et al. [3.57] the use of a molecular beam and of carefully controlled lasers led to reduction of the observed linewidth to 350 kHz (HWHM). Both the argon-ion laser at 514.5 nm and dye lasers have been used and continuously improved since the first publication of EZEKIEL and WEISS in 1968 which described for the first time use of a molecular or atomic beam in laser spectroscopy. In one of the most recent papers a jet-stream continuous dye laser was used and extreme care had been taken to reduce the frequency jitter; the tuning range of this laser was 1.5 GHz. The collimation ratio was 2000/1, giving a residual Doppler width of 200 kHz. The observed linewidth, 350 kHz, is in good agreement with this residual Doppler width combined with the natural width of about 100 kHz and the laser jitter. An example of a hyperfine transition recording is given in Fig. 3.19.

Quite recently the same group reported a still higher performance [3.58]. By using a collimation ratio as high as 10^4 and a slow scanning, in order to have a good signal/noise ratio in spite of the low density of the beam, the linewidth was reduced to 85 kHz which is close to the natural

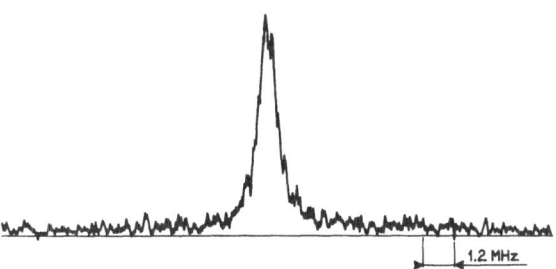

Fig. 3.19. Hyperfine transition in a molecular beam of $^{127}I_2$ (scan rate 1.4 MHz/s). Redrawn from [3.57]

width predicted from lifetime and Hanle effect measurements. In order to achieve a slow scanning, a 514.5 nm argon laser was locked to a transition in a molecular beam of I_2, giving a long-term stability of a few parts in 10^{14} and the frequency of this laser was shifted by an acousto-optical effect. Such high performance is possible only in molecular spectroscopy because the states involved have much longer lifetimes than do atomic states.

As an example of application of this technique, results on quadrupole coupling constants and spin-rotation interaction on $P(13)$ and $R(15)$, 43-0, $^3\Pi_{0_u^+} - ^1\Sigma_g^+$ transitions in $^{127}I_2$ have been measured and interpreted by RUBEN et al. [3.56].

3.4.3 Studies of Photoionization

Photoionization is not such a selective phenomenon even in the case of autoionizing levels that it is necessary to eliminate the Doppler broadening. However, atomic beams have been used in several experiments aimed at studying the phenomenon of photoionization itself on excited states [3.43, 59]. The reasons for choosing a beam in preference to a vapor may be of pure convenience. However, if absolute values of cross sections are desired, it is much easier to know the atomic density in the excited state and the interaction volume if an atomic beam is used.

A series of experiments is now being prepared at Laboratoire Aimé Cotton at Orsay on photo and field ionization and we have chosen to use an atomic beam for the above reasons. Moreover the use of an atomic beam becomes a necessity if states with very high principal quantum number (Rydberg states) are studied, because in these states atoms have such large radii (increasing as n^2) and become so fragile that a collisionless medium like an atomic beam is necessary. This is perfectly exemplified in a recent work by DUCAS et al. [3.44] who have studied the field ionization of *ns* and *nd* states of sodium for very high values of *n*. These states were excited stepwise by two pulsed dye lasers, the first one being tuned to the resonance line so as to populate the $3\,^3P_{3/2}$ state. With a low density $(10^8/\text{cm}^3)$ atomic beam, *ns* states up to $n = 37$ have been studied.

3.4.4 Interaction of Atoms with Intense Light Fields

Such effects as power broadening or dynamic Stark effect (Autler-Townes effect) may be observed in the optical range with lasers even if their power is moderate. The problem may be considered from two different points of view. One can use very powerful, non-tunable optical lasers, and observe important dynamic Stark effects even in conditions very far from resonance. On the contrary one can use tunable lasers of much lower power and make a high-resolution analysis of the effects. This approach allows one to work in much better defined conditions, but it demands that effects like Doppler or collision broadening be avoided in order not to mask the small effects expected. This is why atomic beams are ideal tools for this type of study. In the following we examine experiments performed or suggested, using atomic beams and tunable lasers.

A) *Spontaneous Emission of Atoms Illuminated by an Intense Monochromatic Light.* This is an important subject since the existing theories do not permit calculation of the spectral distribution of the emitted light for any value of the incident intensity varying from weak (i.e., corresponding to $\chi \ll 1$, χ being the parameter used above) to very strong. They almost all predict a purely monochromatic emission at vanishing power and a triplet due to the dynamic Stark effect at very high power. But there are still discrepancies concerning intensities and widths of the two sidebands compared to the central one, the width of which is the natural width γ. By a semi-classical treatment MOLLOW [3.60] found a width $3\gamma/2$ and a relative peak intensity $1/3$ for the sidebands. A treatment using quantum electrodynamics has been given by STROUD [3.61], but it has been carried out only for the initial transient regime and does not give the long-term solution. A spectrum with two sidebands was predicted by this treatment, but the ratio of their heights to the central peak is $2/1$. However a steady-state-quantum treatment has been performed by OLIVIER et al. [3.62] who have confirmed by this way MOLLOW's result.

The first attempt to analyze experimentally this fluorescence spectrum was made by SCHUDA et al. [3.63]. An atomic beam of sodium with a collimation ratio 500/1 was illuminated with a dye laser giving an intensity of about $500\,\text{mW/cm}^2$ and the re-emitted light was analyzed by means of a confocal Fabry-Perot interferometer with a limit of resolution of 20 MHz. The component $F = 2 \leftrightarrow 3$ of the D_2 line was chosen because no optical pumping toward any other $|F\rangle$ state is possible with it. The experiments were made at resonance and for different values of the detuning up to 100 MHz. The main results were the following:

– the central peak is at the same frequency as the laser even if this frequency differs from the resonance frequency;

– two sidebands were observed; they are not completely resolved at resonance but for large detunings they are further from the central peak and well separated, but weaker.

These results are still preliminary and a more complete analysis is to follow. The main difficulties in this type of experiment are the need of a very high resolution for analyzing the fluorescence light and the fact that all the atoms must interact with a field of the same intensity in order not to smear out the sidebands.

The above experiment is the only one to have been published on this subject. But WALTHER and coworkers are carrying out two experiments: in one of them [3.64] the atomic beam is placed inside a confocal Fabry-Perot, in the other [3.65] the fluorescence light is analyzed by photon correlation.

B) *Modulation Resonance of the Scattered Light Under Modulated Monochromatic Excitation*. There is another possibility of detecting the Autler-Townes effect in the visible range by exciting a two-level system by a laser beam weakly modulated at low frequency. ARMSTRONG and FENEUILLE [3.66] have made a calculation showing that the scattered light, modulated at the same frequency, has an amplitude of modulation which shows a resonant behavior with a maximum at the frequency of the Autler-Townes splitting. This can be physically interpreted in the following manner. The modulated strong laser beam is equivalent to a triplet of monochromatic waves. The strong central component produces the Autler-Townes splitting of the upper level whereas the two weak sidebands act as a probe which excites coherently the two components of the Autler-Townes doublet if the modulation frequency has the correct value. One then sees the beat note between the two waves emitted from the two components of the doublet.

Like the preceding one, this experiment can be done only with an atomic beam since the Autler-Townes splitting produced by the available tunable lasers is of the order of a few natural linewidths. This experiment has not actually been done.

C) *Observation of the Autler-Townes Effect in a Two-Step Resonance Experiment*. In an experiment like the one described in Subsection 3.4.1 under C it would be possible to observe the Autler-Townes effect by using a powerful laser for the first transition and a weak one for scanning the second transition. The conditions of such an experiment have been studied in a theoretical paper by Feneuille and Schweighofer [3.67]. Although the effect could also be observed in a vapor by using two laser beams of opposite directions, the use of an atomic beam would provide better defined and controllable conditions.

3.4.5 Measurements of the Relativistic Second-Order Doppler Shift

The special theory of relativity predicts a second-order shift $\Delta v = -v \cdot v^2/2c^2$ for atoms moving perpendicular to the direction of observation. This effect has already been measured by optical experiments but with an accuracy limited to a few percent. By laser spectroscopy it should be possible to increase the accuracy. But it is necessary to get rid of any residual first-order effect by using only atoms with an exactly transverse velocity. This is obtained in the experiment undertaken by Snyder et al. [3.68] by using simultaneously a fast atomic beam and the method of saturated absorption. One is thus assured that only atoms traveling perpendicular to the laser beam interact with it. The atomic beam is a slightly divergent high-velocity beam of light atoms or ions in a metastable state, and the transitions observed are

$$Ne(1s_5 - 2p_2) \quad \text{at} \quad 5882 \text{ Å}$$
$$H(2s^2 S_{1/2} - 3p^2 P_{1/2}) \quad \text{at} \quad 6563 \text{ Å}$$
$$Li^+(2^3 S - 2^3 P) \quad \text{at} \quad 5485 \text{ Å} .$$

The beam is intersected by a laser-produced standing wave approximately perpendicular to it, and the fluorescence is observed while tuning the frequency of this wave across the resonance. A dip in the fluorescence appears at the resonant frequency shifted by the transverse Doppler effect.

For the 5882 Å line of Ne this shift is -26 kHz per electron-volt of energy of the atomic beam, and atomic beams of energy up to 50 keV will be used. Under these conditions an accuracy of the order of 10^{-4} to 10^{-5} is expected by the authors.

3.5 Miscellaneous

3.5.1 Velocity Selection by Means of Lasers

In most experiments, the fact that there is a velocity distribution in an atomic beam has no effect since the atoms are observed transversely.

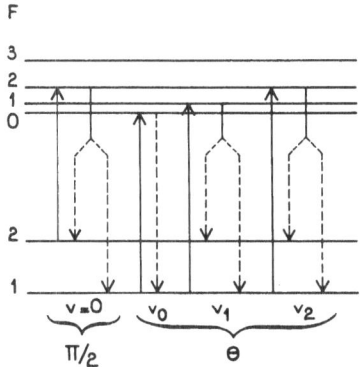

Fig. 3.20. Transition involved in a velocity selection in a Na atomic beam. $\pi/2$ corresponds to a perpendicular laser beam and θ to an oblique laser beam operating downstream. v_0, v_1, v_2 are the atomic velocities [3.10]

However in some of them it may be useful to have a definite velocity or at least a narrower distribution. This is the case if one uses the Doppler effect as a means of shifting frequencies, or if one uses the motion of atoms to observe phenomena depending on time.

Different types of velocity selectors have already been used. Some of them use a system of slotted disks rotating at high speed (see for instance, [3.15]); it is also possible to use the Stern-Gerlach effect [3.69]. In both cases there is an important loss of atomic flux. One may hope to have better characteristics by using the interaction with a laser crossing the atomic beam under oblique incidence.

A) A first method has been proposed by SCHUDA and STROUD [3.10] by which sodium atoms in the hyperfine substate $|F = 2\rangle$ may be selected in a narrow velocity range (Fig. 3.20). This is obtained by two successive optical pumpings. The first one, produced by a laser beam orthogonal to the atomic beam and exactly tuned to the transition $|F = 2\rangle \rightarrow |F' = 2\rangle$, pumps *all the atoms* into the state $|F = 1\rangle$. A second laser beam, issued from the same laser, crosses dowstream the atomic beam at a large angle and in such a direction that the atoms of correct velocities v_0, v_1, and v_2 see the laser frequency upshifted by an amount sufficient to produce the transitions $|F = 1\rangle \rightarrow |F' = 0\rangle$, $|F = 1\rangle \rightarrow |F' = 1\rangle$, and $|F = 1\rangle \rightarrow |F' = 2\rangle$. As seen in the figure, the first transition does not give rise to optical pumping, whereas the two others pumped the atoms of velocities v_2 and v_3 into the $|F = 2\rangle$ state. Since the two levels $|F' = 1\rangle$ and $|F' = 2\rangle$ are very close (35 MHz), v_2 and v_3 are almost equal and the velocity distribution *for atoms* $|F = 2\rangle$ shows two sharp peaks barely resolved. The position of these peaks can be changed by varying the incidence of the second laser beam. In the experiment of SCHUDA and STROUD these peaks were located at $v_1 \simeq v_2 = (1.04 \pm 0.02)10^5$ cm/s, the oven temperature being $400°$ C and the angle of incidence almost $180°$. The collimation ratio was $500/1$.

A refinement of this experiment has been proposed by Abate [3.70] who uses a third optical pumping and a weak magnetic field to pump all the atoms in the state $|F=2, m_F= +2\rangle$. In this way sodium atoms behave as a simple two-level system with non-degenerate ground state.

The above experiments refer to the very particular case of sodium. But the principle is general and can be adapted to many other cases.

B) Another possibility could be the use of the resonant recoil under oblique incidence. Only the atoms tuned to the laser frequency taking account of their velocity would be deflected and selected by a slit at a proper position. In the case of a true two-level system this device should operate rather simply. In the case of more complicated level structures, an analysis of the particular case should be made, as in the preceding experiment. Some transitions could be affected by optical pumping, thus giving small deflections; some others, not affected by optical pumping, could show a much larger deflection as explained in Subsection 3.3.2, and this would be a favorable situation. No attempt has been made to use the recoil effect to produce a velocity selection.

3.5.2 Atomic or Molecular Filters

It has been suggested by Youmans et al. [3.55] that laser-induced transitions in atomic beams could find an application in narrow band filters. Since absorption by atomic beams is usually extremely weak, fluorescence should be used to monitor the tuning of the incident light beam to the atomic transition. Indeed it is the method used for laser stabilization by atomic beams.

Such a "filter" could even be tunable by Doppler effect, by varying the angle between the two beams. But in this case the velocity distribution in the atomic beam tends to broaden the response curve of the filter (see Subsection 3.5.2). The response frequency v is given by $v=v_0+ vv \sin\alpha/c$. For instance, with a thermal sodium beam and $\alpha=45°$, $v-v_0$ is of the order of 3000 MHz. The width of the response is given by

$$\delta v=(v-v_0)\delta v/v \simeq v-v_0 \tag{3.22}$$

since the width δv of the velocity distribution is approximately equal to the mean velocity. This shows that the filter is tunable over only a very small range if a band pass of the order of the natural width is desired. Wide tunability and high selectivity could be reconciled by making a velocity selection of the type indicated above.

3.5.3 Laser Stabilization by Atomic Beam

This has been actually done by several authors. HARTIG and WALTHER [3.47] have locked a dye laser on a hyperfine component of the D_1 line of sodium. The laser was frequency modulated and the modulation of the fluorescence of an atomic beam was used as error signal giving the derivative of the absorption curve versus frequency. This signal was synchronously detected in a lock-in amplifier and fed to the frequency control of the laser. RYAN et al. [3.71] have stabilized a 514.5 nm argon-ion laser with a molecular beam of $^{127}I_2$ by using one of the hyperfine transitions referred to in Subsection 3.4.2, the natural width of which is of the order of 100 kHz. Here also the derivative of the fluorescence signal was used with a servo-loop very similar to the preceding one. Frequency drifts less than 20 kHz for a duration of 20 min were obtained and measured by observing the fluorescence of a second molecular beam identical to the first one.

A somewhat different technique has been used by the same group [3.57, 72, 73]. Here the laser was locked to an external Fabry-Perot cavity and the cavity itself was locked to the atomic beam. By this method a stability of 300 Hz for an integration time of 25 s has been obtained with a jet-stream dye laser, and a stability of 70 Hz for an integration time of 300 s with an argon laser. The last figure corresponds to a stability of 10^{-13}; the reproducibility was 7×10^{-12}.

Stabilization by atomic beam is a very attractive technique since collision shift and broadenings as well as Doppler broadening are avoided. However there may be residual Doppler shift if the two beams are not exactly orthogonal. This can be avoided if the laser beam is exactly reflected back on itself by a mirror; a *small* error of orthogonality gives then a slight broadening of the response, but no shift. Another possibility is to combine saturated absorption with atomic beam, as in the experiment referred to in Subsection 3.4.5.

References

3.1 D. A. JACKSON, H. KUHN: Proc. Roy. Soc. **A167**, 205 (1938)
3.2 D. A. JACKSON, DUONG HONG TUAN: Proc. Roy. Soc. **A280**, 323 (1964)
3.3 K. W. MEISSNER: Rev. Mod. Phys. **14**, 68 (1942)
3.4 R. W. STANLEY: J. Opt. Soc. Am. **56**, 350 (1966)
3.5 A. I. ODINTSOV: Opt. Spectr. **6**, 250 (1959)
3.6 S. EZEKIEL, R. WEISS: Phys. Rev. Letters **20**, 91 (1968)
3.7 N. F. RAMSEY: *Molecular Beams* (Oxford University Press, London 1956)
3.8 H. KOPFERMANN: *Nuclear Moments* (Academic Press, New York 1958)
3.9 PH. CAHUZAC: Physica **67**, 567 (1973)
3.10 F. SCHUDA, C. R. STROUD, JR.: Opt. Commun. **9**, 14 (1973)

3.11 H.J.ANDRÄ, M.GAILLARD, L.HENKE, M.KRAUS, J.MACEK, W.WITTMANN: 4th Intern. Conf. Atomic Physics Heidelberg (July 1974) p. 168

3.12 S.BASHKIN (editor): *Topics in Current Physics*, vol. 1: Beam Foil Spectroscopy (Springer Berlin, Heidelberg, New York 1976)

3.13 H.FIGGER, K.SIOMOS, H.WALTHER: Z. Physik **270**, 371 (1974)

3.14 S.BÜTTGENBACH, G.MEISEL, S.PENSELIN, K.H.SCHNEIDER: Z. Phys. **230**, 329 (1970); S.BÜTTGENBACH, G.MEISEL: Z. Physik **244**, 149 (1971)

3.15 J.P.ANDERSON, R.P.ANDRES, J.B.FENN: Advan. Atomic and Molecular Phys. **1**, 345 (1965)

3.16 S.M.TRUJILLO, P.K.ROLL, E.W.ROTHE: Rev. Sci. Instr. **33**, 841 (1962); H.V.HOSTETTLER, R.B.BERNSTEIN: Rev. Sci. Instr. **31**, 872 (1960)

3.17 J.G.KING, J.R.ZACHARIAS: Advan. Electron. Electron Phys. **8**, 2 (1956)

3.18 J.A.GIORDMAINE, T.C.WANG: J. Appl. Phys. **31**, 463 (1960)

3.19 R.MINKOWSKI, H.BRUCK: Z. Physik **95**, 274 (1935)

3.20 R.CHAMPEAU, J.C.KELLER: J. Phys. (Paris) **36**, L161 (1975)

3.21 R.L.BARGER, T.C.ENGLISH, J.B.WEST: Proc. VIIIth A.M.C.O. Conference, Paris (June 1975)

3.23 S.ISHII, W.OHLENDORF: Rev. Sci. Instr. **43**, 1632 (1972)

3.24 W.LANGE, J.LUTHER, B.NOTTBECK, H.W.SCHRÖDER: Opt. Comm. **8**, 157 (1973)

3.25 J.L.PICQUE: Private communication

3.26 F.Y.CHU, T.OKA: VIIIth Conf. Quantum Electronics, San Francisco (June 1974)

3.27 R.FRISCH: Z. Physik **86**, 42 (1933)

3.28 A.ASHKIN: Phys. Rev. Letters **25**, 1321 (1970)

3.29 J.L.PICQUE, J.L.VIALLE: Opt. Comm. **5**, 402 (1972)

3.30 R.SCHIEDER, H.WALTHER, L.WÖSTE: Opt. Comm. **5**, 337 (1972)

3.31 I.NEBENZAHL, A.SZÖKE: Appl. Phys. Letters **25**, 327 (1974)

3.32 P.JACQUINOT, S.LIBERMAN, J.L.PICQUE, J.PINARD: Opt. Comm. **8**, 163 (1973)

3.33 I.I.RABI: Phys. Rev. **87**, 379 (1952), see also: M.L.PERL, I.I.RABI, B.SENITZKY: Phys. Rev. **98**, 611 (1955)

3.34 R.MARRUS, D.McCOLM: Phys. Rev. Letters **15**, 813 (1965) H.T.DUONG, R.MARRUS, J.YELLIN: Phys. Letters **27B**, 565 (1968)

3.35 H.T.DUONG, P.JACQUINOT, S.LIBERMAN, J.L.PICQUE, J.PINARD, J.L.VIALLE: Opt. Comm. **7**, 371 (1973) U.BRINKMAN, J.GOSELER, A.STEUDEL, H.WALTHER: Z. Physik **228**, 427 (1969)

3.36 H.BUCKA: Z. Physik **191**, 199 (1966); see also: D.ZIMMERMANN: Z. Physik **224**, 403 (1969)

3.37 D.KLEPPNER, H.M.GOLDENBERG, N.F.RAMSEY: Phys. Rev. **126**, 603 (1962)

3.38 J.L.VIALLE: Private communication

3.39 H.T.DUONG, J.L.VIALLE: Opt. Comm. **12**, 71 (1974)

3.40 H.T.DUONG, P.JACQUINOT, S.LIBERMAN, J.PINARD, J.L.VIALLE: C. R. Acad. Sci. (Paris) **B276**, 909 (1973)

3.41 S.A.TUCCIO, J.W.DUBRIN, O.G.PETERSON, B.B.SNAVELY: VIIIth Intern. Quantum Electronics Conf., San Francisco, (June 1974)

3.42 U.BRINKMANN, W.HARTIG, H.TELLE, H.WALTHER: Appl. Phys. **5**, 109 (1974)

3.43 R.F.STEBBINGS, F.B.DUNNING: Phys. Rev. **A8**, 665 (1973), R.SOLARZ, L.CARLSON, C.MAY: 1975 Spring Conf. Opt. Soc. of Am., Anaheim, California

3.44 T.W.DUCAS, M.G.LITTMAN, R.R.FREEMAN, D.KLEPPNER: Private communication (to be published)

3.45 T.W.HÄNSCH, I.S.SHABIN, A.L.SCHAWLOW: Phys. Rev. Letters **27**, 707 (1971)

3.46 F.SCHUDA, M.HERCHER, C.R.STROUD: Appl. Phys. Letters **22**, 360 (1973)

3.47 W.HARTIG, H.WALTHER: Appl. Phys. **1**, 171 (1973)

3.48 R. KLAPISCH, C. THIBAULT, A. M. POSKANGER, R. PRIEELS, C. RIGAUD, E. ROEKL: Phys. Rev. Letters **29**, 1254 (1972)

3.49 M. DEIMLING, R. NEUGARD, H. SCHWEICKERT: Z. Physik **A273**, 15 (1975)

3.50 G. HUBER, C. THIBAULT, R. KLAPISCH, H. T. DUONG, J. L. VIALLE, J. PINARD, P. JUNCAR, P. JACQUINOT: Phys. Rev. Letters **34**, 1209 (1975)

3.51 J. H. BROADHURST, M. E. CAGE, D. L. CLARK, G. W. GREENLEES, J. A. R. GRIFFITH, G. R. ISAAK: J. Phys. B7, L513 (1974)

3.52 D. E. ROBERTS, E. N. FORTSON: Phys. Rev. Letters **31**, 1539 (1973)

3.53 H. T. DUONG, S. LIBERMAN, J. PINARD, J. L. VIALLE: Phys. Rev. Letters **33**, 339 (1974)

3.54 W. HARTIG, V. WIEKE, H. WALTHER: Opt. Comm., in press

3.55 D. G. YOUMANS, L. A. HACKEL, S. EZEKIEL: J. Appl. Phys. **44**, 2319 (1973)

3.56 D. J. RUBEN, S. G. KUKOLICH, L. A. HACKEL, D. G. YOUMANS, S. EZEKIEL: Chem. Phys. Letters **22**, 326 (1973)

3.57 F. Y. WU, R. E. GROVE, S. EZEKIEL: Appl. Phys. Letters **25**, 73 (1974)

3.58 R. E. GROVE, L. A. HACKEL, F. Y. WU, D. G. YOUMANS, S. EZEKIEL: VIIIth Intern. Conf. Quantum Electronics, San Francisco, (June 1974)

3.59 K. J. NYGAARD, J. D. JONES, R. E. HEBNER: VIIIth Intern. Conf. Phys. of Electronic and Atomic Collisions, Belgrade (1973)

3.60 B. R. MOLLOW: Phys. Rev. **188**, 1969 (1969)

3.61 C. R. STROUD, JR.: Phys. Rev. **A3**, 1044 (1971)

3.62 G. OLIVIER, E. RESSAYRE, A. TALLET: Lettere al Nuovo Cimento **2**, 777 (1971)

3.63 F. SCHUDA, C. R. STROUD, M. HERCHER: J. Phys. B7, L198 (1974)

3.64 H. WALTHER: Private communication

3.65 W. RASMUSSEN, R. SCHIEDER, H. WALTHER: VIIIth Intern. Conf. Quantum Electronics, San Francisco (June 1974)

3.66 L. ARMSTRONG, JR., S. FENEUILLE: J. Phys. **B8**, 546 (1975)

3.67 S. FENEUILLE, M. G. SCHWEIGHOFER: J. Phys. (Paris), in press

3.68 J. J. SNYDER, J. L. HALL, M. S. SOREM: VIIIth Intern. Conf. Quantum Electronics, San Francisco (June 1974)

3.69 V. W. COHEN, A. ELLETT: Phys. Rev. **52**, 502 (1937)

3.70 J. A. ABATE: Opt. Commun. **10**, 269 (1974)

3.71 T. J. RYAN, D. G. YOUMANS, L. A. HACKEL, S. EZEKIEL: Appl. Phys. Letters **21**, 320 (1972)

3.72 L. A. HACKEL, D. G. YOUMANS, S. EZEKIEL: Proc. Vth A.M.C.O. Conf., Paris (June 1975)

3.73 R. E. GROVE, F. Y. WU, S. EZEKIEL: Opt. Eng. **13**, 531 (1974)

4. Saturation Spectroscopy

V. S. LETOKHOV

With 38 Figures

In this chapter we consider the principles and methods of laser saturation spectroscopy for Doppler-broadened transitions, as well as the basic information obtained by this method.

4.1 Background Material

4.1.1 Historical Remarks

The discovery of saturation spectroscopy was connected with the first experiments of studying physical effects when the laser radiation interacted with the amplifying medium of the first gas laser created by JAVAN et al. [4.1]. Among them we should mention the works by BENNETT [4.2] and by LAMB [4.3]. The laser light burns a "hole" in the Doppler-broadened amplification line, and the laser output power decreases resonantly, when the laser frequency is tuned to the centre of the Doppler-broadened line. This effect was termed "the Lamb dip". Experimentally the Lamb dip was revealed in works of two independent groups at MIT [4.4] and by Yale [4.5]. The saturation method was further elaborated by three laboratories in the USSR and USA [4.6–8] that started a wide use of absorption saturation spectroscopy. They proposed to put a resonantly absorbing low-pressure gas cell into the laser cavity. Saturation of absorption in a standing wave laser field results in a narrow Lamb dip at the centre of the Doppler-broadened absorption line. Thus the laser output power has a narrow peak at the centre of the absorption line, termed often as "the inverted Lamb dip". The early experimental works in observing the narrow inverted Lamb dip were reported in Refs. [4.7–9]. In the subsequent works, other methods were suggested to improve considerably the usefulness of saturation spectroscopy.

4.1.2 Saturation Approach

Saturation spectroscopy, free of Doppler broadening, is an example of the great improvements in the methods of atomic and molecular spectroscopy which became practicable with the advent of lasers. This method of laser spectroscopy is one of the most efficient and promising in regards to both fundamental and applied works. The basis for saturation spectroscopy is a change in the velocity distribution of particles at the levels n and m when a coherent light wave acts upon the Doppler-broadened transition $n - m$. This approach gives the foundation for most experiments of laser spectroscopy inside the Doppler contour conducted in the last ten years. There are three main methods for obtaining narrow resonances: 1) saturated absorption resonances in a two-level transition; 2) absorption and emission resonances in transitions connected to either level m or n of the transition under saturation; 3) resonances observed in the total number of atoms (or molecules) in the levels n or m which interact with the laser field.

This chapter presents principles of saturation spectroscopy. Anyone who wants to familiarize himself in more detail with the methods and the theory of saturation spectroscopy may use the original papers, which are referred to below, as well as a monograph [4.10] and more comprehensive reviews [4.11, 12]. Yet the ideas and methods of saturation spectroscopy have been set forth in a more popular and accessible form in Refs. [4.13–15].

4.2 Interaction of a Laser Wave with a Doppler-Broadened Transition

In the present section we shall list in brief and give final formulas for basic resonance effects resulting from the interaction of the laser field (a running wave, a standing wave, a combination of a strong running and a weak counterrunning waves) with the Doppler-broadened transition. Also we shall consider both the case of simple two-level transition and resonance effects in two coupled transitions with a common level.

4.2.1 Hole in the Velocity Distribution Induced by a Traveling Wave

Assume that the Doppler-broadened transition between two levels interacts with a traveling light wave which has the form:

$$E(t, r) = \mathscr{E} \cos(\omega t + \varphi - kr) . \tag{4.1}$$

The field interacts most effectively with atoms (or molecules) which have a velocity v (see Subsect. 2.2.2):

$$|kv - \omega + \omega_0| \gtrsim \Gamma_B , \tag{4.2}$$

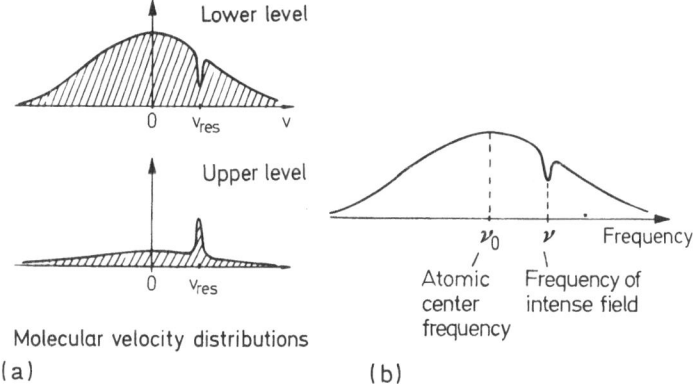

Molecular velocity distributions

(a)   (b)

Fig. 4.1a and b. Distribution of the projection of atomic velocities on the light wave direction in the lower and upper levels: $v_{res} = (\omega - \omega_0)/k$ is the projection of the velocity of atoms resonantly interacting with the laser wave of frequency $\omega = 2\pi\nu$, and $\omega_0 = 2\pi\nu_0$ is the atomic center frequency

where Γ_B is the resonance half-width at half-maximum (HWHM of the Bennett hole)

$$\Gamma_B = \Gamma(1 + G)^{1/2} , \qquad\qquad (4.3)$$

which increases with the saturation parameter G.

In the notation of Chapter 2 the parameter $G = (\tau/\gamma)|x|^2$, and $\Gamma = \gamma$. The same parameter determines the decrease in the absorption coefficient of the Doppler-broadened absorption line $(\Delta\omega_D \gg 2\Gamma_B)$ in the strong field of a running wave:

$$\kappa(\omega) = \kappa_0(\omega)/(1 + G)^{1/2} , \qquad\qquad (4.4)$$

where $\kappa_0(\omega)$ is the absorption coefficient per unit length for the weak field. Relation (4.4) can be used for a direct experimental estimation of the saturation parameter G without determining the constants, $x = \mu_{ab}\mathscr{E}/\hbar$, Γ, and τ, which determine the value of G. As mentioned in Subsection 2.2.2, the saturation of absorption results in the following: in the lower level there is a shortage of atoms which comply with the resonance condition (4.2), i.e., "hole burning", while in the upper level there is a surplus of atoms with the same velocity, i.e., a peak in the velocity distribution (Fig. 4.1a). As a result, the velocity distribution of the population difference can be written in the form:

$$n_1(v) - n_2(v) = [n_1^0(v) - n_2^0(v)]\left[1 + \frac{G\Gamma^2}{(\Omega - kv)^2 + \Gamma^2}\right]^{-1} , \qquad (4.5)$$

where $\Omega=\omega-\omega_0$, $\omega=2\pi v$, $\omega_0=2\pi v_0$, $n_i^0(v)=N_i^0 W(v)$ is the velocity distribution of population for the i-th level in the absence of field, N_i^0 is the total density of particles on the level in the absence of field, $W(v)$ is the Maxwell distribution of velocity projection on wave vector $\boldsymbol{k}$ which is given by (2.27). There is therefore a "hole" in the distribution of the population difference (4.5) for the atoms complying with the resonance condition (4.2). This corresponds to the "hole burning" in the Doppler contour (Fig. 4.1b) which was described by BENNETT [4.2].

4.2.2 Narrow Resonance of Saturated Absorption

During absorption only a small part of the atoms is excited at a resonance velocity. The light wave seems to set up a beam of excited particles with $kv=\omega-\omega_0$ in the gas. Just as the spectral line of a particle beam, if observed perpendicular, has no Doppler broadening (see Chapt. 3), so "an excited atomic beam" in gas induced by a strong running wave can be observed through the use of a properly oriented probe wave; information on the spectrum of such particles without Doppler broadening can thus be obtained. Most widely used cases of observation of narrow resonances by saturated absorption are discussed below.

1) Lamb Dip in the Standing Wave

Assume that the laser field is a standing plane wave which can be represented as a superposition of two oppositely propagating waves of the same frequencies:

$$
\begin{aligned}
E &= \mathscr{E} \cos(\omega t + \varphi - \boldsymbol{kr}) + \mathscr{E} \cos(\omega t + \varphi + \boldsymbol{kr}) \\
&= \mathscr{E}_s \cos(\omega t + \varphi) \cos\boldsymbol{kr},
\end{aligned}
\tag{4.6}
$$

where $\mathscr{E}_s=2\mathscr{E}$ is the amplitude of the standing wave. This field interacts with two groups of atoms with velocities which comply with one of the resonance conditions:

$$
\omega-\omega_0\pm\boldsymbol{kv}=0.
\tag{4.7}
$$

In the velocity distribution, and on the Doppler contour, these two groups occupy symmetric positions about the centre. If the detuning $\Omega=\omega-\omega_0$ is somewhat larger than the resonance half-width Γ_B, each running wave burns its "hole" independently from the other (Fig. 4.2a). The parameters of each hole and the saturated absorption of each running wave are described by the equations of Subsections 2.2.2 and 4.1.1, where the amplitude of the field in the saturation parameter G is to be taken as the amplitude of one running wave.

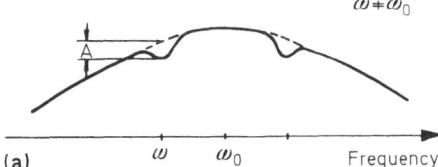

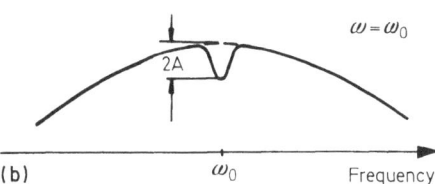

Fig. 4.2a and b. The shape of the Doppler contour in a standing light wave, when the frequency is shifted from the centre of the line (a), and in the case of exact resonance (b)

When the laser frequency lies at the centre of the Doppler line $(|\omega - \omega_0| \gtrsim \Gamma_B)$ the holes begin to overlap each other, and the same group of atoms interacts with two running waves (Fig. 4.2b). In the centre of mass system of the atoms the light waves have different frequencies $\omega \pm kv$. This corresponds to the fact that in the laboratory coordinate system any atom moves in a space-modulated standing light wave. The non-monochromaticity (in the centre of mass system) or the inhomogeneity (in the laboratory system) complicates greatly the study of the nonlinear resonant interaction. At the same time, the principal effect occurring in the standing wave, i.e., the occurrence of a resonance dip in the Doppler line centre (Lamb dip) of the saturated absorption coefficient for the standing wave, can be understood simply in terms of hole burning. BENNETT explained the Lamb dip in laser output this way in 1962. In fact, when the laser frequency is tuned to the line centre, the effective field acting upon the atoms with $kv = 0$ becomes twofold. Consequently the saturation parameter increases also by a factor of two and the absorption coefficient drops resonantly. This corresponds to the merging of two holes at $\Omega = 0$ and the formation of one deeper hole in the centre of the Doppler contour (Fig. 4.2b).

This effect was first investigated by LAMB [4.3] in the weak saturation approximation, where a perturbation method could be used. The saturated absorption coefficient of a standing wave with frequency ω has the form:

$$\kappa(\omega) = \kappa_0(\omega) \left[1 - \frac{G}{2} \left(1 + \frac{\Gamma^2}{\Gamma^2 + \Omega^2} \right) \right], \quad G \ll 1, \tag{4.8}$$

where G is the saturation parameter for one running wave. The degree of absorption saturation is equal to G at the Doppler line centre and it is equal to $G/2$ far from the resonance. The full width of the dip at the line centre corresponds to 2Γ.

In studying saturated absorption, the strong field case is of great importance. Saturation in the strong field of a standing wave was theoretically investigated by a number of authors: RAUTIAN and SOBELMAN [4.16], RAUTIAN [4.17], GREENSTEIN [4.18], STENHOLM and LAMB [4.19], FELDMAN and FELD [4.20], SHIMODA and UEHARA [4.21,22], BAKLANOV and CHEBOTAYEV [4.23]. In the case of an arbitrary degree of saturation, and of arbitrary detuning and relaxation constants, the problem can be solved only with the help of computer. It is possible to get an analytic solution in the particular case of exact resonance ($\omega_0 = \omega$) and equal relaxation constants ($\gamma_1 = \gamma_2 = \Gamma$). However, approximate methods enable us to get some idea of the intense standing wave interaction and to answer questions of practical importance. The complications in solving problems of this type can be explained by changes in the line shape of atomic emission and the level populations in the strong field. These phenomena cannot be treated separately. When two fields with frequencies ω_1 and ω_2 interact simultaneously, the induced polarization contains combination frequencies $\omega_1 \pm n (\omega_1 - \omega_2)$, where $n = 1, 2 \dots$. The polarizations at these frequencies in turn result in a modulation of the population difference. Equations (2.12) are interconnected by time dependent off-diagonal and diagonal elements of the density matrix, which are related directly to the polarization and the population of the levels, respectively.

In the rate-equation approximation one can ignore the well-known changes of the absorption—or emission—line shape of a particle which take place under the action of a strong field (oscillation of the probability amplitudes between the two levels at the Rabi frequency, $x = \mu\mathcal{E}/\hbar$). When $x \ll \Gamma$, we may neglect oscillations. But, the condition $x \ll \Gamma$ does not all mean that no saturation effects show up. If $\Gamma \gg 1/\tau$ (or $T_2 \ll T_1$, where $T_2 = 1/\Gamma$ is the transversal relaxation time, $T_1 = \tau$ is the longitudinal relaxation time), then, nevertheless, the saturation parameter may be rather large and saturation of the level population difference would occur. Ignoring the spatial inhomogeneity the absorption of a standing wave was studied by a number of authors [4.18, 21, 23]. They found expressions which are identical and differ only in their form. The shape of the Lamb dip when no coherence effects are taken into account can be expressed by

$$\kappa/\kappa_0 = (a_+ + a_-)^{-1}[1 + (\delta^2 + 1)^{1/2}/(1 + \delta^2 + 2G)^{1/2}] \qquad (4.9)$$

where

$$a_\pm = \{1 + G - \delta^2 \pm [G^2 - 4\delta^2(1 + G)]^{1/2}\}^{1/2}$$

and the parameter $\delta = \Omega/\Gamma$ is the frequency detuning. This expression (4.9) can be rewritten in the form given by UEHARA and SHIMODA [4.21]

$$\kappa(\omega) = \kappa_0(\omega)\Gamma(1 + A/B)/[(A + B)^2 - 4\Omega^2]^{1/2} \tag{4.10}$$

where $A = (\Omega^2 + \Gamma^2)^{1/2}$ and $B = [\Omega^2 + \Gamma^2(1 + 2G)]^{1/2}$.

At frequencies far from resonance, the absorption coefficient is approximated by

$$\kappa(\omega) = \kappa_0(\omega)/(1 + G)^{1/2}, \qquad \Omega \gg \Gamma_B, \tag{4.11}$$

which is in agreement with the absorption coefficient of the strong traveling wave (4.4). This corresponds to independent propagation of the traveling waves through the gas medium. In the case of exact resonance the absorption coefficient is

$$\kappa(\omega_0) = \kappa_0(\omega_0)/(1 + 2G)^{1/2}, \qquad \Omega \ll \Gamma_B. \tag{4.12}$$

At the centre of the Doppler line the saturated absorption coefficient decreases because of an increase of the saturation parameter. Figure 4.3 shows curves characterizing the shape of the Lamb dip for various values of the saturation parameter G. The FWHM of the dip is shown in Fig. 4.4 as a function of the saturation parameter G (solid line). The width by rate-equation approximation is also shown (dotted line). For large saturation the shape of the Lamb dip is a function of the parameter Ω/Γ_B. It is close to a Lorentzian function with half-width Γ_B. In this approximation the depth of the dip depends on G in a simple way

$$H = \frac{\Delta\kappa}{\kappa_0} = (1 + G)^{1/2} - (1 + 2G)^{1/2}. \tag{4.13}$$

The dip depth is maximum for $G = 1.42$ being $H_m = 0.133$.

Neglect of the spatial inhomogeneity of the standing wave field and of coherence effects when we solve the equations results in a loss of some results. A method of approximation for the calculation of the contribution from coherent processes was developed by BAKLANOV and CHEBOTA-YEV [4.23]. The main idea of their approach is that they find coherence corrections which depend on the parameter $(\tilde{\gamma}/\Gamma) \cdot G$, where

$$\frac{2}{\tilde{\gamma}} = \frac{1}{\gamma_1} + \frac{1}{\gamma_2}. \tag{4.14}$$

In the optical region the relaxation constants of levels γ_i differ greatly as a rule. Therefore the parameter $\tilde{\gamma}/\Gamma \ll 1$ and hence the condition $(\tilde{\gamma}/\Gamma) \cdot G \ll 1$ can be met, even with high G. The presence of collisions,

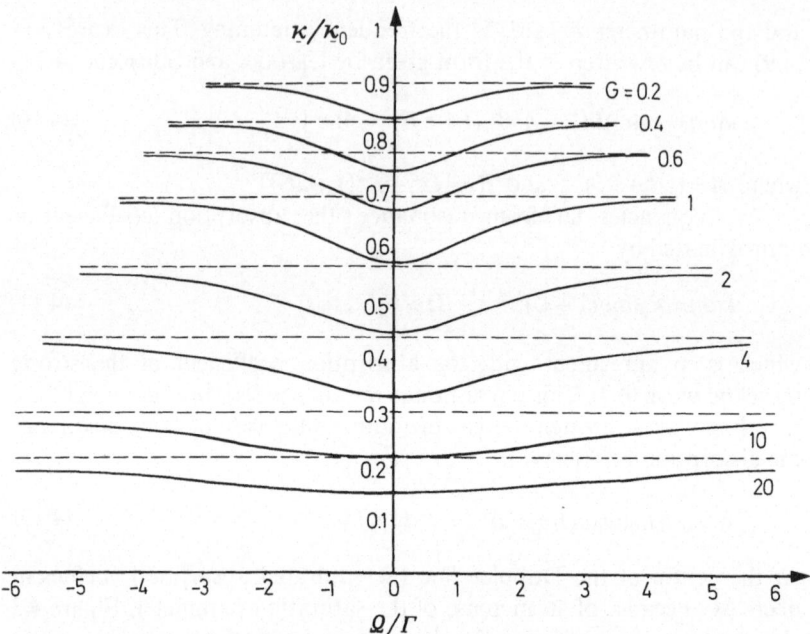

Fig. 4.3. The shape of the Lamb dip with various degrees of saturation G in the rate equation approximation

which result in a phase shift but not a change of the level lifetime, also decreases the ratio $\tilde{\gamma}/\Gamma$.

The quantitative contribution of coherence processes to the standing wave absorption is not so large compared with that obtained from the rate equation. But, when we calculate the velocity distribution and determine the weak wave absorption in the presence of the standing wave on the same levels, it is, in essence, important to take into account the coherence effects which result in qualitatively new results (BAKLANOV and TITOV [4.24]). Figure 4.4 shows the peak width vs. saturation parameter estimated with regard to coherence effects. As seen, the coherence correction is quite small. The contribution of the spatial non-uniformity effect and of coherence effects has been illustrated well by FELDMAN and FELD [4.20] through numerical integration of the equations on a computer. The solution of the problem by a computer has shown that there are no qualitative changes in the Lamb dip structure as compared with the rate-equation approximation. In the strong field regime, the dip depth, with equal relaxation constants $\gamma_1 = \gamma_2 = \Gamma$, decreases by some 20% compared with the result of exact calculation [4.21, 23]. Figure 4.5 gives the results of a numerical calculation for the imaginary part of the polarizability of Doppler-broadened absorption

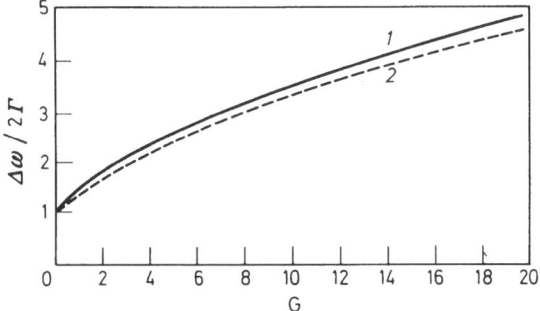

Fig. 4.4. The dependence of the width of the Lamb dip $\Delta\omega$ on the saturation parameter G in the rate equation approximation (dotted line), and exact calculation (solid line) for $\gamma_1 = \gamma_2$

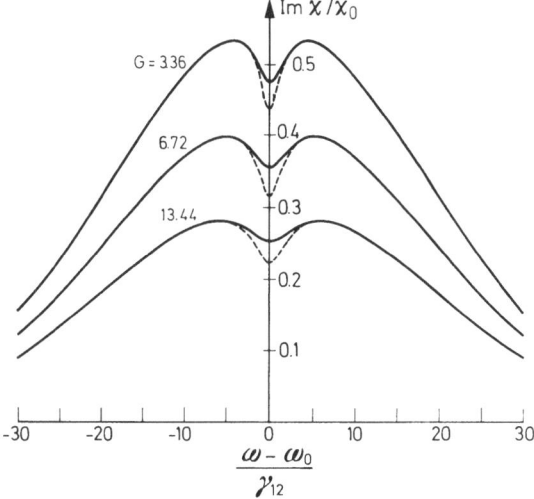

Fig. 4.5. The frequency dependence of the imaginary part of the susceptibility $\bar{\chi}/\bar{\chi}_0$ of the two-level Doppler-broadened transition for various values of saturation parameter. The calculation is performed for the case $ku = 25\gamma_{12}$, $\gamma_1 = \gamma_2$, $\gamma_{12} = \frac{1}{2}(\gamma_1 + \gamma_2)$. The dotted line shows the result of calculation in the rate equation approximation (FELDMAN and FELD [4.20])

under different degrees of saturation. The dotted line shows the results of the computation in the rate-equation approximation.

2) Dip for the Counter Probe Wave

Suppose that the light field having at least one strong traveling wave saturates an absorption. To observe the resonant distortion of the Doppler contour, the second wave should be used as a probe. The probe wave

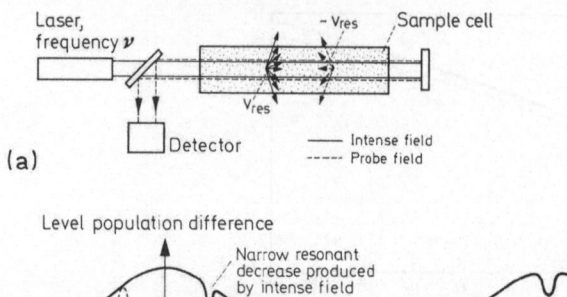

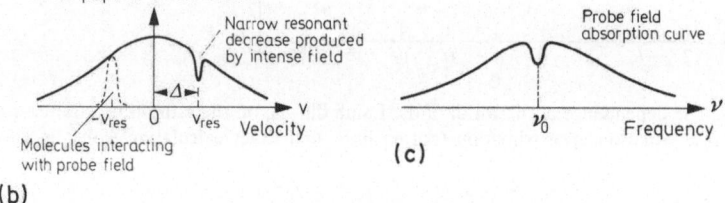

(a)

(b)   (c)

Fig. 4.6a–c. The narrow resonances of saturated absorption obtained by the method of an oppositely directed probe wave: (a) the scheme of experiment; (b) the shape of absorption line in a strong traveling wave field; (c) dependence of absorption on the weak probe frequency

may have the same frequency as the strong wave; then they should propagate in opposite directions (Fig. 4.6). Clearly when the laser frequency coincides with the line centre ω_0 the weak backward wave interacts with the atoms saturated by the strong forward wave. As a result, the absorption of the weak probe wave decreases sharply at the line centre. Thus, there is a resonant dip in the absorption of the weak probe wave (LETOKHOV and CHEBOTAYEV [4.25]).

Assume that a light field consists of two oppositely propagating waves of the same frequency but different in amplitude:

$$E(t, r) = \mathscr{E}_0 \cos(\omega t + kr) + \mathscr{E} \cos(\omega t - kr). \tag{4.15}$$

The traveling wave with the amplitude $\mathscr{E}$ is strong and able to saturate the absorption. The oppositely directed wave $\mathscr{E}_0$ is weak and does not induce saturation. The weak probe wave interacts linearly with the atoms inside the homogeneous width at the mirror-image frequency $\omega_0 + (\omega_0 - \omega)$. The transmission for the probe wave can be easily calculated in the approximation where only the change in the velocity distributions of the populations is taken into account and the effects of coherent interactions are disregarded. The change in the velocity distribution of the population difference $n(v) = n_1(v) - n_2(v)$ under the action of a strong forward wave is determined by the relation (4.5). The linear absorption coefficient of a weaker backward wave is determined by

$$\kappa(\omega) = \int \sigma(v, \omega) dv, \tag{4.16}$$

where $\sigma(v, \omega)$ represents the absorption cross section of the atoms at the velocity v in a field $\mathscr{E}_0 \cos(\omega t + kr)$, which is given by the expression:

$$\sigma(v, \omega) = \sigma_0 \frac{\Gamma^2}{(\omega_0 - \omega + kv)^2 + \Gamma^2}, \tag{4.17}$$

where σ_0 is the absorption cross section at resonance. Substituting the expression for the distribution $n(v)$ from (4.5) into expression (4.15) we obtain (BASOV et al. [4.26, 48], MATIUGIN et al. [4.27])

$$\kappa(\omega) = \kappa_0(\omega) \left\{ 1 - [1 - (1 + G)^{-1/2}] \mathscr{L} \left(2\frac{\Omega}{\Delta\omega} \right) \right\}, \tag{4.18}$$

where $\Delta\omega$ is the dip width (FWHM) determined by

$$\Delta\omega = \Gamma[1 + (1 + G)^{1/2}] = \Gamma + \Gamma_B, \tag{4.19}$$

$\mathscr{L}(y) = (1 + y^2)^{-1}$ is the Lorentzian contour. This shows that the dip width is equal to half the sum of the width $2\Gamma_B$ of the hole burnt by the strong wave and the homogeneous width 2Γ which corresponds to the range of frequencies interacting with the weak probe wave.

The calculated shape of the absorption line (4.18) for the weak oppositely directed wave is not exact, because we considered only the change in the population difference. The strong field also changes the shape of the absorption line for individual atoms. This problem has been solved without any limitation of the field amplitude and the relaxation constants in two-level approximation by BAKLANOV and CHEBOTAYEV [4.28, 29]. HAROCHE and HARTMANN [4.30] have obtained similar results for the case when both levels have equal relaxation constants. The absorption coefficient for the probe wave consists of two terms. The first term corresponds exactly to the absorption coefficient found in the rate-equation approximation. Then the second term can be interpreted as a contribution from coherence effects. In the Doppler limit ($\Delta\omega_D \gg \Gamma_B$) the absorption line shape for the probe wave is given by [4.28]

$$\frac{\kappa(\omega)}{\kappa_0(\omega)} = \frac{\kappa^{(1)}(\omega)}{\kappa_0(\omega)} + x^2 \frac{\Gamma_B - \Gamma}{\Gamma_B} \mathrm{Re} \left[-\frac{f(\Omega - i\Gamma_B)}{2\Omega + i(\Gamma_B + \Gamma)} \right], \tag{4.20}$$

where

$$f(z) = \frac{(3z - \Omega + i\Gamma)(z + i\Gamma)(2z + i\Gamma)}{(3z - \Omega + i\Gamma)(z + \Omega + i\Gamma)(2z + i\gamma_1)(2z + i\gamma_2) + x^2(2z + i\Gamma)^2}. \tag{4.21}$$

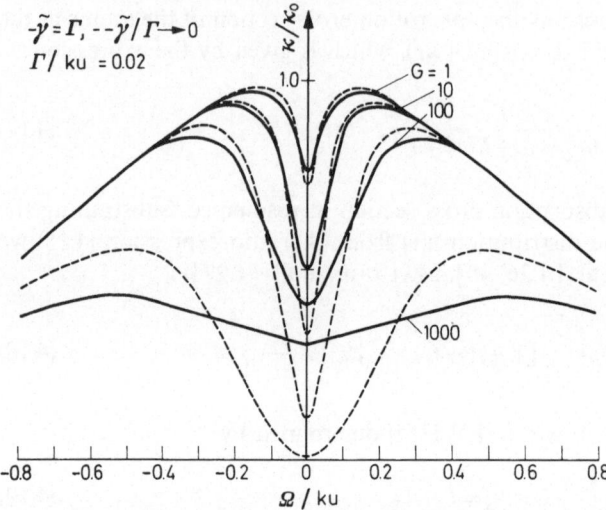

Fig. 4.7. The shape of the absorption line for a weak probe wave in the presence of a strong counter-running wave at a different saturation parameter G, when coherence effects are taken into account (the solid curves), and are neglected (the dotted curves) (BAKLANOV and CHEBOTAYEV [4.28])

The term $\kappa^{(1)}(\omega)/\kappa_0(\omega)$ is given by Eq. (4.18), $\Omega = \omega - \omega_0$, and x is the Rabi frequency. Figure 4.7 shows the absorption line shape for the probe wave with coherence effects taken into account. This line shape is determined by expression (4.20).

Coherence effects produce an additional broadening for the narrow dip at the line centre. From the physical point of view this is naturally due to coherent oscillations in the two-level system (optical nutations) under a strong field. Figure 4.8 gives the results for the dip width in probe wave absorption estimated by formulas (4.20) and (4.21) for various values of the parameter $\varrho = \tilde{\gamma}/\Gamma$, where the constant $\tilde{\gamma}$ is determined by (4.14). The case of $\varrho = 0$ corresponds to a very large difference between the decay rates γ_1 and γ_2, when the rate-equation approximation holds. The case of $\varrho = 1$ corresponds to $\gamma_1 = \gamma_2 = \Gamma$, and coherence effects make the largest contribution. Thus, an increase in the dip width with the parameter ϱ is evidently the effect of coherent interaction of the strong field.

The contribution of coherence effects is proportional to the parameter ϱ and appears only in the even orders of the saturation parameter. Coherence effects result in a number of important features in the absorption line shape of the probe wave. Firstly, the absorption coefficient of the probe wave is always larger than the saturated absorption of the strong

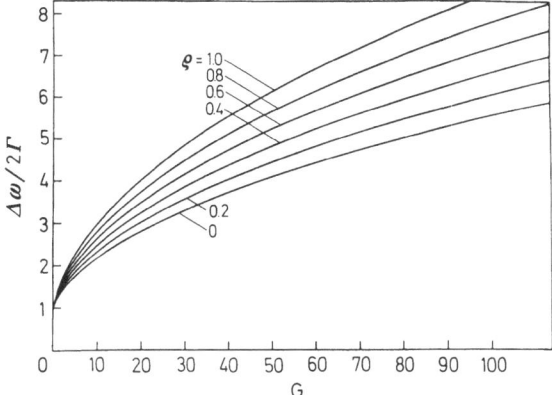

Fig. 4.8. Dependence of the dip width $\Delta\omega$, observed by the method of counter-running probe wave, on the value of the parameter G for absorption saturation by a strong wave at different values of the parameter $\varrho = \tilde{\gamma}/\Gamma$, with allowance made for coherence effects in a strong wave

wave. For example, in a very strong field $(G \gg 1)$ at the centre of the Doppler contour $(\Omega = 0)$, we have

$$\frac{\kappa}{\kappa_0} = \frac{3\varrho}{2(3+\varrho)} \exp\left[-\frac{1}{3}\varrho\left(\frac{\Gamma_B}{ku}\right)^2\right], \tag{4.22}$$

where $\varrho = \tilde{\gamma}/\Gamma$, $2\tilde{\gamma} = 1/\gamma_1 + 1/\gamma_2$. The relationship between the absorption of the weak wave at the line centre and the intensity of the strong wave, determined by expression (4.20), is given in Fig. 4.9. Secondly, with an increase in the intensity of the strong wave, the absorption of the weak wave approaches a constant value which is determined by (4.22) and depends only on the ratio between the relaxation constants. For example, with equal relaxation constants for the levels γ_1 and γ_2, the absorption of the weak wave approaches the constant value $(3/8)\kappa_0$. When the relaxation constants differ greatly, or if there are dephasing collisions acting on the absorbing particles, the contribution of coherence effects is small and we can use expression (4.18) obtained from the rate equations with fairly good accuracy.

3) Narrow Resonance for the Unidirectional Probe Wave

A probe wave can propagate in the same direction as the strong wave. In this case its frequency ω_2 should be scanned (Fig. 4.10) to reveal a resonance in the Doppler contour. The resonance dip occurs in this case at the frequency of the strong wave ω_1, but not at the line centre.

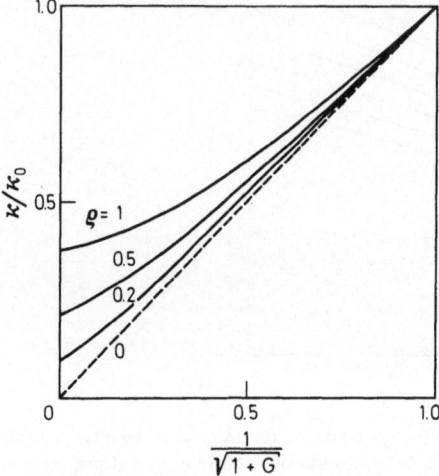

Fig. 4.9. The relation between the absorption of the weak wave at the line centre ($\omega = \omega_0$) and the intensity of the strong wave for various values of the parameter ϱ. The case $\varrho = 1$ corresponds to $\gamma_1 = \gamma_2 = \Gamma$, i.e., maximal contribution of coherence effects, while $\varrho = 0$ corresponds to the greatly different γ_1 and γ_2, which is the case of incoherent saturation (BAKLANOV and CHEBOTAYEV [4.28])

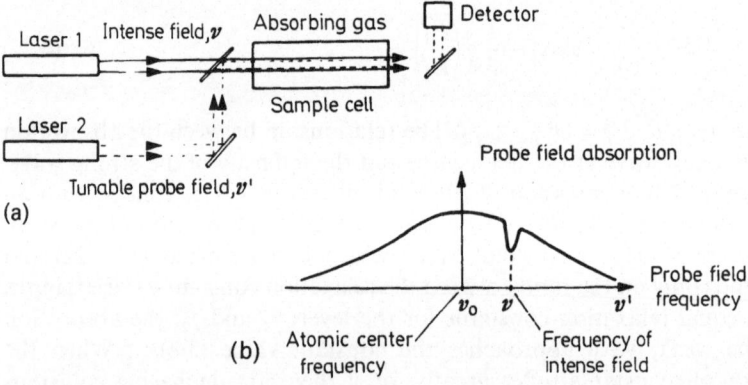

Fig. 4.10a and b. The narrow resonances of saturated absorption obtained by the method of a unidirectional probe wave: (a) the scheme of experiment; (b) the absorption line for the probe wave

When unidirectional waves interact in a gas, new features of the absorption line appear. Apart from the "Bennett hole", caused by a decrease of the population difference, additional resonances appear in the line shape with their widths equal to the decay constants γ_1 and γ_2. These resonances, which give information about the rates of decay of the individual levels γ_1 and γ_2, are characteristic of the interaction of unidirectional waves and are absent in the interaction of oppositely directed waves. The physical essence of this phenomenon can be understood from qualitative explanation below.

Two unidirectional waves with close frequencies ω_1 and ω_2 create at each point a composite field with an amplitude which varies at the difference frequency $\Delta = (\omega_1 - \omega_2)$. If the field is sufficiently strong, it may significantly change the populations. The time-dependent amplitude of the field induces a modulation of the population difference which gives rise to a corresponding modulation of the absorption coefficient and, hence, to amplitude modulation of the fields. Additional frequency components, appearing as sidebands due to the amplitude modulation can be regarded as a decrease in absorption of the initial waves. The depth of modulation of the population difference depends on the modulation frequency Δ and the decay constants γ_1 and γ_2. If $\Delta \ll \gamma_1$ and γ_2, the population follows the change in the amplitude of the composite field, and the amplitude-modulation effect is maximal. When $\Delta \gg \gamma_1, \gamma_2$, the medium has no time to respond to the change in the instantaneous amplitude of the composite field. In this region only a change in the average population is essential. Thus, the additional resonances are associated primarily with the temporal modulation of the population.

When the relaxation constants differ greatly ($\gamma_1 \ll \gamma_2$), and with a limitation on the field ($\varrho G \ll 1$) of the strong wave, the absorption coefficient of the probe wave has been found by RAUTIAN [4.17]. The absorption coefficient of the probe wave in a gas of two-level atoms has been found by BAKLANOV and CHEBOTAYEV [4.29], with no limitations on the strong wave amplitude and the relaxation constants in the presence of collisions which quench and shift the phase of emission. In the general case the formula for the absorption of the probe wave is very lengthy. For a weak probe field the formula for the absorption coefficient becomes comparatively simple:

$$
\frac{\kappa(\omega)}{\kappa_0(\omega)} = 1 - \frac{G}{2} \mathscr{L}\left(\frac{\Delta}{2\Gamma}\right)\left[1 + \frac{\tilde{\gamma}}{2}\left(\frac{\gamma_1}{\gamma_1^2 + \Delta^2} + \frac{\gamma_2}{\gamma_2^2 + \Delta^2}\right)\right.
$$
$$
\left. + \frac{\tilde{\gamma}}{4\Gamma}\left(\frac{\Delta^2}{\Delta^2 + \gamma_1^2} + \frac{\Delta^2}{\Delta^2 + \gamma_2^2}\right)\right] \tag{4.23}
$$

where (Δ, Γ and $|\omega_1 - \omega_0|) \ll \Delta\omega_D$, $\mathscr{L}(y) = (1 + y^2)^{-1}$. Equation (4.23) gives dips with widths determined by the relaxation constants γ_1, γ_2, and Γ. In the presence of phase-changing collisions and under the condition $\Gamma \gg \gamma_1, \gamma_2$, the line shape consists of the sum of three Lorentzian-type dips, with half-widths 2Γ, γ_1, and γ_2 and depths

$$
\frac{G}{2}, \frac{G}{2}\frac{\gamma_2}{\gamma_1 + \gamma_2} \quad \text{and} \quad \frac{G}{2}\frac{\gamma_1}{\gamma_1 + \gamma_2}
$$

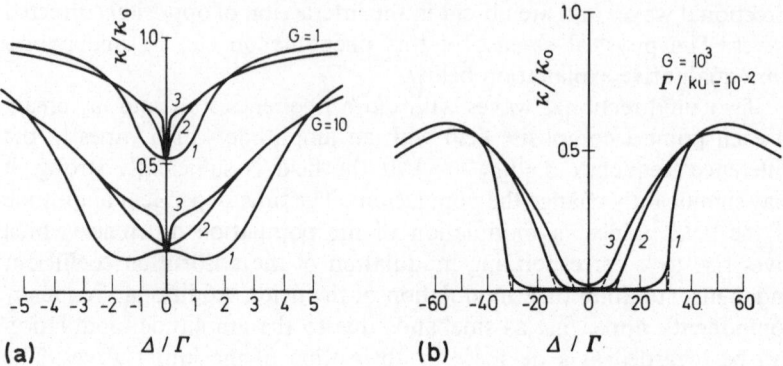

Fig. 4.11a and b. The shape of the absorption line of a probe wave in the presence of a strong unidirectional wave for the saturation parameters $G=1$ and 10 (a) and $G=10^3$ (b). Curves *1* correspond to $\gamma_1/\gamma_2=1$, 2 to $\gamma_1/\gamma_2=10$, 3 to $\gamma_1/\gamma_2=10^2$; $\Gamma/ku=10^{-2}$ (BAKLANOV and CHEBOTAYEV [4.29])

respectively, on the background of a Doppler contour. This fact is of particular interest for saturation spectroscopy, since every resonance carries direct information on the damping of the off-diagonal and diagonal elements of the density matrix (the line width and the lifetimes of the levels).

The first Lorentzian term gives the saturated absorption line, which depends only on the saturation of the population difference in the strong wave. The subsequent terms in (4.23) determine the contribution of coherence effects, which in contrast to the case of oppositely directed waves show up in the first order of the saturation parameter. With an increase of the field, there are changes in the line shape. The width and depth of the dip, depending on changes in the populations of the levels, increase. The relative width of the sharp dips with widths $2\gamma_1$ and $2\gamma_2$ grows and then begins to decline. Their width depends in a complicated way on the field. In very strong fields ($x \gg \Gamma$), the sharp structure of the lines almost disappears and the absorption coefficient tends to zero over a wide frequency range. The absorption line shape, calculated for various relaxation constants, is illustrated in Fig. 4.11. In strong fields the expression for the absorption coefficient becomes simple (accurate to $1/G^{1/2}$) to have the form [4.29]

$$\frac{\kappa}{\kappa_0} = \begin{cases} 0, & \Delta < x \\ \dfrac{|\Delta|(\Delta^2 - x^2)^{1/2}}{\Delta^2 + \Gamma_B^2(1-\varrho)}, & \Delta \geq x. \end{cases} \tag{4.24}$$

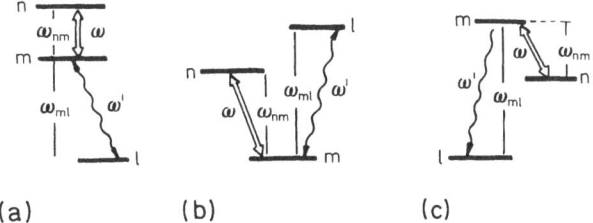

Fig. 4.12a–c. Energy level configuration for two coupled transitions: (a) is a cascade configuration: (b, c) are folded configurations. The arrow points to the transitions *m-n* acted upon by a strong coherent light wave, the wavy line shows the transition where a spontaneous radiation is observed or a probe wave acts

These peculiarities can be explained by the Stark effect in a light field acting on the Doppler-broadened transition (BAKLANOV and CHEBOTAYEV [4.28, 29], HAROCHE and HARTMANN [4.30]).

4.2.3 Narrow Resonances on Coupled Transitions

Narrow "hole" and "peak" in the atomic velocity distribution at two levels of the transition *n-m*, which is acted upon by a coherent light wave, may also occur at its connected transitions. Figure 4.12 shows all possible profiles of two coupled transitions *n-m* and *m-l* with their common level *m*, where the transition *m-n* is saturated by a strong coherent light wave. The narrow-band saturation in the Doppler-broadened transition *m-n* induces narrow resonances at the coupled transition *m-l*. These induced narrow resonances are related not only to the change of the velocity distribution but also to coherence effects resulting from two-quantum transitions in the three-level system. Both processes may be observed simultaneously, and therefore the nature of these processes is more complicated in the three-level system than in the two-level system.

It should be noted that such a simple structure of narrow resonances appears when $k' > k$. When the frequency ratio of the coupled transitions is reversed ($k' < k$), the narrow resonance Γ_- is superposed by an additional structure which involves level splitting in a strong light field on the transition *m-n* [4.31].

The structure of narrow resonances in two coupled transitions has been studied in detail by many workers. In addition to the qualitative discussion we shall list some principal works. SCHLOSSBERG and JAVAN [4.32] were the first to study the resonance structure at two closely spaced transitions with the common level *m* in a two-frequency field. They revealed that for unidirectional waves the resonance width is determined only by the initial and final level widths, while the intermediate level width is ruled out because of the two-quantum nature of transitions (the case $k' = k$ in (4.29) for folded transition scheme). NOTKIN et al. [4.33] predicted the line shape anisotropy of spontaneous emission at the transition m-l, in the presence of a weak saturating wave at the transition m-n, and estimated the widths of narrow and broad resonances. An analogous conclusion from simple considerations was drawn by HOLT [4.34]. FELD and JAVAN [4.35] pointed out a relation between these conclusions. They further studied in detail [4.32, 36] the structure of resonances in the stimulated emission of the transition m-l when the transition m-n is highly saturated. For the case of weak saturation the results of this work agree with those of Ref. [4.33]. POPOVA et al. [4.37] and POPOV [4.38] showed that there may occur a more complex structure of narrow resonances in the coupled transition owing to level splitting in the strong light field of the transition m-n. An analogous problem, as applied to the problem of the three-level gas amplifier, collisions in the gas being taken into account, was investigated by HÄNSCH and TOSCHEK [4.39]. Other details of resonance structure in coupled transitions for the case of a standing wave at the transition m-n were considered by FELDMAN and FELD [4.40]. The case of two frequency-coincident coupled transitions in a single-frequency field (level crossing) has been discussed by SHIMODA [4.41] and by FELD et al. [4.42]. This case is a stimulated version of the well-known Hanlé effect in spontaneous emission. Many arising effects have been discussed in detail by FELD [4.31], in the review by BETEROV and CHEBOTAYEV [4.12] and in a book [4.10] and Chapter 6. Below we shall describe only the physical nature of the effects without discussion of too complicated formulae.

Let us consider at first the spectral line shape at the coupled transition m-l, taking into account only the change in atomic velocity distribution on the common level m. The strong laser field of the absorbing transition m-n forms either a Bennett hole (for the schemes in Fig. 4.12a, b) or a peak (for the scheme in Fig. 4.12c) in the velocity distribution, their half-width being $\Delta v = \Gamma_B/k = (\Gamma_{mn}/k)(1 + G)^{1/2}$, where $2\Gamma_B$ is the Bennett hole width, $2\Gamma_{mn}$ is the homogeneous width of the transition m-n, G is the saturation parameter for the transition m-n, k is the wave vector of the strong wave. The Doppler-broadened line of the transition m-l has a hole or a peak, respectively, with their width $(k'/k)\Gamma_B$, where k' is the wave vector for the probe wave on the transition m-l. Since the transition m-l has the homogeneous width $2\Gamma_{ml}$, the resultant half-width of the hole or peak at the coupled transition will be

$$\Gamma_0 = \Gamma_{ml} + \frac{k'}{k} \Gamma_{mn}(1 + G)^{1/2} . \tag{4.25}$$

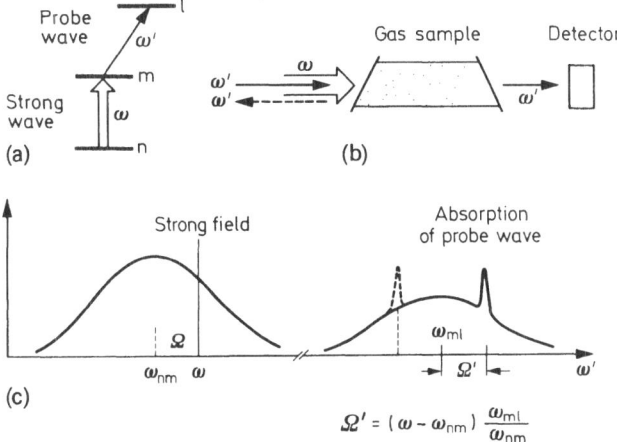

Fig. 4.13a–c. Spectroscopy of saturated absorption of two coupled transitions: (a) scheme of transitions; (b) scheme of observation; (c) absorption line shapes at a coupled transition with a narrow resonance peak, when observed in the forward (solid curve) and in the backward direction (dashed curve), caused by stepwise transitions

The position of this hole (peak) at the Doppler profile depends on the direction of observation of the coupled transition (Fig. 4.13).

    This simple picture presupposes independence of quantum processes of photon absorption and emission of the transitions n-m and m-l. Actually these two processes can occur simultaneously and, hence, can be correlated [4.43]. In other words, two-quantum processes are possible for the transition from the state n to l, while the intermediate state m is bypassed. For the scheme in Fig. 4.12c this resembles the process of resonance Raman scattering where a photon is absorbed by the transition n-m and another photon is emitted at the same time by the transition m-l. However, the presence of a real level makes single-quantum cascade transitions $n \rightarrow m \rightarrow l$ possible, which, unlike two-quantum transitions, are successive independent single-quantum processes of absorption and emission. In the case, when relaxation constants of levels differ markedly and when the intermediate state lifetime is much shorter than that of the initial state $(\gamma_m \gg \gamma_n)$, the two-quantum process of resonance Raman scattering type is dominant. In the other case, when the intermediate state lifetime is much longer than that of the initial state $(\gamma_m \ll \gamma_n)$, a two-step process occurs. The above-described process with a change in population holds true in this case. The relaxation constants being equal $(\gamma_m \cong \gamma_n)$, the both processes come out together, and it is impossible to specify the way of transition from the state n to l. In this case, as a general rule of quantum mechanics, the two processes are subject to interference.

The importance of two-quantum processes in nonlinear three-level spectroscopy has been pointed out by JAVAN and SCHLOSSBERG [4.32].

In the limiting case of $\gamma_m \gg \gamma_n$, when the two-quantum process prevails, two new effects show up which do not take place in the two-step process $n \to m \to l$ where only the change in velocity distribution of population is considered [4.34, 35]: 1) the narrow resonance width of the coupled transition m-l differs between observations in forward and backward directions with respect to the strong wave direction k of the transition m-n; 2) the narrow resonance width at the coupled transition m-l may be much smaller than the natural width of the transition Γ_{ml}.

These effects can be easily explained. Let us consider, say, a two-quantum process in the cascade transition scheme given in Fig. 4.12a. For the two-quantum transition in such a scheme the frequency condition is

$$\omega_{nm} + \omega_{ml} = \omega + \omega', \tag{4.26}$$

where we see that the total energy of two photons equals the difference in energies between the initial and final states. This condition is not so rigid as that for two successive single-quantum transitions ($\omega_{nm} = \omega$, and $\omega_{ml} = \omega'$). Let us consider atoms moving at the velocity v. In the coordinate system connected with the moving atom the light frequency on the transition n-m will be $\omega - kv$. If an atom emits a photon by the transition m-l, which satisfies the double-resonance condition (4.26), its frequency in the moving coordinate system will be $\omega_{nm} + \omega_{ml} - (\omega - kv)$. The frequency of these photons in the laboratory coordinate system depends on the direction of their propagation k' as

$$\omega'(v) = \omega_{nm} + \omega_{ml} - \omega + (k + k')v, \tag{4.27}$$

or

$$\omega'_{\pm} = \omega_{nm} + \omega_{ml} - \omega + \left(1 \pm \frac{k'}{k}\right) kv \tag{4.28}$$

where the sign "$+$" corresponds to the observation in the direction k, while "$-$" corresponds to the observation in the opposite direction to k. Relation (4.28) shows that the photons at the transition m-l, emitted during the two-quantum process in the direction opposite to k, are less sensitive to the velocity than those emitted in the forward direction. When $k = -k'$, the Doppler effect has no influence at all (see Chapt. 8). Therefore in the opposite direction, the narrow peak resulting from two-quantum processes is observed to have a width given by

$$\Gamma_{\text{nar}} = \Gamma_{ln} + \left(\frac{k'}{k} - 1\right) \Gamma_{\text{B}}, \tag{4.29}$$

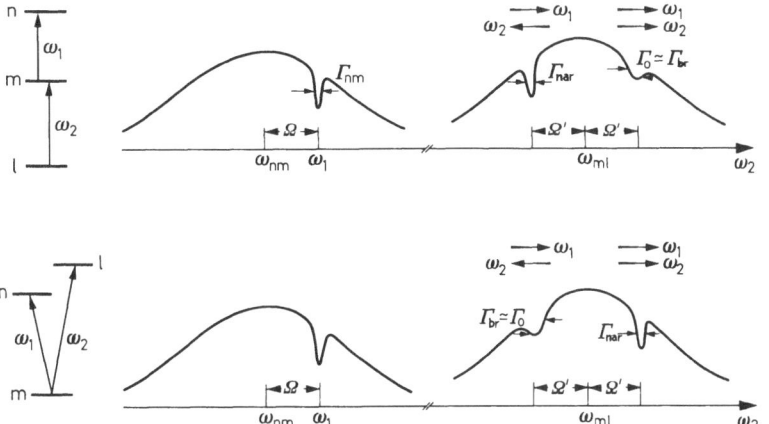

Fig. 4.14. Line shape for spontaneous emission or absorption of the probe signal at the Doppler-broadened transition $m-l$, coupled to the Doppler-broadened transition $m-n$ acted upon by a strong running light wave, with allowance made for two-quantum transitions $n-l$. The relative orientation of running waves at the coupled transition is shown with arrows. The simplest case $\omega_2 > \omega_1$ is considered, when the narrow resonance shape at the coupled transition is bell-shaped

while the line width in the forward direction is

$$\Gamma_{\text{bro}} = \Gamma_{lm} + \left(\frac{k'}{k} + 1\right)\Gamma_{\text{B}} \simeq \Gamma_0 \, . \tag{4.30}$$

The Doppler contour for the transition $m-l$ in the presence of a strong wave on the transition $m-n$ is shown in Fig. 4.14 for two directions of observation. It is essential that for the wide intermediate level, when $\Gamma_{lm} \gg \Gamma_{ln}$, the observed narrow peak width may be much narrower than the homogeneous width of the transition $m-l$ if $k' \simeq k$, since it depends on the width of the initial (n) and final (l) levels of the two-quantum transition. When $k = k'$, the narrow peak width is equal to the natural width of the "forbidden" transition $n-l$, and in this case we can speak of nonlinear spectroscopy inside the natural width. This effect was first observed experimentally in Ref. [4.35].

For the coupled transition schemes illustrated in Fig. 4.12b, c, the two-quantum frequency condition is determined not by (4.26) but by

$$\omega_{ml} - \omega_{mn} = \omega' - \omega \, . \tag{4.31}$$

For such schemes all the above-said things hold, with a difference that a narrow peak (4.29) is observed in the forward direction, and a broad

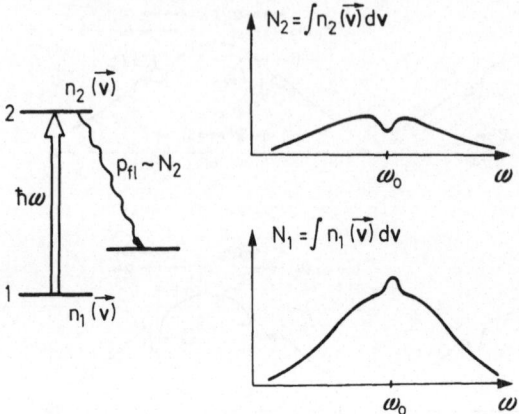

Fig. 4.15. Dip formation in the fluorescence intensity P_{fl} under a strong standing wave due to the total number of excited particles N_2 as a function of the frequency ω

peak (4.30) in the opposite direction. Therefore, for the folded schemes of coupled transitions in Fig. 4.14 (lower part) the positions of narrow and broad peaks are interchanged.

4.2.4 Resonances of the Total Level Population in a Standing Wave

In tuning the standing wave frequency to the centre of a Doppler-broadened line, a resonant reduction in the saturated absorption is accompanied also by a resonant change of the total number of atoms on each level of the transition regardless of their velocity (Fig. 4.15). There is a resonance minimum in the total population of the upper level and a resonance peak in that of the lower level.

The population of excited atoms is related to the saturated absorption coefficient $\kappa(\omega)$ by

$$N_2(\Omega) = \kappa(\Omega, P) P \tau_2 , \tag{4.32}$$

where $\Omega = \omega - \omega_0$, ω is the standing wave frequency, P is the intensity of the wave, τ_2 is the lifetime of the excited level. For example, in the weak saturation approximation, the dependence of $\kappa(\Omega, P)$ on the frequency is given by (4.8), and N_2 is

$$N_2(\Omega) = \kappa_0(\Omega) P \tau_2 \left[1 - \frac{G}{2} \left(1 + \frac{\Gamma^2}{\Gamma^2 + \Omega^2} \right) \right]. \tag{4.33}$$

Thus, against the background of the gaussian curve $\kappa_0(\omega) P \tau_2$ there is a narrow resonance minimum in the excited atomic population with a width 2Γ. Since the saturation parameter $G \sim P$, the magnitude of dip in $N_2(\omega)$ increases with the square of the intensity for low saturations $(G \ll 1)$. Its relative value is given by the same expression as for the Lamb dip.

The effect of resonant change in $N_2(\omega)$ can be obtained from a simple consideration of hole burning (LETOKHOV [4.44]). In the standing wave at the off-resonance frequency $(|\Omega| \gg \Gamma)$, two holes occur in the velocity distribution of the population difference (Fig. 4.2a). The total number of particles in the excited level N_2 is proportional to the total area of the holes $(S = S_1 + S_2)$. In the case of a true resonance $(|\Omega| \ll \Gamma)$ both holes unite and the number of excited particles becomes proportional to the area of the common hole S_0. A resonant change of the total number of excited particles may occur when $S_0 \neq S_1 + S_2$.

From rate equations (2.38) and (2.39) one can easily obtain an expression for the density of excited atoms in the stationary state without coherence effects as

$$N_2 = N_2^0 + (N_1^0 - N_2^0) \frac{G}{1 + (\gamma_2/\gamma_1)} f(G, \Omega) \tag{4.34}$$

in which

$$f(G, \Omega) = \frac{1}{2} \int_{-\infty}^{\infty} dv\, W(v) \frac{\mathscr{L}(\Omega + kv) + \mathscr{L}(\Omega - kv)}{1 + G/2[\mathscr{L}(\Omega + kv) + \mathscr{L}(\Omega - kv)]} \tag{4.35}$$

The resonance effect in the population of excited atoms is given by the integral function $f(G, \Omega)$. The general analysis can be carried out by calculating the integral $f(G, \Omega)$ and expressing it in terms of the plasma dispersion function (see STENHOLM and LAMB [4.19]). We restrict ourselves to consider two limiting cases here.

In the case of true resonance $(|\Omega| \ll \Gamma \ll ku)$, the integral (4.35) can be evaluated exactly:

$$f(G, 0) = f_0/(1 + G)^{1/2}, \quad f_0 = \sqrt{\pi} \Gamma/ku. \tag{4.36}$$

Far from resonance $(\Gamma \ll |\Omega| \ll ku)$ the integral (4.35) can be broken into the sum of two integrals; the main contribution to each of them is made by particles with velocities near $v = \pm \Omega/k$. As a result we obtain

$$f(G, |\Omega| \gg \Gamma) = f_0/(1 + 2G)^{1/2}. \tag{4.37}$$

Comparing the expressions (4.36) and (4.37) we find that the total number of atoms on the upper level decreases resonantly when the frequency passes through the Doppler line centre. When $G \ll 1$ the depth of the narrow dip is proportional to G^2, and when $G \gg 1$, the rise is proportional to $\sqrt{G}$. The relative depth of the dip (dip contrast) approaches the constant value $H = (1 - 1/\sqrt{2}) = 0.29$ for $G \gg 1$. The optimum value of the saturation parameter is at $G \simeq 1$. In this case the dip contrast is $H = 0.16$, that is, about one half the maximum value, and the factor of resonance broadening by the strong wave is $\sqrt{2}$.

The resonant change in the number of atoms of the lower level is

$$(N_1 - N_1^0) = -(N_2 - N_2^0). \tag{4.38}$$

The resonant change in density of excited molecules in a strong standing wave, which acts on one rotational-vibrational absorption line, was estimated by LETOKHOV and PAVLIK [4.45]. They calculated that, owing to rotational relaxation, it is possible to attain a more intense resonance by accumulating excited molecules of many rotational levels.

4.3 Methods of Laser Saturation Spectroscopy

The resonance effects of saturation of a Doppler-broadened absorption line, described in Section 4.2, may be used as the basis for many methods of saturation spectroscopy. Such methods and their realization of different schemes of nonlinear spectrometers are considered in this section.

4.3.1 Spectroscopy of Unconnected Transitions

There are some rather efficient methods suggested for and used in saturation spectroscopy of overlapping independent spectral lines (lines with no common level). They are classified in Fig. 4.16; a) a Lamb dip in the standing wave field inside or outside the laser cavity; b) a dip in the total number of excited atoms in the standing wave field observed through changes in fluorescence intensity (4.2.4); c) a dip in absorption of the probe wave when atoms are saturated by a strong counter-running wave; d) a dip in absorption of the probe wave when atoms are saturated by a strong unidirectional wave; e) a dip in absorption of the probe wave when atoms are saturated by a strong counter-running wave with a different frequency. Let us consider each of the above methods in more detail.

1) Absorption Cell Inside the Laser Cavity

The Lamb dip in a standing wave can be observed if the absorption cell is placed inside the laser cavity. In this case it is necessary that, firstly, the absorption line should coincide with the amplification line of the

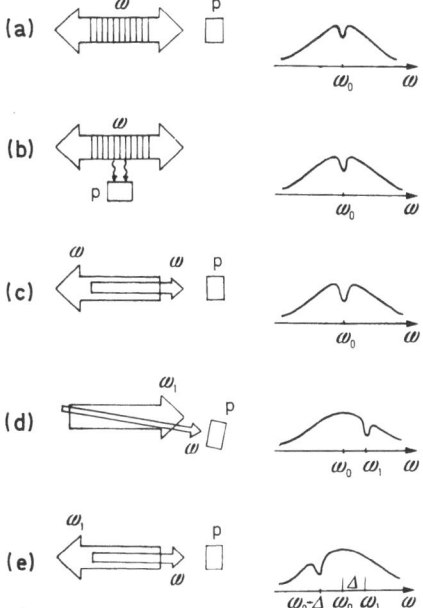

Fig. 4.16a–e. Classification of saturation spectroscopic methods for unconnected transitions: on the left (a, b) denote the standing wave: (c, d, e) denote the strong and weak running waves, on the right: absorption or fluorescence for signals is recorded by the photodetector P, when the wave frequency ω is being scanned (with a fixed frequency of the strong wave ω_1 for (d, e))

active medium or lie inside the amplification line width. Secondly, the field intensity in the cavity should suffice to saturate the absorption. Then a Lamb dip arises at the centre of the absorption line (Fig. 4.17b), and the effective amplification inside the cavity develops a resonance peak. This gives rise to a resonance peak in the output power (Fig. 4.17c) which is often termed as a inverted Lamb dip. It is this method by which the first experiments in saturation spectroscopy have been realized (Ne-absorbing cell in He$-$Ne laser at 6328 Å [4.7, 8], CH$_4$-absorbing cell in He$-$Ne laser at 3.39 μm [4.9], etc.).

The power peak depends in a complicated way on the characteristics of the amplifying and absorbing media; the shape of the peak does not resemble exactly the shape of the Lamb dip in absorption. This presents certain problems when the power peak is used for spectroscopic measurements. Yet there is a range of parameters where the power peak takes a nearly Lorentzian shape. When the saturation parameters of the amplifying media G_a and absorbing media G_b are both small

$$G_a, G_b \ll 1, \tag{4.39}$$

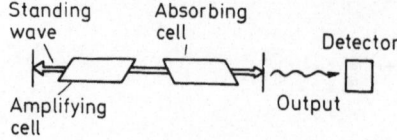

(a)   Experiment

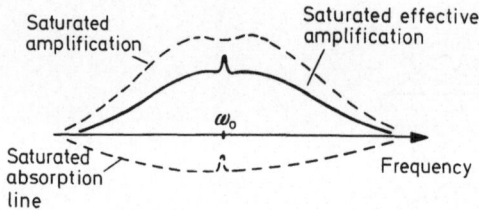

(b)   Amplification of two-component media

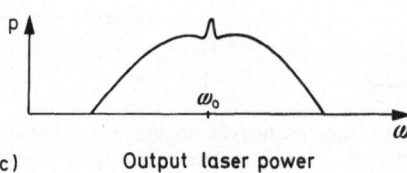

(c)   Output laser power

Fig. 4.17a–c. Formation of the inverted Lamb dip in a gas laser with a saturated absorption cell: (a) scheme of experiment: (b) dependence of saturated amplification of the two-component medium at the standing wave frequency; (c) laser output power as a function of oscillation frequency

denoting a low degree in saturation of the amplification and absorption, the relative output power is given by

$$P(\omega) \sim \frac{(\eta - 1)}{(2 - \beta) - \beta \mathscr{L}(\omega - \omega_b / \Gamma_b)}, \tag{4.40}$$

where $\mathscr{L}(x) = (1 + x^2)^{-1}$ represents the Lorentzian contour, $\eta = (\alpha_{\rm eff} - \gamma_0)/\gamma_0$ is the excess of initial effective gain $\alpha_{\rm eff} = \kappa_{a0} - \kappa_{b0}$ over linear losses γ_0, $\beta = \kappa_{a0} G_a / \kappa_{b0} G_b$ is a parameter characterizing the change in the effective gain $\alpha_{\rm eff}$ when $P = 0$ (when $\beta > 1$ the value $d\alpha_{\rm eff}/dP < 0$, and when $\beta < 1$, $d\alpha_{\rm eff}/dP > 0$).

The power peak, determined by the Lorentzian term in the denominator, has a Lorentzian shape with the FWHM of

$$\Delta \omega_{\rm res} = 2\Gamma_b \left\{ \frac{2(1 - \beta)}{(2 - \beta)} \right\}^{1/2}, \tag{4.41}$$

where $2\Gamma_b$ is the homogeneous full width of absorption. Thus, one should remember that by extrapolation the experimental data of peak width into the weak-intensity region, we obtain the peak width $\Delta\omega_{res}$, which is not generally equal to the homogeneous width of absorption. The peak width coincides with the homogeneous width only under the additional condition of $\beta \ll 1$. In practice this condition can be met with the use of very low absorption.

The relative amplitude or peak contrast may be defined by

$$H = \frac{1}{P_0}(P_{max} - P_0).$$ (4.42)

Under the condition of (4.39), it becomes

$$H = \beta/2(1-\beta).$$ (4.43)

Thus, the shape of the power peak coincides with that of the Lamb dip when the peak has a very small contrast. A large contrast can be obtained with an increase of the parameter β by increasing, for example, the absorption coefficient κ_{b0}. For a large peak contrast H of strongly saturated absorption, the shape of the power peak differs greatly from a Lorentzian shape. In the general case of large saturation, the broadening can be written in the form

$$\Delta\omega_{res} = f_P \cdot 2\Gamma_b,$$ (4.44)

where f_P represents the factor of broadening for the power peak. Calculations of H and f_P for large saturations for a saturation spectrometer with internal absorption cell have been performed by GREENSTEIN [4.46].

The power peak was observed in a large number of investigations with various lasers and the absorbers. The list of these studies is given in Table 4.1. There are two types of nonlinear absorbers used in the experiments: 1) atoms and molecules used in the amplifying medium but under conditions when there is absorption and not amplification on the working transition; 2) molecules with an absorption frequency which coincides accidentally with the laser line.

Not all kinds of absorbing gas, when inserted into a laser cavity, are suitable for the inverted Lamb dip. It is desirable that the parameter of saturation of the absorption should be of the same order of or larger than that of amplification. Yet, when they differ greatly as $G_b \gg G_a$, very high saturation of the absorption occurs, which results in disappearance of the Lamb dip and hence of the power peak. A typical example of such an unsuitable pair is a CO_2 laser with an SF_6 nonlinear absorption cell

Table 4.1. Inverted lamb dip experiments with a saturated absorption cell inside the laser cavity

Laser	Wavelength (μm)	Absorber	Reference
He–Ne	0.6328	^{20}Ne[a]	4.7, 8, 145, 177–180
		^{127}I$_2$	4.123, 158–161, 181–183
		^{129}I$_2$	4.161, 183
		^{79}Br$_2$	4.184
He–^{22}Ne	0.6328	^{22}Ne[a]	4.185
		^{81}Br$_2$	4.184
He–Ne	1.52	^{20}Ne[a]	4.186
	3.39	^{12}CH$_4$	4.9, 93, 94, 96, 154, 155
		^{12}CH$_4$[b]	4.139
		^{12}CH$_4$[c]	4.79, 137, 138
		^{12}CH$_4$[d]	4.191
He–Ne[b]	3.39	^{13}CH$_4$, CH$_3$OH, C$_2$H$_6$	4.100
		C$_2$H$_4$, C$_2$H$_6$, C$_3$H$_8$, C$_4$H$_{10}$	4.99
		CH$_3$F[c]	4.135
He–Xe	3.507	H$_2$CO	4.67
CO$_2$	10.6	CO$_2$	4.47, 148, 192, 193

[a] In discharge.   [b] In magnetic field.   [c] In electric field.   [d] In molecular beam.

inside the cavity. Another unfavourable situation might occur for molecules with a long relaxation time of vibration in the specific pressure range, where the length of free path of molecules due to collisions is much shorter than the diameter of the laser beam. In this case there is an accumulation of excited molecules with different velocities, and strong saturation of the whole vibrational band appears. As a result the amplitude of the Lamb dip decreases markedly and the power peak practically disappears despite the fact that the homogeneous width of the absorption line is much narrower than the Doppler width. It is difficult to observe the power peak under conditions of continuous oscillation in such a laser. The peak arises easily in the transient time when the laser is suddenly switched on, and exists until the excited molecules are accumulated and they diffuse in velocity space [4.47].

2) Absorption Cell Outside the Laser Cavity

Observation of the Lamb dip in a nonlinear absorber inside the laser cavity calls for a special selection of saturation parameters and absorption and amplification coefficients. In many cases this is impracticable and, of course, one should saturate absorption by the laser field outside the

Table 4.2. The experiments on saturation resonances in external absorbing cell

Laser	Wavelength (μm)	Absorbing particle	References
Ar	0.5017	$^{127}I_2$	4.122
	0.5145	$^{127}I_2, ^{129}I_2$	4.56, 122
	0.5208	$^{127}I_2, ^{129}I_2$	4.122
Kr	0.5682	$^{127}I_2, ^{129}I_2$	4.121, 122
Dye laser	0.5890	Na	4.107
He–^{20}Ne	0.6328	^{20}Ne [a]	4.27, 50, 149
	0.6563	H [a]	4.108, 120
		D [a]	4.120
	3.39	CH_4	4.93, 96
		CH_4 [b]	4.80
He–Ne [b]	3.39	$CH_3OH, CH_3Br,$ $CH_3F, ^{13}CH_4,$	4.94
Spin-flip-laser	5.3	H_2O	4.115
CO	5.714	NH_3, H_2CO [c]	4.158
CO_2–N_2–He	9.6	PF_5, CF_2Cl_2	4.195
		$CH_3F^{3)}$	4.66, 89, 136, 158, 176
	10.6	SF_6	4.26, 48, 105, 152, 195–197
		NH_2D [c]	4.132–134
		SiF_4	4.198, 199
		OsO_4	4.200
		$^{189-192}OsO_4$	4.125
		$CH_3^{35}Cl$	4.124
	9.6–10.6	CO_2	4.54
N_2O–N_2–He	10.8	C_2H_4	4.195
		NH_3	4.58, 195
		N_2O	4.140
Microwave oscillator	8.2×10^3	OCS, CH_3CN	4.201
	$(1-3) \times 10^3$	$OCS, CH_3F,$ $^{35}ClCN$	4.202

[a] In discharge.   [b] In magnetic field.   [c] In electric field.

cavity and observe the Lamb dip in the standing wave. The use of an external absorption cell eliminates any influence of the absorber on the amplitude and frequency of the laser and permits us to determine and to change easily the spatial shape of the light field. At present this technique is generally accepted. Table 4.2 lists atoms and molecules in which narrow resonances have been observed with saturated absorption in an external cell, by using the lasers given in the table. The recent progress of tunable lasers with a narrow laser line width (dye visible lasers and infrared spin-flip Raman lasers) enables us to extend this list over and over.

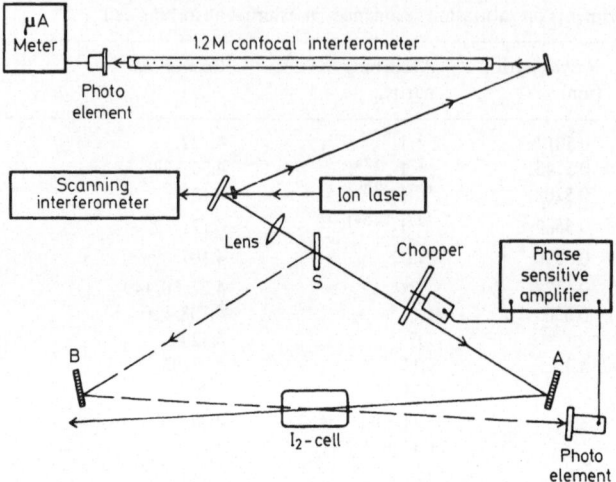

Fig. 4.18. Saturation spectrometer using the intensity modulation of strong wave and synchronous detection of induced modulation of backward probe wave (measurement of hyperfine structure I_2). (From Ref. [4.121])

The method of the counter-running probe wave illustrated in Fig. 4.6 is the most convenient for experiments with an absorption cell outside the cavity. The first studies of saturation resonances in the absorption of an oppositely directed probe wave were carried out in SF_6 by BASOV et al. [4.26] and Ne by MATIUGIN et al. [4.27]. The parameters of narrow resonances have been studied in detail in Refs. [4.48, 49].

As seen from Eq. (4.18), the dip amplitude H in the absorption line is

$$H = \frac{\Delta \kappa}{\kappa_0} = 1 - (1 + G)^{-1/2} . \tag{4.45}$$

It is important to note that the amplitude of the transmission peak grows monotonically as the strong wave intensity increases in contrast to the case of saturation by a standing wave inside a cavity. The resonance amplitude can be substantially increased with the use of an optically dense absorption cell, for which $\kappa_0 L \gg 1$ [4.48].

Figure 4.18 shows a very effective and most generally employed scheme of saturation spectrometer with the method of counter-running probe wave. The scheme employs the amplitude modulation of the strong running wave, which induces absorption saturation [4.50, 51]. The counter-running probe wave will be modulated only when the probe wave is responsive to the hole burnt in the Doppler contour by the strong modulated light wave. In this case the modulated intensity of the probe wave recorded by a phase-sensitive amplifier depends only on saturated absorption and is not sensitive to the unsaturated part of the Doppler contour. Such a scheme is highly sensitive in detecting the spectrum of saturated absorption.

Usually the weak probe wave is directed towards the strong wave at a small angle θ in order not to enter the laser cavity. This gives an additional (geometrical) broadening of the resonance by the value of

$$\Delta v_{\mathrm{g}} = \frac{1}{2\pi}\,\Delta\omega_{\mathrm{g}} \simeq \theta\frac{u}{\lambda}, \tag{4.46}$$

where $u = (2\kappa T/M)^{1/2}$. The influence of weak reflected wave decreases markedly with the use of an optical isolation element. The isolation may be accomplished with non-reciprocal devices such as Faraday isolators. It is possible to use the effect of optical isolation in a Doppler-broadened molecular absorber [4.52]. In such a unidirectional device, the velocity selection by the interaction between a monochromatic wave and a Doppler-broadened resonance is utilized. A strongly saturating wave propagates through the medium with little attenuation, whereas a weak oppositely directed wave suffers large attenuation, provided the laser frequency is detuned from the Doppler peak by an amount larger than the homogeneous (pressure-broadened or power-broadened) line width.

3) Fluorescence Cell in a Standing Wave Field

The narrow resonance of excited atoms by a standing light wave (Subsect. 4.2.4) can be detected by recording their fluorescence intensity (Fig. 4.16b). The main advantage of the fluorescence method (or the observation of population of the excited atoms) lies in the possibility of studying weakly absorbing transition for which $\kappa_0 l_b \ll 10^{-2}$, where κ_0 is the initial absorption coefficient and l_b is the length of the absorption cell. The characteristic feature of the nonlinear fluorescence method is that there is no direct effect of the strong wave saturating the absorption on the photodetector. It is precisely this feature that provides the high sensitivity of the method (see Sect. 2.4). This technique was proposed by BASOV and LETOKHOV [4.53] and realized independently by FREED and JAVAN [4.54]. They observed sharp fluorescence peaks of CO_2 molecules at a low pressure in the 4.3 μm transition when one of the rotational-vibrational lines of the 10.6 μm band was saturated by the CO_2 laser radiation.

The sensitivity of the method is so high that it permits observation of transitions between the excited levels of the CO_2-molecule having an

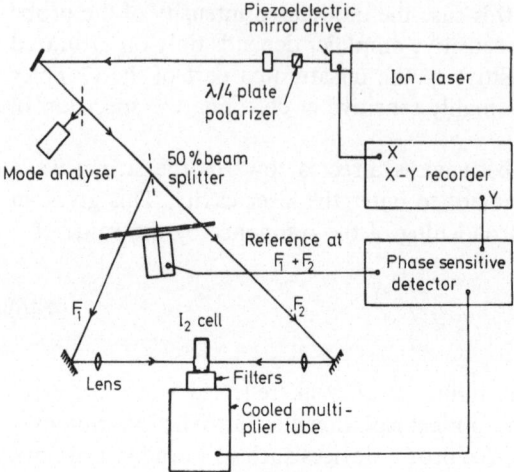

Fig. 4.19. Saturation spectrometer with detection of fluorescence from the saturated cell, and independent intensity-modulation of two strong laser waves (SOREM and SCHAWLOW [4.56]

absorption coefficient at 300 K of 1.5×10^{-6} cm^{-1}. Using a highly sensitive IR photodetector (Ge:Cu) with a large receiving area, it is possible to detect resonances in CO_2 and N_2O molecules (with CO_2 and N_2O lasers, respectively) at a pressure of $10^{-3} - 10^{-4}$ Torr [4.55].

To increase the sensitivity it is convenient to employ saturation by two strong oppositely directed waves which are intensity-modulated at different frequencies F_1 and F_2. The scheme of such an experiment is shown in Fig. 4.19. In this case the information about the dip is contained in the fluorescence signal which is modulated at the sum or difference. This method is important in that it allows the elimination of the continuous background of fluorescence and the parasitic background light. SOREM and SCHAWLOW [4.56] used this technique to resolve the hyperfine structure of the lines P(13), R(45), (43−0) of I_2 with the aid of argon ion laser at 514.5 nm.

Apart from detection of the spontaneous emission of excited atoms, there are other methods (LETOKHOV [4.44, 57]) of observing resonant change in the total number of atoms in the upper level. For example, it is possible to measure the absorption coefficient for the coupled transition between the upper level of the saturated transition and a higher state of atoms (or molecules) including transitions to the continuous spectrum (Fig. 4.20, II). Also we can add another gas that has a level which is close to the excited level of the saturated transition and which decays radiatively with a short lifetime (Fig. 4.20, III). In this case there is a narrow fluores-

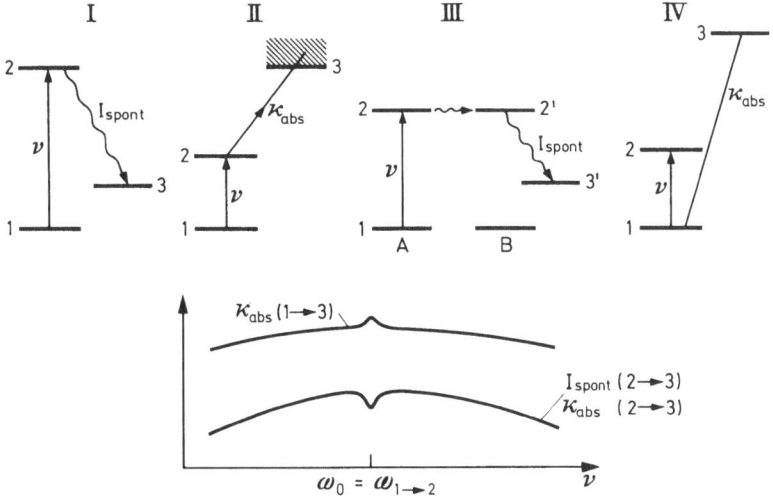

Fig. 4.20. Various methods for detection of the resonant change in the total population of the upper (2) and lower (1) levels of the saturated transition (LETOKHOV [4.57])

cence resonance of the added molecules. It is possible to detect the resonant change in the number of molecules of the lower level by a change of the absorption coefficient of the transition (Fig. 4.20, IV). In this case instead of a resonance dip we can observe a resonance peak.

The saturated fluorescence method can be used to investigate absorption lines 10^4 times as weak as those studied by the method of saturated absorption. That is why the former method is effective for weakly absorbing molecular transitions: molecular transitions between excited levels, transitions at vibrational overtones, quadrupole vibrational transitions of homonuclear molecules, etc.

4) Absorption Cell in a Two-Frequency Field

An interesting modification which enables us to obtain narrow resonance at any point of a Doppler-broadened absorption line was suggested [4.58]. The method is based on saturation by a running light wave at frequency ω_1 and on probing the resultant hole by a counter-running light wave at frequency ω_2 (Fig. 4.16e). When the frequency of the two running waves are tuned symmetrically on opposite sides of the center frequency ω_0, a narrow resonance similar to the Lamb dip will appear. In contrast to the Lamb dip, however, the atoms which give rise to a resonant dip are those with a non-zero velocity component v_{res} along the laser beam given by $v_{res} = [(\omega_i - \omega_0)/\omega_0]c$, where c is the speed of light, ω_i is ω_1 or ω_2. For $\omega_0 - \omega_1 = \omega_2 - \omega_0$ only one velocity group of atoms interacts with both traveling waves. The narrow absorption resonance centered at $\omega_0 - \omega_1 = \omega_2 - \omega_0$ appears when the saturation by at least one wave becomes appreciable (Fig. 4.16e). The expression for the narrow resonance with a two-frequency field (one strong ω_1 wave and second backward probe wave ω_2) is similar to (4.18).

In the regime where collision broadening exceeds the natural broadening and other homogeneous broadening, the width of the above resonance observed for ω_1 (not equal to ω_2) can differ from that of the Lamb dip, owing to the dependence of the collision-broadening cross section on the atomic velocity.

This method has been applied to the observation of the velocity dependence of collision broadening of an infrared transition of NH_3 [4.58]. A cw N_2O laser on the P(13) line at 10.8 μm, which is in close coincidence with the v_2 [asQ(8, 7)] transition of $^{14}NH_3$, is utilized in first experiments. Part of the laser output is sent to a standing-wave Ge acoustic-optic modulator which produces light symmetrically shifted above and below the laser frequency. The frequency shift is about 75 MHz, which at room temperature is $1.5\Delta\omega_D$, corresponding to $v_{res} = 1.5u$. The spatially separated frequency-shifted radiation is split into a strong saturating wave and a weaker probe wave which are sent in opposite directions through an NH_3 absorption cell.

4.3.2 Spectroscopy of Coupled Transitions

It is possible to apply a larger variation of certain experimental methods and spectrometer schemes to two-coupled Doppler-broadened transitions. Below we consider the most widely used and popular methods and schemes.

1) Narrow Resonance in Spontaneous Emission

Narrow resonances in the Doppler-broadened line of spontaneous emission at the level m in the presence of a strong wave on the coupled transition m-n were observed in the first experiments of saturation spectroscopy (BENNETT et al. [4.59], CORDOVER et al. [4.60], SCHWEITZER et al. [4.61], HOLT [4.62]). In all these experiments He–Ne lasers at $\lambda = 3.39$ μm [4.59] or at $\lambda = 1.15$ μm [4.60–62] were used. These radiations saturated the amplification of the corresponding transition of ^{20}Ne or ^{22}Ne. Spontaneous emission was observed either at $\lambda = 0.6328$ μm [4.59] or at $\lambda = 0.6096$ μm [4.60–62]. Figure 4.21 shows a typical scheme of the spectrometer based on this method of three-level spectroscopy. As explained in Subsection 4.2.3, the forward and backward change signals

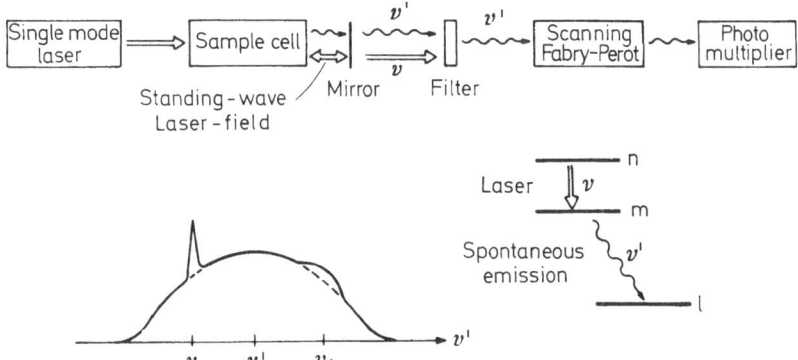

Fig. 4.21. Three-level saturation spectrometer for observation of line-narrowing effects in spontaneous emission. The laser field is put in the form of a standing wave by means of a partially reflecting mirror. (From Ref. [4.31])

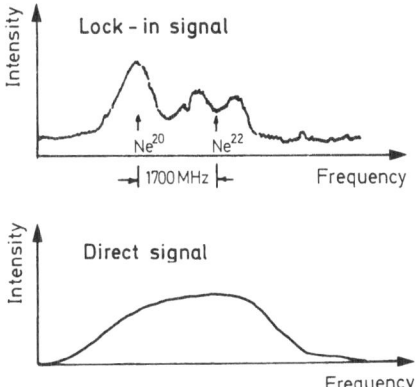

Fig. 4.22. Experimentally observed neon isotope shifts by a three-level saturation spectrometer which is shown in Fig. 4.21. Wavelength of the laser standing field $\lambda_{las} = 1.15\ \mu m$, wavelength of spontaneous emission $\lambda_{sp} = 0.6096\ \mu m$. The lower trace shows a direct signal from the detector. The upper trace shows a modulation signal of saturation effect. (From Ref. [4.60])

are symmetrically located on opposite sides of the Doppler profile. Therefore, by studying forward and backward change signals together, it is possible to determine the atomic center frequency of the coupled transition with an accuracy limited only by the homogeneous line width. By utilizing a standing wave laser field, as shown in Fig. 4.21, it is actually possible to make both change signals appear together at frequencies v_+ and v_-, respectively, symmetrically located about v_0'.

Figure 4.22 shows an experimental trace in which the sample cell contains a mixture of ^{20}Ne and ^{22}Ne (CORDOVER et al. [4.60]). The lower trace is the direct output of the photomultiplier and shows the normal broad Doppler-broadened spontaneous emission spectrum of the ^{20}Ne and ^{22}Ne which overlap closely. The upper trace shows the lock-in signal from the saturation spectrometer when the laser frequency is tuned to the center of the Doppler profile of the ^{20}Ne laser transition and is somewhat detuned to the high-frequency side of the ^{22}Ne laser transition. Therefore, at the coupled transition the forward and backward ^{20}Ne resonances coincide, but the ^{22}Ne resonances are split. The aniso-tropic effect of spontaneous emission in the presence of a strong wave on the coupled transition, treated qualitatively in Subsection 4.2.3, was first observed in Ref. [4.62]. The relationship between the spontaneous emission line shape and the direction of observation was revealed also by DUCAS et al. [4.63], when they studied the hyperfine structure of ^{21}Ne at $\lambda = 1.15\ \mu m$ (laser transition) and $\lambda = 0.6096\ \mu m$ (spontaneous transi-tion). The allowance made for anisotropy in this paper permitted measur-ing the hyperfine structure parameters and estimating the quadrupole moment of ^{21}Ne.

Since the detection of weak spontaneous emission signals is a severe problem, this method of saturation spectroscopy is rather difficult. Therefore, the method of weak probe wave is used more often to study the line shape of coupled transitions.

2) Optical Double Resonance

The line shape of stimulated emission by a weak probe wave agrees with that of spontaneous emission. There may be some difference only due to the fact that the spontaneous emission probability is proportional to the population of the common level m, while that of stimulated emission is proportional to the difference in population between the levels m and l.

Figure 4.23 shows schemes of narrow resonances observation on the coupled transition in spontaneous and stimulated emissions. There are two principal differences between them as far as the possibilities of the experiment are concerned. Firstly, in the spontaneous version the resolu-tion is determined by the Fabry-Perot scanning interferometer and the angle at which the spontaneous emission is detected from the laser beam. A higher resolution decreases the sensitivity. In the stimulated version this problem is ruled out because the probe beam can be easily directed in parallel to the strong wave, and a high spectral resolution is realized by scanning the single-mode laser. Secondly, spontaneous emission can be observed only in the case of allowed atomic transition and rarely in the case of vibrational molecular transitions. In the stimulated version we may study either allowed or forbidden transitions in the visible and infrared regions of atoms and molecules.

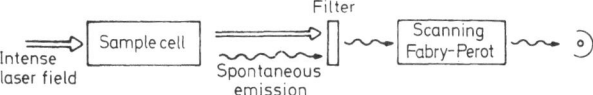

(a) Spontaneous emission version

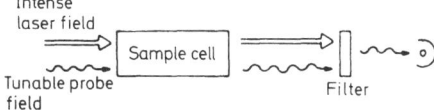

(b) Stimulated emission version

Fig. 4.23. Experimental arrangements for observation of laser-induced line-narrowing effects in coupled systems. a) Spontaneous emission version; b) stimulated emission version. (From Ref. [4.31])

By now the line shape of stimulated emission in the presence of the strong wave on the coupled transition has been studied in a few experiments by He–Ne laser in the Ne transitions (BETEROV and CHEBOTAYEV [4.64, 43], HÄNSCH et al. [4.65]). They revealed anisotropy of Ne line at $\lambda_{ml} = 1.15\,\mu m$ with a strong wave at $\lambda_{mn} = 0.63\,\mu m$. The distinguishing feature of this case is $k_{ml} < k_{mn}$, when the non-Lorentzian structure of narrow resonance is theoretically more complex than that described in Subsection 4.2.3. Figure 4.24 gives calculated relationships between the spectral line shape for the probe wave at λ_{ml} and the varied population differences of the levels m and l for the case when the strong wave is detuned with respect to the central frequency ω_{mn} (HÄNSCH and TOSCHEK [4.39]). It is seen that, if the weak and strong waves are unidirectional, the narrow resonance shape may have a splitting, interpreted as dynamic Stark effect. Figure 4.24b shows experimental relationships [4.65] of amplification of a weak probe signal at $1.15\,\mu m$ in the presence of wave at $0.63\,\mu m$, which is also detuned about the line centre. As in Fig. 4.24a, the narrow resonances on the left and on the right correspond to the cases of counter-running and unidirectional waves for different values of amplification (absorption) at the transition m-l. One can see here clearly that the experimental and theoretical data are in agreement. Other details of quantitative comparison between theory and experiment for this method of three-level saturation spectroscopy are obtained in Ref. [4.43].

The method of probe wave at a coupled transition is often termed as the double-resonance method. It is widely used in molecular saturation spectroscopy. When both waves act on coupled vibrational molecular transitions, a double IR–IR resonance occurs. Such a method was used, for example, by BREWER [4.66] for spectroscopy of CH_3F molecule by

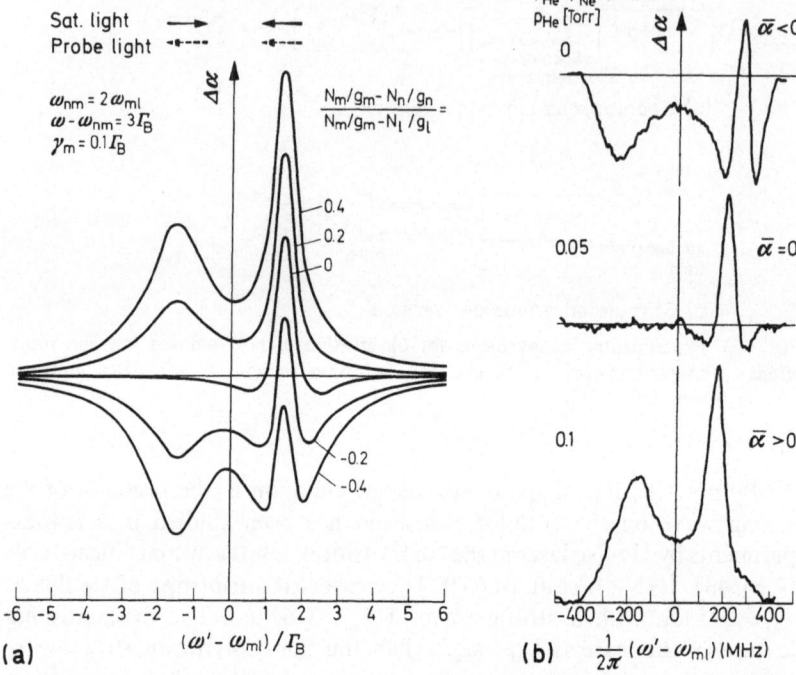

Fig. 4.24a and b. Calculated (a) and experimental (b) spectra of the probe wave amplification on the 1.15 μm Ne transition in the presence of a strong standing wave at 0.63 μm which is detuned from exact resonance (theoretical curves from paper [4.39], experimental curves from [4.65]). The explanation is in text

means of two CO_2 lasers. The experiment is discussed in Section 4.6. When one wave acts on the vibrational molecular transition and the other acts on its connected rotational transition, a double IR-microwave resonance takes place. This method was employed by TAKAMI and SHIMODA [4.67] for spectroscopy of H_2CO molecule. They managed to observe a narrow resonance splitting of the vibrational molecular transition caused by the dynamic Stark effect in a strong microwave field at the rotational transition. A more comprehensive discussion of this method is given by SHIMODA [4.68], and all the methods of double resonance are described in review by SHIMODA and SHIMIZU [4.69].

3) Mode-Crossing

The method of mode-crossing is very similar to the above-discussed method of optical double resonance. The only difference is that two coupled transitions are so closely spaced that the Doppler profiles of

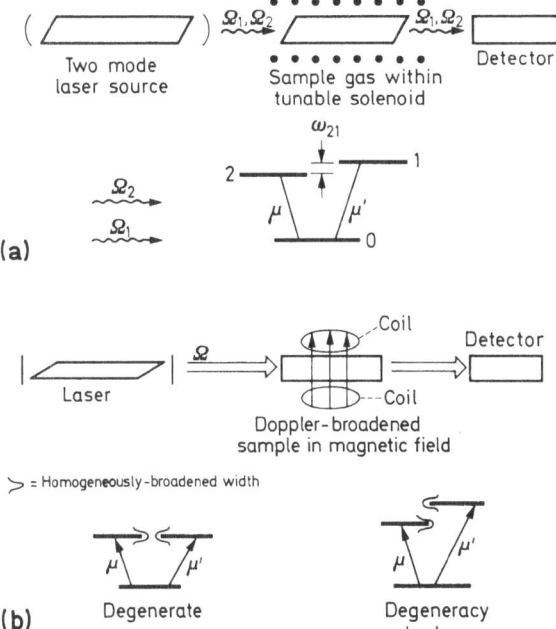

(a)

(b)

Fig. 4.25a and b. Saturation spectrometers using the mode-crossing effect (a), and level-crossing effect (b). (From Ref. [4.35] and [4.42])

their lines overlap with each other. In this case we can use the waves of two adjacent modes of the same laser for saturation spectroscopy of such overlapping transitions. Figure 4.25a shows a saturation spectrometer based on mode-crossing.

The effect manifests itself as resonant change in the absorption induced by the applied field. The resonance occurs when the separation between the closely spaced levels equals the frequency separation between the two laser modes

$$\Omega_1 - \Omega_2 = \omega_{12} . \tag{4.47}$$

In this case an atom will only interact with both waves if their Doppler shifts are equal or $\Omega_1 - \omega_{01} = \Omega_2 - \omega_{02}$. Thus, if one of the laser frequencies is varied or if the level splitting ω_{21} is tuned by the Zeeman or Stark effect, a resonant response of saturation will appear at the resonance condition (4.47). An important experimental consequence of the mode-crossing frequency condition is that it depends only upon the frequency separation between the laser modes, and not on their absolute frequencies. This makes the method easily realizable with a free-running multimode laser, where the mode spacing remains controlled although the absolute frequencies may change over the broad Doppler profile during the time of measurement. This advantage is not general and holds only for the unidirectional waves and closely spaced levels. It does not occur for the antiparallel waves with different frequencies (see (4) in Subsect. 4.3.1). In the case of a Doppler-broadened laser medium with closely spaced tunable structure, the mode-crossing effect is also observable as a resonant change in the output signal of the laser itself. But in this case only the forward signal is stable against the frequency instability.

The theory of mode-crossing effect was developed by SCHLOSSBERG and JAVAN [4.32]. They have shown that the width of mode-crossing resonance is determined only by the width of initial and final levels $\gamma_1 + \gamma_2$. If the optical field for one transition, say $1-0$, is strong, the resonance width is subjected to power broadening in such a way as $\gamma_1\sqrt{1+G}+\gamma_2$.

The effect of mode-crossing in a magnetic field has been experimentally observed by SCHLOSSBERG and JAVAN [4.70]. They used it in measuring the hyperfine structure of the 3.37 μm xenon line [4.70] and g-factors of a few oxygen transitions [4.71, 31]. In these works they have proved experimentally that the common level width does not contribute to the width of the resonance signal under observation. For instance, at $\lambda = 3.37$ μm the observable resonance width is 0.5 MHz, that is 30 times less than the common level radiation width. The effect of mode-crossing on saturation of an atomic transition in a magnetic field has been recently discussed in detail by DUMONT [4.72, 73].

4) Level Crossing

Level crossing in saturated absorption may be regarded as a particular case of mode crossing when $\omega_{01} = \omega_{02}$, that is one of the levels is doubly generated. Then one running wave, which connects two sublevels with transitions via the common level, will suffice for resonance condition (4.48) to be fulfilled. There may be a resonance in the change of running wave absorption when degeneration is removed by an external electric or magnetic field. Crossing resonances may occur either at high fields ("level crossing") or at zero field ("Hanle effect"). The width of level-crossing resonance is determined by the homogeneous broadening of degenerate (crossing) levels.

A typical experimental arrangement of a level-crossing saturation spectrometer is shown in Fig. 4.25b. A sample cell containing a low-pressure Doppler-broadened gas is subjected to the traveling laser wave. It should be noted that an experiment of this type does not require the high stability of laser frequency. It is enough to keep laser frequency inside the Doppler profile and not the homogeneous width. This is in contrast to the narrow resonance of the Lamb-dip type, which is disturbed under small frequency variations.

The experimental scheme illustrated in Fig. 4.25b corresponds to the observation of level crossing in stimulated transitions. It is well known that mode-crossing signals appear in spontaneous emission as a resonant change in angular distribution and polarization characteristics of fluorescence from an atom or a molecule at the instant of level crossing (see reviews [4.74, 75]). FELD et al. have shown in their work [4.42] that the same physical processes are responsible for both stimulated and spontaneous emission versions of the effect.

The level crossings induced by the laser radiation have been observed in both stimulated and spontaneous emission. Stimulated level crossings were initially observed in the splitting of the Zeeman levels in fine structure of oxygen at zero magnetic field by FELD [4.76] and in crossings of hyperfine levels of xenon in a high magnetic field [4.71]. Manifestations of stimulated level crossings were also present in experimental [4.77] and theoretical [4.78] studies of the output characteristics of a Zeeman-tuned He–Ne laser near zero magnetic field. The stimulated level crossings in CH_4 with Stark tuning of levels were observed by LUNTZ et al. [4.79] and with Zeeman tuning by UZGIRIS et al. [4.80]. Level crossings in the spontaneous emission of sidelight of a Zeeman-tuned laser have also been studied [4.81–85] and recently for molecular iodine as well [4.86].

5) Collision-Induced Optical Double Resonance

So far we have considered the three-level cases in which a strong wave on one transition gives use to a sharp resonance in the other coupled transition. That is why an atom can transfer from the initial state n to the final state l at a definite resonance velocity v_{res}. A narrow resonance may also arise at the transition k-l owing to saturation at the transition m-n which has no common level with k-l. For the first time the effect was observed on transitions in Ne atoms [4.87]. It is conditioned by the transition of atoms, which are colliding with electrons, from one of the levels of m-n to the levels of k-l. Owing to the mass difference between the electron and the atom, the atomic velocity remains almost constant during collisions. Therefore, the holes and peaks in the atomic velocity distribution of the m and n levels may be transferred by collisions to the distributions of the k and l levels, hence producing a narrow resonance on the Doppler-broadened transition k-l.

An analogous effect can be observed in molecular collisions. The quantitative measurements [4.88] show that the average velocity change in CH_3F-type molecular collisions may be as low as 200 cm/s. At the same time, a molecular transition from the state of fixed angular mo-

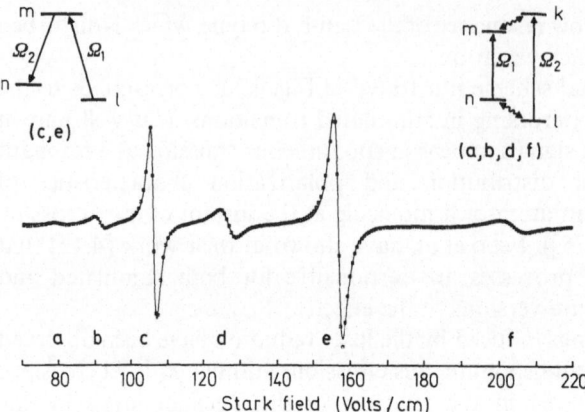

Fig. 4.26. Optical double resonance spectrum for $^{13}CH_3F$. Lines (c) and (e) correspond to the left configuration (with a common level) and others to the right configuration (collision-induced optical double resonance). (From BREWER et al. [4.89])

mentum projection M to another state M' needs a small number of collisions in which the molecular velocity does not change markedly. A collision-induced double-resonance on CH_3F molecule was observed by BREWER et al. [4.89]. The results of this experiment are given in Fig. 4.26. Two light fields from two CO_2-lasers with the fixed frequency difference $\Omega_1 - \Omega_2 = 30$ MHz were passed through a cell filled with $^{13}CH_3F$ gas at a pressure of 3 mTorr. To detect double resonance, the scanning was achieved by Stark frequency tuning of $^{13}CH_3F$ transitions. Apart from the two resonances, conditioned by coupled transitions with a common level, there were resonances observed at transitions connected by collisions only.

Collision-induced double resonance can be observed at any pair of transitions, provided that the collisions connecting transitions conserve the molecular velocity. Among such collisions are those which give rise to molecular inversion [4.90] or angular momentum transfer [4.91, 92]. In these cases double resonances at transitions with no common level have been also observed.

4.4 Resolution of Saturation Spectroscopy

The limit of the resolving power of saturation spectroscopy is determined by the narrow dip width $\Delta\omega$. In most general case the dip width is determined by the contribution of the two effects:

1) Atomic spectral line broadening caused by radiative level decay, collisions, etc. (see Chapt. 2).

2) Instrumental broadening of resonance caused by light-field interaction.

4.4.1 Contribution of Various Effects

The instrumental resolution is determined by several effects: broadening by saturation, geometric broadening caused by the wave front curvature, transit broadening due to the finite time of interaction, and the width of the laser line (see Chapt. 2).

The ultimate task of saturation spectroscopy is to achieve the natural width of a spectral line without contribution made by instrumental effects. Let us discuss their contribution to resolving power in detail.

1) Power Broadening

When the absorption is saturated by a standing wave field the resonance width is

$$\Delta\omega = 2\Gamma \cdot f_P, \tag{4.48}$$

where 2Γ represents the homogeneous full width at half-height of the transition and f_P is the factor of power broadening of the dip, the value of which is given in Fig. 4.4 (standing wave saturation) and Fig. 4.8 (weak probe wave). In the simplest approximation f_P can be taken as

$$f_P = (1+G)^{1/2} \tag{4.49}$$

When the method of a counter-running probe wave is used, the resonance width without coherent effects is determined by expression (4.19), i.e.

$$f_P = \tfrac{1}{2}\{1 + (1+G)^{1/2}\}. \tag{4.50}$$

For strong saturation ($G \gg 1$) the broadening caused by the field is half of that in the case with a standing wave.

2) Transit-Time Broadening

The contribution made by the flight time of an atom through a light beam that saturates absorption was discussed by HALL [4.93, 94], since in his experiments with a record resolution this effect has to be allowed for. For simplicity, let us take a Gaussian light beam and look upon the atoms flying in parallel with the wave front at the radial velocity into the region of the beam waist where there is no distortion in the wave front. Then the broadening owing to the flight time is determined only by the effect of amplitude modulation when atoms move in the field. Hall used the following expression to evaluate approximately the broadening by this effect:

$$\Delta v_{tr} \simeq \frac{1}{4} \frac{\langle v_r \rangle}{r_{1/e}}, \tag{4.51}$$

where Δv_{tr} is the FWHM, $r_{1/e}$ is the Gaussian beam radius on the level $1/e$ of amplitude, $\langle v_r \rangle$ is the average radial velocity which is given by the expression

$$\langle v_r \rangle = (\sqrt{\pi}/2)u\,, \quad u = (2kT/M)^{1/2}\,. \tag{4.52}$$

Recently BAKLANOV et al. [4.95] have carried out a rigorous estimation for the narrow dip width in a low-pressure gas, with allowance made for final cross section of a light beam with a Gaussian profile of intensity and velocity distribution of atoms. Their expression for the FWHM has the form

$$\Delta v_{tr} = \frac{1}{2\pi} \Delta \omega_{tr} = \frac{0.58}{\pi} \frac{u}{r_{1/e}}\,. \tag{4.53}$$

The value of (4.53) is just 16% less than the approximate value of (4.52).

The transit-time broadening is very small. For instance, it comes to only 140 kHz for $r_{1/e} = 1$ mm [4.94] for the 3.39 µm transition of CH_4 at 300 K. In most experiments with $r_{1/e}$ of several millimeters the contribution by transit-time broadening may be neglected. To reach a high possible resolution, however, as in the experiment by HALL and BORDE [4.96] in measuring the magnetic hyperfine structure of CH_4, the line width at 3.39 µm must be less than 10 kHz, and the beam diameter has to be increased up to 5 cm.

3) Geometrical Broadening

The directional alignment of two counter-running plane waves through an angle θ broadens the resonance by a value given by relation (4.46). This relation can be written as

$$\Delta \omega_{geom} = 2\pi \Delta v_{geom} \simeq 0.6\theta \Delta \omega_D\,. \tag{4.54}$$

This broadening can be termed as the residual Doppler effect caused by light wave non-parallelism.

An analogous broadening due to a spherical wave front was estimated by LETOKHOV [4.97] to be

$$\Delta \omega_{sph} \simeq ku/(kR)^{1/2}\,, \tag{4.55}$$

where R denotes the radius of curvature. It should be noted that, even though we have at first a plane light wave of diameter a, the wave front becomes distorted owing to diffraction. The radius of curvature for a

Gaussian beam takes a minimum value of R_{dif} at the distance of $ka^2/2$, so that we may assume the approximate value as

$$R_{dif} \simeq ka^2 . \tag{4.56}$$

From (4.55) and (4.56) the maximum broadening due to wave front curvature will be

$$\Delta\omega_{sph}^{min} \simeq \frac{u}{d} \simeq \Delta\omega_{tr} , \tag{4.57}$$

that coincides with the transit-time broadening. The coincidence of these values is not accidental, since both types of broadening are characterized by the beam diameter.

4) Laser Linewidth and Others

If the geometrical broadening is eliminated, the resolution of saturation spectroscopy is limited by the linewidth of the laser used, that is by the short-term frequency stability. The theoretical relative linewidth of a cw gas laser determined by spontaneous emission and thermal noise is as small as $10^{-13} - 10^{-15}$ (JAVAN et al. [4.98]). But practically a very great contribution to linewidth can be made by acoustic and other fast disturbances of the laser cavity which are hardly compensated by a servo-system stabilizing the oscillation frequency. In a very carefully constructed arrangement it is possible to obtain a linewidth of about 1 kHz (He–Ne laser at 3.39 μm [4.94]).

Narrowing the laser radiation linewidth, in order to obtain a resolution much better than 10^{11}, requires consideration of other fundamental effects that limit the resolution of the saturation method. There is, for instance, a thermal distribution of atomic transition frequencies in gas due to the second-order Doppler effect. The value of such broadening is

$$\Delta\nu_{S.D.}/\nu_0 \simeq \frac{kT}{Mc^2} . \tag{4.58}$$

This value, say, for CH_4 at 300 K amounts to 1.8×10^{-12}.

4.4.2 Some Peculiarities of Saturation Spectroscopy

The saturation spectroscopy of overlapping complex (three or more levels) transitions has some special features (additional crossing effects and possibility of elimination of intermediate level width for coupled

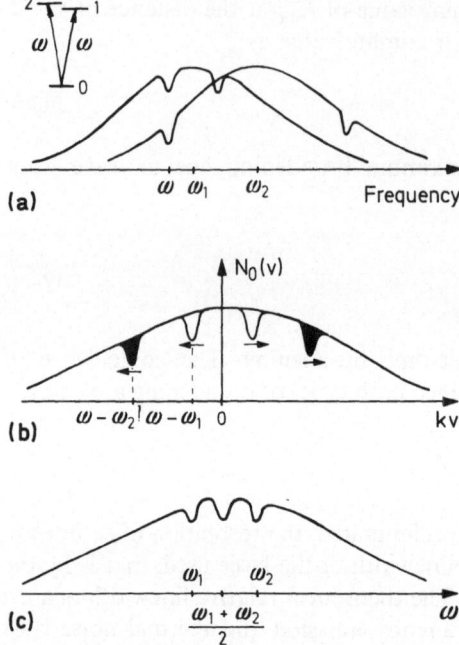

(a)

(b)

(c)

Fig. 4.27a–c. The occurrence of additional cross-resonances when one is saturating the absorption of overlapping transitions with a common level in the field of a standing wave with a frequency ω: (a) the Doppler profile of two lines; (b) velocity distribution on the common level "0": (c) the saturation spectrum

transitions), which are absent in linear spectroscopy. In the case of spectroscopy of unknown structures of transitions these effects should be taken into account.

1) Additional Cross-Resonances

If close-lying spectral lines inside the Doppler width belong to transitions with a common level, the saturation spectrum will contain additional resonances (crossings) which one should take into account in analysing the line structure. Assume that a standing wave saturates absorption of the spectral lines formed by an overlap of two Doppler-broadened lines with a common level (Fig. 4.27). Two holes are burnt in each line due to the interaction with atoms having velocity components $kv = \pm(\omega - \omega_1)$ and $\pm(\omega - \omega_2)$. When the wave frequency is varied, the two dips at the frequencies ω_1 and ω_2 appear due to hole overlapping in the line centres and further a cross-dip at the frequency $(\omega_1 + \omega_2)/2$ appears due to overlapping of the right-hand hole on the ω_1 line with the left-hand hole on the $\omega_2(\omega_1 < \omega_2)$ line. This effect was considered by SCHLOSSBERG and JAVAN [4.32] (see Subsect. 4.2.3) and has been observed in experiments, for example, by UZGIRIS et al. [4.80] using the Zeeman splitting of the absorption line of CH_4 and by HALL and BORDE [4.96] in saturation spectrum of the hyperfine structure of CH_4.

Cross-resonances also occur when a transition is saturated by one running wave and is probed by an oppositely directed weak wave. Also the saturation spectrum of a fluorescent cell might contain additional crossings.

2) On Spectroscopy Inside the Natural Width

In a number of cases the width of saturation resonances turns out to be smaller than the homogeneous width of the transition 2Γ and, consequently, less than the radiative (natural) width $\gamma_{rad} = \gamma_1 + \gamma_2$ provided the radiative decay makes a main contribution to the homogeneous width. Therefore it is advantageous to look into the possibilities of spectroscopy inside the radiative width by methods of nonlinear optical resonances.

Narrow resonances by the unidirectional wave method ((3) in Subsect. 4.2.2) are observed with widths close to the decay rates of the initial and final levels. If the travelling wave at frequency ω_1 is strong and that at a scanned probe frequency ω_2 is weak, then the absorption coefficient of the probe wave has a resonance minimum at $\omega_2 \approx \omega_1$ which is complicated in structure. To cite an example, when $\gamma_1 \ll \gamma_2$, the shape of a complex resonance represents the sum of three dispersion dips with half-widths $2\Gamma = \gamma_{rad}$ (in the absence of collisions), γ_1 and γ_2, against the Doppler profile. Thus, observation of narrow resonances by this method permits one to obtain information on the lifetimes of the levels, not by direct measurement of lifetimes but by methods of saturation spectroscopy. It should be stressed that the occurrence of a narrow resonances with a width $\gamma_1 \ll \gamma_{rad}$ by no means provides the possibility of resolving spectral lines with the same resolving power. If the spectral line consists of several overlapping lines, a resonance occurs only at one frequency of the strong wave.

In the case of saturation spectroscopy of coupled transitions the narrow resonance width can be determined (Subsect. 4.2.3 and Sect. 6.2) by the widths of the initial and final levels, and the intermediate level makes no contribution to the broadening. If the width of the intermediate level $\gamma_2 \gg \gamma_1$ and γ_3, the width of the narrow resonance observed at the frequency $\omega_2 \simeq \omega_{32}$ is nearly $\sim (\gamma_1 + \gamma_3)$, that is, much smaller than the radiative width of the probe transition $\gamma_{rad} = \gamma_2 + \gamma_3$ (Fig. 4.28). Such a narrow resonance arises at the frequency $\omega_2 = \omega_1 - \omega_{13}$, where ω_{13}

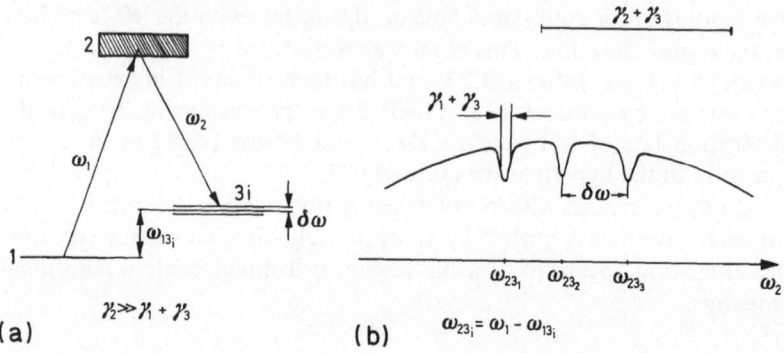

Fig. 4.28a and b. The potentialities of spectroscopy inside the natural width of the 2−3 transition when the three-level system (a) lies in the field of two unidirectional light waves (ω_1 is a strong wave with a fixed frequency; ω_2 is a weak probe wave with a scanned frequency). The absorption coefficient of the probe wave (b) has narrow resonances of the transition

represents the frequency of the "forbidden" transition 1−3. If, for example, level 3 has a complicated structure with the level separation $\delta\omega$, and

$$\gamma_1 + \gamma_2 > \delta\omega > \gamma_1 + \gamma_3 , \tag{4.59}$$

this structure will be resolved when the probe field frequency ω_2 (Fig. 4.28b) is scanned.

4.4.3 Information Capacity of Saturation Spectroscopy

A saturation laser spectrometer used for atomic and molecular spectroscopy without Doppler broadening has at least two important advantages for applied spectroscopy. Firstly, its information capacity of the spectral range will be

$$\mathscr{P} = p_0 R \frac{\Delta v}{v_0} \, [\text{bit}] , \tag{4.60}$$

where p_0 is the number of information units (bits) obtained within the resolved spectral width v_0/R owing to the spectral intensity measurement at a given frequency. For example, with a resolution $R = 10^8$, the spectral range $\Delta v = 1 \text{ cm}^{-1}$ at $v_0 = 3 \times 10^3 \text{ cm}^{-1}$ contains 3×10^4 bits, that is, with $p_0 = 10$ the value is $\mathscr{P} = 3 \times 10^5$ bits. For a common high-quality infrared spectrometer with a resolution of 0.1 cm^{-1} there are only about 10^2 bits in the same range. Secondly, complex molecules are characterized by overlapping of a large number of rotational-vibrational lines,

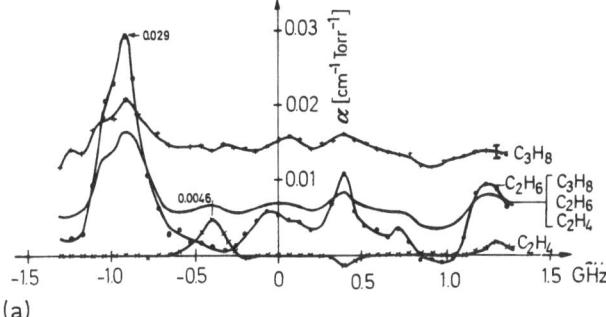

(a)

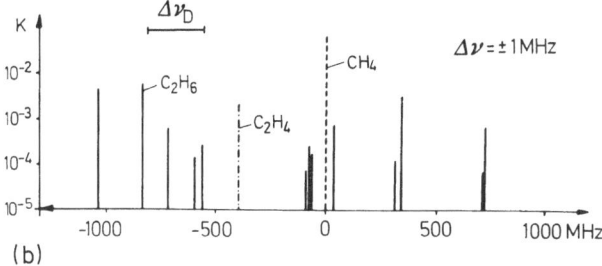

(b)

Fig. 4.29a and b. Laser spectroscopy of hydro-carbon compounds: (a) linear spectrum for C_2H_6, C_2H_4, C_3H_8; (b) saturation spectrum for CH_4, C_2H_4 and C_2H_6 (from [4.99])

the distance between them being less than the Doppler width. In this case the maximum number of information bits in one infrared octave is

$$\mathscr{P}_{max} \simeq p_0 \frac{v_0}{\delta v_{vib}}, \qquad (4.61)$$

where δv_{vib} is the vibrational bandwidth. This greatly limits the potentialities of IR molecular spectroscopy. This limitation is absent in saturation spectroscopy, and information on complex molecules sufficient for quantitative and qualitative spectral analyses can be obtained from a rather narrow spectral range.

The first advantageous experiments using the saturation spectrometer for such purposes are reported by RADLOFF and BELOW [4.99]. The C–H stretching vibration lies in the range of 3.4 μm. When the frequency of the 3.39 μm He–Ne laser is scanned by a magnetic field, we may find in the region of only 0.2 cm^{-1} narrow resonances of 0.5 MHz in width caused by rotational-vibrational lines of CH_4, C_2H_4, C_2H_6 and other molecules. Figure 4.29 shows an experimentally observed absorption saturation spectrum for the gas mixture of some hydrocarbons. The detection sensitivity of molecular admixtures by this method ranges from 10^{-2} to 10^{-3}. High-resolution studies of methyl halides by saturation spectroscopy are described in detail in Chapter 5.

4.5 Lasers for Saturation Spectroscopy

For saturation spectroscopy without Doppler broadening we may use only lasers with a very high temporal and spatial coherence. The time coherence or laser linewidth Δv_{las} should be much less than the desired resolution Δv_{resolv}. Depending on the method used, we need a long-term frequency stability. For example, in experiments with the method of inverted Lamb dip the frequency drift during scanning should not exceed Δv_{resolv}. In some other methods, say, in mode-crossing, the laser frequency drift is not so severe, but the difference between laser frequencies of two modes must be stable. The diameter of the laser beam and its angular divergence in a gas cell must provide smaller values of transit and geometrical broadenings determined by (4.51) and (4.54), than Δv_{resolv}. Besides, the laser power should suffice to saturate absorption of the atomic or molecular transition in gas under study.

The whole complex of requirements for the laser source in saturation spectroscopy is a problem of great concern, and so far they have been met only at a small number of laser lines of cw gas lasers which have a power density inside the cavity of the order of 1 W/cm^2 under a single-mode operation. Therefore most experiments in saturation spectroscopy in the period of 1967–74 have been conducted with atomic and molecular transitions, the absorption frequencies of which accidentally coincide with the oscillation frequencies of such lasers. Such coincidences are listed in Tables 4.1 and 4.2. It was evident from the very beginning that the progress in tunable lasers with narrow and frequency-tunable lines would widen the scope of saturation spectroscopy. But only recently has it enabled us to start realizing the potentialities of tunable lasers.

4.5.1 Lasers with Narrow Frequency Tuning

There are two types of narrow-range tunable lasers now used successfully: He-Ne lasers tuned by a magnetic field and CO$_2$ lasers with pressure-broadened lines.

MAGYAR and HALL [4.100] used a 3.39 μm He-Ne laser with frequency tuning in a magnetic field by ± 3.2 GHz and observed about 30 resonances of absorption saturation in hydrocarbon molecules (CH$_3$OH, C$_2$H$_6$), in methyl halides (CH$_3$Br, CH$_3$F, CH$_3$Cl, CH$_3$I) and in the $P(6)$ component of the ^{13}CH$_4$ molecule. They measured the frequencies of these resonances accurately using a He-Ne laser stabilized on the component $F_2^{(2)}$ of the $P(7)$ line of $^{12}CH_4$. The absolute frequency of this laser was measured by EVENSON et al. [4.101] accurate to 50 kHz (!). This method and the observed results are fully described in Chapter 5. It should be said that the great progress achieved by JAVAN et al. at MIT [4.102] and by EVENSON et al. at NBS [4.103] in measuring absolute frequencies of infrared lasers makes the method of heterodyning very promising for precision saturation spectroscopy over a wider spectral range.

The other approach is based on pressure broadening; PROVOROV and CHEBOTAYEV [4.104] suggested development of cw CO_2-lasers at a pressure of up to 1 atm isotope mixtures ($^{12}C^{16}O_2$, $^{12}C^{18}O_2$, $^{13}C^{16}O_2$), that allows, in principle, a continuous tuning within a range up to $50-100$ cm^{-1}. The gain band-width of CO_2-laser transitions increases at a rate of ~ 5 MHz/Torr, the Doppler width being 53 MHz. Therefore, at a pressure of 300 Torr, we can obtain a frequency tuning range of about 1.5 GHz; that is wider than the typical Doppler broadening of molecular transitions in the region of 10 μm. The first experiments with such a laser in saturation spectroscopy (for the SF_6 molecule) were discussed by BETEROV et al. [4.105]. This approach is very promising, especially in the light of the recent progress achieved in elaborated waveguide CO_2-lasers. Now it is possible to build a simple, small, sealed-off cw CO_2-laser at a pressure of about 300 Torr and with an output power of about 0.1 W (ABRAMS [4.106]).

4.5.2 Tunable Lasers

In saturation spectroscopy tunable dye lasers are used in the visible and tunable "spin-flip" lasers in the infrared.

The first successful application of a pulsed dye laser to saturation spectroscopy without Doppler broadening of atomic lines was carried out by HÄNSCH et al. [4.107]. They studied the hyperfine structure of resonance Na lines with a tunable pulsed dye laser pumped by an N_2-laser at 3371 Å. The tunable laser had a line width of 300 MHz and it was further narrowed to 7 MHz by using an external confocal interferometer as a narrow-band filter. This experiment made it possible to resolve hyperfine splittings of the ground and excited $^3P_{1/2}$ states of Na and to measure the broadening of resonances only 50 MHz in width with a Doppler broadening of 1350 MHz. The experiment has shown that a rather simple pulsed tunable dye laser may widen the use of saturation spectroscopy and remove the restriction of accidental coincidences between the laser and absorption lines. This method was used properly in the next work of HÄNSCH et al. [4.108] to measure the Lamb dip at the H_α-line (6563 Å) of a hydrogen atom. The laser linewidth was reduced down to 30 MHz approximately [4.108]. The experiment is discussed in more detail in Subsection 4.6.2. The tunable laser used in these experiments has been described by HÄNSCH [4.109].

Much narrower lines, enabling us to conduct experiments with a resolution of $R \gtrsim 10^{10}$, can be obtained in cw dye lasers. With a wide-band servosystem (0–100 kHz) we can stabilize the cw dye laser frequency with respect to a transmission peak of a high-quality Fabry-Perot interferometer (BARGER et al. [4.110]). By this method the dye laser frequency drift, which without servosystem is ±50 MHz, can be reduced down to $1-2$ MHz/min with an averaging time of one second. Further work produced a dye laser linewidth of about 0.1 MHz [4.111, 112]. The first experiments in saturation spectroscopy (the I_2 lines at 5957 Å) were carried out by BARGER et al. [4.110]. Recently LETOKHOV and PAVLIK [4.113] have studied the possibility of narrowing the dye laser spectrum at the expense of frequency autostabilization of the internal saturated absorption cell with vapour of atoms (Sr, Mg, Ca).

The cw spin-flip InSb laser tuned by a magnetic field and pumped by a cw CO laser is a very promising laser source for saturation spectroscopy in the infrared. The frequency characteristics of such a laser have been studied in detail by BRUECK and MOORADIAN [4.114]. They have proved that the linewidth of such a laser may be reduced down to 30 kHz (with a time of observation being several minutes) and tuned over a range of $5-6$ μm by the use of a wide-band (0–30 MHz) servosystem which controls the oscillation frequency. The first experiment on saturation spectroscopy (the line of H_2O vapour near 5.29 μm) with such a laser was conducted by PATEL [4.115]. The Lamb dip in his experiment was 200 kHz in width; that was 10^3 times narrower than the Doppler width. The cw spin-flip laser frequency in the range of $5-6$ μm can be measured accurately with respect to the known frequencies of the CO-laser lines. This opens up a real possibility for precision spectroscopy of molecular absorption with absolute measurement of transition frequencies.

4.6 Applications of Laser Saturation Spectroscopy

4.6.1 Spectroscopic Data

Experiments with a resolution of 10^8 have become not uncommon at present. In the best experiments with the method of saturated absorption, a resolution of about 5×10^{10} has been attained [4.116] (see Chapt. 5 in this book), that is 10^6 times as high as that of the best classical spectrometers and 10^5 times that of linear laser spectroscopy. The comparison

Table 4.3. The spectroscopic atomic effects

Effect	Resolution
1. Fine structure of excited levels	10^5–10^7
2. Isotopic structure	10^5–10^7
3. Hyperfine structure, including atoms with isometric nuclei	10^5–10^8
4. Relativistic effects (Lamb shift)	10^6–10^8
5. Radiative broadening of spectral lines	10^6–10^9
6. Collisional broadening of spectral lines (at 1 Torr pressure)	10^7–10^9

of the methods of nonlinear laser spectroscopy discussed above shows that a resolution of the order of $10^{13} - 10^{15}$ is expected to be attained. It is determined by a fundamental limit, that is by the natural width. Therefore another question is to be posed: what new spectroscopic information becomes accessible with the methods of nonlinear laser spectroscopy of atoms and molecules without Doppler broadening?

Table 4.3 lists some effects in atomic spectra which must be measured by the methods of spectroscopy without Doppler broadening. A multiplet of fine structure is usually well resolved by classical methods, but for highly excited states the fine structure splitting decreases in proportion to n^{-3} (n is the principal quantum number) and it is masked by the Doppler broadening. For the study of isotopic and hyperfine structures, arising from spin and quadrupole moment of nuclei as well as excited nuclei (isometric hyperfine structure), it is necessary that the resolution should range within $10^5 - 10^8$. Many components of isotopic and hyperfine structures can be resolved by classical high-resolution devices (Fabry-Perot interferometer), but it is necessary to work into the Doppler contour to investigate the structure fully. Before the nonlinear laser spectroscopy was discovered, it was possible to do this by narrowing spectral lines in an atomic beam or by microwave spectroscopy of the ground and some excited states. It should be specially stressed that the methods of nonlinear spectroscopy enable us by rather simple means to measure the broadening of the spectral lines inside the Doppler contour caused by radiative decay and collisions. The value and power of the methods of nonlinear spectroscopy in solving the problems of atomic spectroscopy, listed in Table 4.3, reside above all in the fact that with the help of a number of methods they permit us to obtain systematically the entire spectroscopic information of atoms and ions with an unprecedented accuracy.

In molecular spectroscopy, and in the infrared especially, the classical methods do not assure any reasonable resolution. Therefore, only with

Table 4.4. The spectroscopic molecular effects in infrared

Effect	Resolution
1. Hyperfine structure due to quadrupole interaction	10^6–10^8
2. Collisional broadening of spectral lines (at 1 Torr pressure)	10^7–10^8
3. Magnetic hyperfine structure	10^9–10^{11}
4. Isomeric shift due to nuclear excitation	10^8–10^{10}
5. Difference of energy levels between left-hand and right-hand molecules due to weak interaction	10^{13}–10^{15}

the advent of lasers was the way opened for the infrared molecular spectroscopy with its resolution better than 10^5. Table 4.4 lists certain molecular spectroscopic effects which must be measured by the methods of spectroscopy without Doppler broadening. First of all there is the hyperfine structure of vibrational-rotational transitions due to quadrupole and magnetic interactions.

A magnetic interaction between the molecular angular momentum and nuclear spins brings about a splitting from 10^3 to 10^5 Hz. To detect it the resolution must be of the order of $10^9 - 10^{11}$.

A nuclear excitation in a molecule must bring about variations in the molecular vibrational frequencies since the nuclear excitation energy is equivalent to a decrease in the nuclear mass by $\Delta m = \Delta E_{\text{nucl}} c^{-2}$ (isomeric shift) [4.117]. This effect offers possibilities of measuring energies of metastable nuclei with a high accuracy, irrespective of the type of radiative decay by methods of infrared molecular spectroscopy.

There is another very fine effect in molecular spectra lying beyond the up-to-date experimental scope, but in the future it may be detected by methods of nonlinear laser spectroscopy. This effect consists of a small difference between the energy levels of two molecules which are mirror images of each other due to parity violation in weak interactions between electrons and nucleons in a molecule [4.118]. Physically the effect shows itself when there is a small admixture of odd potential of interaction between electrons and nucleons forming the molecule. The odd interaction removes the energy-level degeneracy of left- and right-hand molecules, so that their vibrational energies, for example, become different from one another by an extremely small value of the order of $\Delta E \sim 10^{-15}$ eV.

Let us enumerate briefly the main experiments which demonstrate the spectroscopic information stated above by the methods of saturation spectroscopy.

4.6.2 Measurements of Isotopic and Hyperfine Structures

1) Atomic Transition

Reference [4.60] was one of the first to demonstrate the efficiency of nonlinear laser spectroscopy. Narrow resonances in the line of spontaneous emission from the cavity of the 1.15 μm He-Ne laser were used for precise measurements of isotope shifts of two optical transitions in neon. The results of this work are treated in Subsection 4.3.2 and in Fig. 4.22 where the method of detecting narrow resonance in spontaneous emission is described.

The hyperfine structure of spectral lines for the odd isotope ^{21}Ne was investigated and the nuclear quadrupole moment was measured by FELD et al. [4.31, 63]. The radiation of the 1.15 μm He-Ne laser was directed to an external gas-discharge cell of low pressure (~ 0.1 Torr). The spectrum of spontaneous emission at 6096 Å was studied in the forward and backward directions with respect to the incident wave. The two different spectra must be described by the same set of parameters. The fitting of parameters, with the difference of linewidths in the three-level system in two opposite directions taken into account (Subsect. 4.3.2), has made it possible to determine the quadrupole moment of ^{21}Ne nucleus. The observed value of $Q = (+0.1029 \pm 0.0075)$ barn agrees with the value obtained less precisely before.

In Ref. [4.119] the hyperfine and isotopic structures of some excited states of several Xe isotopes (129, 134, 136) were studied. The particular feature of this experiment is the saturation effect of Xe transitions in an external amplifying cell of very low pressure (10^{-3} Torr) excited by an electric discharge. In this experiment the resolution limit determined by the natural width of the 3.51 μm line ($\Delta\nu_{nat} = 4.6 \pm 0.7$ MHz) has been obtained. The isotope shift for the 3.51 μm lines of ^{136}Xe and ^{134}Xe ($\delta\nu_{136-134} = 36.0 \pm 0.6$ MHz) has been measured, and at last the *hfs* constant A has been measured for two levels of the $5p^3 5d[5/2]_2 \rightarrow 5p^5 6p[3/2]_1$ transition at $\lambda = 3.36$ μm ^{129}Xe.

The progress of tunable lasers has allowed systematic studies of atomic transitions, the frequencies of which have no accidental coincidence with those of narrow band lasers. The first experiments of the kind were carried out at Stanford University on the D-line of Na [4.107] and H_α-line of hydrogen [4.108]. The lasers used in these experiments are outlined in Subsection 4.5.2. The experimental technique stated in Subsection 4.3.1 was based in observing saturated absorption by a probe wave in the presence of a strong counter-running wave. Under the conditions providing a maximum resolution of about 25 MHz it was possible to measure the hyperfine splitting ($\Delta\nu = 177$ MHz) of the $2S_{1/2}$ and $2P_{1/2}$ states and the Lamb shift of the $2\,^2S_{1/2}$ state (1058 MHz).

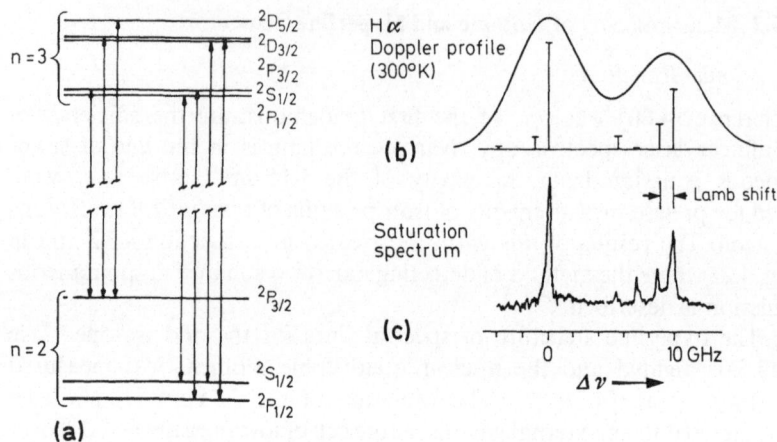

Fig. 4.30a–c. Lamb shift in H_α line: (a) the quantum levels and transition involved in hfs; (b) Doppler profile at room temperature; (c) the spectrum of saturated absorption attained by a tunable dye laser (HÄNSCH et al. [4.108])

Figure 4.30 is a diagram of the energy levels and transitions responsible for the hyperfine structure of the H_α-line and the saturated absorption spectrum. Similar experiments were conducted by HÄNSCH et al. [4.120] for the D_α-line of deuterium to determine the Rydberg constant.

2) Molecular Transition

The hyperfine structure of a large number of optical transitions in the $^{127}I_2$ and $^{129}I_2$ molecules was studied by HÄNSCH et al. [4.121] and later in more detail by LEVENSON and SCHAWLOW [4.122]. In these experiments they used the argon ion (514.5 nm, 501.7 nm) and krypton ion (568.2 nm, 530.8 nm, 520.8 nm) lasers. All the lines under study correspond to the transitions between the $^1\Sigma_g^+(X)$ and $^2\Pi_u^+(B)$ electronic states, but the rotational and vibrational quantum numbers are different. The studies have revealed two causes of the hyperfine structure: 1) the nuclear electric quadrupole interaction, which is almost constant for the different lines, and 2) magnetic spin-orbit interaction, which greatly depends on the vibrational energy in the excited electronic state. The hfs of $^{127}I_2$ has also been studied by HANES et al. [4.123] using the He-Ne laser at 0.63 μm.

The quadrupole hyperfine structure of infrared molecular rotational-vibrational lines has been observed for $^{12}CH_3{}^{35}Cl$ by MEYER et al. [4.124] and for $^{189}OsO_4$ by KOMPANETZ et al. [4.125]. The quadrupole hfs of the $v_6{}^RQ_3(6)$ transition was studied by the P(26) 9.4 μm CO_2 laser line [4.124]. The observed splitting shows a good agreement with

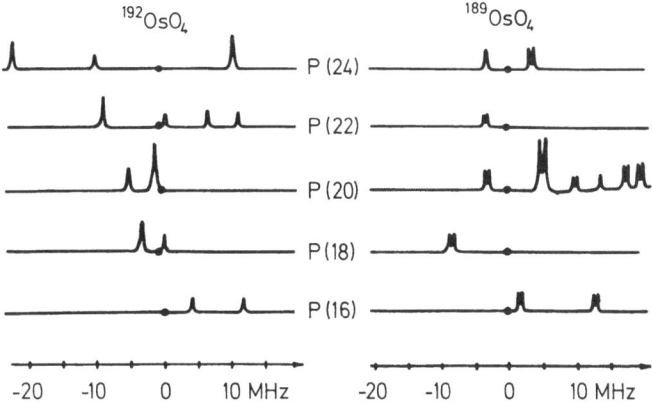

Fig. 4.31. Spectra of saturated absorption of $^{192}OsO_4$ and $^{189}OsO_4$ molecules measured through some lines of the P-branch of CO_2 laser at 10.6 μm (small circles note the centres of amplification lines of the CO_2 laser) (KOMPANETZ et al. [4.125])

the theoretical value. Manifestation of three effects (quadrupole, magnetic and isometric) in hfs is considered in the vibrational-rotational spectrum of monoisotopic molecules of OsO_4 [4.124]. For this aim the masked structures in the Doppler-broadened absorption lines of the molecules of $^{187}OsO_4$, $^{189}OsO_4$, $^{190}OsO_4$ and $^{192}OsO_4$ were studied by the P- and R-branch CO_2 laser lines. The electric quadrupole structure was observed in the $^{189}OsO_4$ spectrum. Figure 4.31 shows the saturation spectrum of $^{192}OsO_4$ and $^{189}OsO_4$ for several lines of the CO_2 laser. The $^{189}OsO_4$ spectrum has a character of double lines (Fig. 4.31b). It is manifested more brightly on the $P(20)$ CO_2 laser line. A few individual resonances in the $^{189}OsO_4$ spectrum can be ascribed to an admixture of other isotopic molecules. The ^{189}Os nucleus in contrast to other isotopes possesses a comparatively large magnetic moment (0.65004 nuclear magneton) and an electric quadrupole moment (0.8 barn). The magnetic hfs is considered on the infrared spectrum of $^{187}OsO_4$ and $^{189}OsO_4$. On the basis of these experiments and calculation [4.125] the possibility of measuring the nuclear energy of isomeric state is discussed for $^{189}OsO_4$ by the method of saturation spectroscopy.

The highest resolution (better than 10^{10}) has been attained in the study of the magnetic hfs of the $F_2^{(2)}$ component of the P(7) v_3 line of methane by HALL and BORDE [4.94, 96] (see Chapt. 5). The magnetic hfs consists of three similar components of different intensities. The distances between the lines measured experimentally were close to the values theoretically deduced. Resolved observations of the hyperfine structure are of great importance, since these resonances are used as an optical wavelength standard [4.126].

4.6.3 Zeeman and Stark Effects

In conventional optical spectroscopy, large magnetic and electric fields are required to study these effects, since level splitting must exceed the Doppler width. The use of saturation allows us to eliminate the Doppler broadening and, hence, to study the Zeeman and Stark effects inside the Doppler width. An external electric or magnetic field can play a double role. For a non-degenerate two-level transition, the external field changes the single transition frequency, and it can be used for tuning the central frequency of the Doppler-broadened line. Usually the atomic and molecular energy levels are degenerate, and the external field causes a splitting of these levels. As a result, the Doppler-broadened line becomes a set of overlapped lines. The merits of saturation spectroscopy are that one can record this splitting before the Doppler-broadened line splits. That is to say, it is possible to detect the splitting inside the Doppler width. This effect has been used in spectroscopy of atomic transitions in a magnetic field (the Hanle effect) where the effect of splitting the Doppler width shows itself as changes in the polarization of spontaneous emission. In the case of saturation spectroscopy this would be manifested by changes in absorption, that is, in a stimulated rather than in a spontaneous process. That is why the methods of saturation spectroscopy are of special importance for transitions with a fast non-radiative relaxation when the spontaneous emission is very weak in, say, molecular rotational-vibrational transitions.

1) Atomic Transitions

Narrow resonances in saturation of atomic amplification or absorption in an external magnetic field have been observed in a number of experiments, mainly to measure the g-factors of atomic levels as well as the hyperfine level splitting. Such an experiment was first conducted in Ref. [4.70]. In this experiment the amplifying transition of ^{129}Xe at $\lambda = 3.37$ µm was acted upon by two traveling waves of a Xe-laser with the frequency difference $\Delta = c/2L$ (two axial modes), while the constant magnetic field set up several pairs of coupled transitions with their frequency splitting depending on the magnetic field strength. Well-resolved resonances of narrow half-width of $\Gamma = 0.6$ MHz were observed, as the splittings of the appropriate pairs of coupled transitions were tuned into Δ ("mode-crossing" method, in Subsect. 4.3.2). These resonances were used to obtain an accurate value of the zero field hyperfine splitting. Figure 4.32a gives the signal observed by JAVAN [4.127]. The $\Gamma_{\text{nat}} = 0.6$ MHz half-width of resonances is in full agreement with (4.29):

$$\Gamma_{\text{nat}} = \frac{\gamma_l}{2} + \frac{\gamma_n}{2}, \tag{4.62}$$

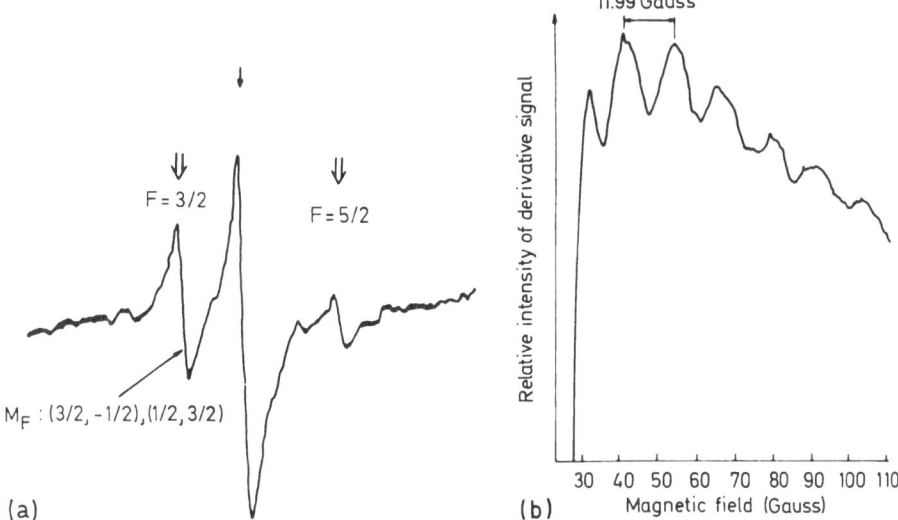

Fig. 4.32a and b. Saturation mode-crossing resonances, obtained via Zeeman tuning: (a) 3.37 μm transition ^{129}Xe (JAVAN [4.127]); (b) 8446 Å transition OI (FELD [4.31])

where $\gamma_l = \gamma_n = 0.6$ MHz. The width of the common level γ_m is at least one order of magnitude larger and does not contribute to broadening of the narrow resonance.

The similar experiments for atomic oxygen laser lines have been done by FELD [4.31, 76]. A typical trace for the 8446 Å transition is shown in Fig. 4.32b. A mode-crossing signal is observed each time the splitting between a pair of upper levels connected to a common lower level approaches the spacing between any two axial modes of the multimode laser. For perpendicular orientation of the laser polarization and the magnetic field ($e \perp B$) the selection rule is $\Delta M = 2$. The g-values can be obtained from the mode-crossing signals using the relationship

$$2\mu_0 g \Delta B = \frac{c}{2L}, \tag{4.63}$$

in which ΔB is the separation between resonances in units of magnetic field, μ_0 is the Bohr magneton. The measured g-factor of the upper level of 8446 Å ($3p^3P - 3s^3S$) transition is $g = 1.51 \pm 2\%$. It is in good agreement with the value predicted by the L-S coupling scheme for the upper level ($g_{LS} = 3/2$). Mode-crossings of the lower levels were not observed.

This rather general method of measurement for atomic g-factors may be used for any pair of coupled (not necessarily close) transitions where coherent radiation can be produced. For instance, in Ref. [4.43] this technique was applied to study the Zeeman effect on the $2s_2 - 2p_1$ transition of Ne (1.52 μm, a strong running wave) and $2s_2 - 2p_4$ (1.19 μm, a weak probe running wave). Line splitting in the magnetic field was used to determine accurately the ratio of g-factors of the $2s_2$ and $2p_4$ levels (the $2p_1$ level has the total angular momentum $J = 0$ and is not split in the magnetic field). The measured values of g-factors are $g_{2p_4}/g_{2s_2} = 1.035 \pm 0.02$, $g_{2p_4} = 1.30 \pm 0.03$ and $g_{2s_2} = 1.26 \pm 0.03$.

In the last few years the Zeeman effect in narrow resonance of saturated absorption has been studied by French scientists [4.82, 85, 86]. In a multimode laser operation, saturation resonances have been obtained in zero magnetic field [4.85, 128, 129] and in nonzero magnetic field, where the beat frequency between modes is equal to the Zeeman splitting [4.130]. See Chapter 5 in Vol. 2 of *Topics in Applied Physics* [4.131].

2) Molecular Transitions

Electric field. The experiment by BREWER et al. [4.132] is a classic example of the precise measurement of a Stark spectrum by the use of saturation. In this work they investigated the $4_{04}(a) \rightarrow 5_{14}(s)$, ν_2 transition of NH_2D in a static electric field by the 10.6 μm $P(20)$ CO_2 laser line. Usually this line of the CO_2 laser is not absorbed in NH_2D, but in a uniform electric field some rotational-vibrational lines can be resonantly tuned to coincide with the CO_2 laser line. When the line centre is tuned exactly to the laser frequency, a narrow saturation resonance arises for each line, which permits us to obtain very precise measurements. The accuracy of such measurements is sufficient to determine the second-order Stark shift in weak electric fields (few kV/cm). Measurements of this kind were conducted with NH_2D molecules and CO_2 laser lines (BREWER and SWALEN [4.133]; KELLY et al. [4.134]), CH_3F molecules and the 3.39 μm He-Ne laser with magnetic tuning (LUNTZ et al. [4.135]) or the 9.4 μm CO_2 laser (FREUND et al. [4.136]). A few examples of Stark-Lamb dips in CH_3F are shown in Fig. 4.33. The Lamb dips of the Q-branch lines of CH_3F with $\Delta M = \pm 1$ are composed of $2J$ components.

Brewer improved the technique for the precise measurement of Stark shift by observing the narrow resonance at two coupled Stark transitions in a two-frequency laser field [4.13, 66]. The accuracy of dipole moment measurements in the ground and excited vibrational states of the CH_3F molecule was about one part in 2000. This accuracy is comparable with that of the method of microwave spectroscopy in a molecular beam, but the latter is applicable only to the vibrational ground state. At the same time the method of saturated absorption is applicable, in principle, to the measurement of the molecular dipole moment in any stable quantum state.

Q (1,1)   Q (2,2)   Q (3,3) ^{13}C

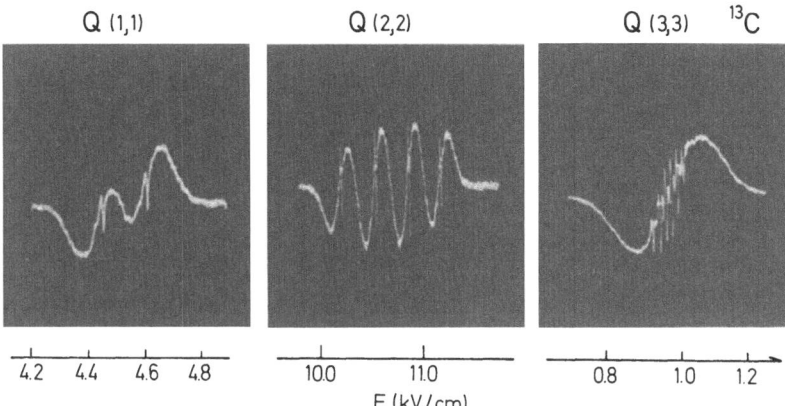

4.2 4.4 4.6 4.8   10.0   11.0   0.8   1.0   1.2

E (kV/cm)

Fig. 4.33. Some examples of Stark-Lamb dips. The sharp features are the Stark-Lamb dips while the broader features are normal Doppler-broadened Stark resonances. For the $Q(1,1)$ and $Q(2,2)$ lines of $^{12}CH_3F$ the CO_2, $P(18)$ line was used. For the $Q(3,3)$ transition of $^{13}CH_3F$ which has studied using CO_2, $P(40)$ lines, the different M lines are resolved only by Stark-Lamb dips (FREUND et al. [4.136])

Some interesting experiments on saturation Stark spectroscopy were conducted with the CH_4 molecule and the 3.39 µm He-Ne laser. The excited state of the $P(7)$ transition of the v_3 band of CH_4 has six Coriolis sublevels. The E-sublevel has the first-order Stark effect, while the four F-sublevels show only the second-order Stark effect. The $P(7)$, $F_2^{(2)}$ transition is in almost exact resonance with the 3.39 µm He-Ne laser, and the $P(7)$, E transition is displaced from it by 0.096 cm^{-1} to the long wavelength side. The linear Stark effect of the E transition has been studied by LUNTZ et al. [4.79, 137]. They detuned the frequency of the He-Ne laser by 3 GHz using an axial magnetic field and placed the Stark CH_4 cell inside the laser cavity. Owing to a small electric dipole moment of 0.0200 ± 0.0001 Debye in the excited vibrational state, the spectral line is split into $2J + 1 = 13$ equidistant components, which can be clearly seen in the saturation spectrum. The quadratic Stark effect of the $F_2^{(2)}$ transition was observed by UEHARA [4.138]. The Stark splitting could not be resolved even in a field of 40 kV/cm but it showed itself as asymmetric broadening. The small value of the Stark shift for this line is of importance in obtaining a high reproducibility of the He-Ne laser stabilized with CH_4.

Magnetic Field. The use of saturation spectroscopy in studying the Zeeman effect is of most importance for molecules in the ground electronic state $^1\Sigma$ for which the magnetic moment is due to molecular rotation and the nuclear magnetic moments. The study of the Zeeman effect of a

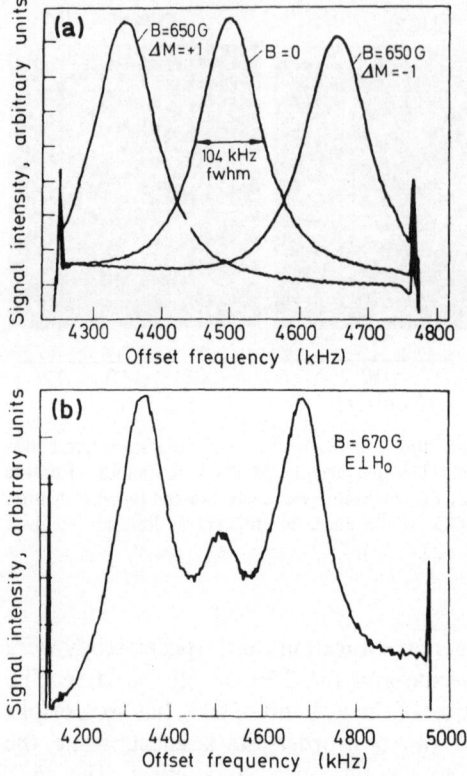

Fig. 4.34a and b. The Zeeman structure of $F_2^{(2)}$ component of the $P(7)$ line CH_4 observed inside the Doppler profile by saturated absorption: (a) resonances in a circularly polarized light for two different polarizations; (b) resonances in a linearly polarized light (UZGIRIS et al. [4.80])

Doppler-broadened line of such molecules by the methods of linear spectroscopy requires a magnetic field of several tens of teslas. Narrow saturated absorption allows the study of Zeeman effect in a magnetic field of hundreds of times weaker. Such experiments were conducted by LUNTZ and BREWER [4.139] and UZGIRIS et al. [4.80] on the $F_2^{(2)}$ component of the $P(7)$ CH_4 line. The observed effects are determined by the occurrence of two saturation resonances due to transitions with the selection rules $\Delta M_J = \pm 1$ for laser waves having left and right circular polarizations. In a circularly polarized light, one narrow observed resonance is shifted in energy by $\pm \mu_N g_J B$ (Fig. 4.34a). The value of the g_J factor is found from the magnetic shift of the resonance frequency. It has been found for CH_4 to be $g_J = +0.311 \pm 0.006$, which agrees well with the value of $|g_J| = 0.3133 \pm 0.0002$ obtained from molecular beam experiments. Not only the absolute value but also the sign of g_J is measured, which is a characteristic feature of laser spectroscopy. Using linearly polarized light and the axial magnetic field, both transitions $\Delta M_J = \pm 1$ have been resolved, and therefore two circularly polarized light waves interact with two coupled transitions. The saturation resonances for this case, observed also by UZGIRIS et al. [4.80], are shown in Fig. 4.34b. In this case one can observe simultaneously two narrow resonances due to the Zeeman effect, and an additional cross-resonance (crossing), arising from the common level as discussed above in Subsection 4.4.2.

Recently JAVAN [4.140] successfully applied the fluorescence method of narrow resonance observation [see (3) in Subsect. 4.3.1] to rotational-vibrational transitions $10^0 0 - 00^0 1$ of CO_2 and N_2O in a magnetic field by using CO_2 and N_2O-lasers, respectively. The specific feature of his experiments was the use of a large-diamter fluorescence cell (the beam diameter of up to 6 cm) at a pressure of up to 0.5×10^{-3} Torr. The fluorescence resonances were 70 kHz in width and their relative amplitude was up to 20%. He studied the anomalous Zeeman effect and measured the g-factors for the upper ($00^0 1$) and lower ($10^0 0$) vibrational states of laser transitions, the sign of g-factor included. For CO_2 the following values were obtained: $g_u = -0.042$ and $g_l = -0.043$; and for N_2O: $g_u = -0.061$, $g_l = -0.062$. High sensitivity of the fluorescence method allows undoubtedly a larger increase in this accurately.

4.6.4 Collision Effects

The methods of high resolution saturation spectroscopy have provided a new approach to the study of collisional effects on the spectral line shape. Narrow resonances inside the Doppler profile make it possible to observe collisions at a very low gas pressure when Doppler width significantly exceeds the collisional broadening. Under such conditions the shape of the narrow resonances can be substantially affected by elastic collisions.

1) Elastic Collisions

Atomic Transitions. In many early works detailed studies on the collision broadening of the Lamb dip by foreign gas (helium) have been carried out with a He-Ne laser operating at $\lambda = 0.63\,\mu m$ [4.141–143] and $\lambda = 1.15\,\mu m$ [4.4, 144]. Within the limits of experimental error a linear dependence was obtained for the collisional broadening versus helium pressure with a slope of 60 MHz/Torr. The collisional shift of the Ne lines at 0.63 μm caused by helium atoms has been measured with the use of Ne-He lasers stabilized by the Lamb dip [4.145, 146] and a He-Ne laser with an Ne-arsorbing cell [4.7, 8]. The shift measured by the latter method was 21 ± 3 MHz/Torr. Collisional broadening and shift of the Lamb dip by foreign gases (He and Ne) have been observed with the Hg laser on $\lambda = 1.5$ μm [4.147]. The measured results of broadening and shift of the Lamb dip by collisions with foreign atoms are rather reliable.

However, the results of numerous experiments on the natural broadening and line shift of Ne atoms (Ne-Ne collisions) [4.148, 149, 43] disagree. For example, the data on shift of the 0.63 μm Ne line range from 0 to -25 MHz/Torr (red shift). But this can be explained not by the shortcoming of the method of saturation spectroscopy but by the difficulty in control of experimental conditions, since the shift depends on such effects as radiation trapping [4.150] and collision with electrons [4.51].

Note that not all potentialities of saturation spectroscopy have been brought into use for purposeful studies in atomic collisions. Just one experiment [4.147] has been carried out in which a great difference in masses of colliding atoms (Hg with He or Ne) has permitted determination of the interaction potential from the measurements of broadening and shift.

Molecular Transitions. The very first experiments with low-pressure molecular gases [4.9] revealed an abnormally small shift of the frequency of the vibrational-rotational transitions caused by collisions. To cite an example, for the $P(7)$ line of CH_4 molecule, the shift (<0.1 MHz/Torr) was at least one hundred times less than the collosion broadening (16.3 ± 0.6 MHz/Torr). Very similar results were obtained for some other molecules (SF_6 [4.26, 152], CO_2 [4.153], OsO_4 [4.125]). In studying collisional broadening and the shift of vibrational-rotational molecular transitions by saturation spectroscopy, a particular emphasis should be placed upon the important features in the limiting case of low pressure (<0.01 Torr).

At low pressures the elastic scattering of atoms in collisions is substantial which results in a nonlinear relation between broadening and shift as functions of the gas pressure. The nonlinear dependence manifests itself primarily in the pressure region where the Doppler shift by an elastic scattering $ku\theta$ (θ is the characteristic scattering angle) is of the magnitude of the homogeneous (collisional) line width. These effects have been found and qualitatively explained by BAGAEV et al. [4.154, 155]. The theoretical description of the collisional effect for saturation resonances at low pressure has been developed by ALEKSEEV et al. [4.156]. Experimental observations of the nonlinear dependence of width and shift of the Lamb dip have been carried out in CH_4 with the 3.39 μm He-Ne laser [4.154] and the measured dependences are given in Fig. 4.35. Within a pressure range of $1-5$ mTorr the width-pressure relation is

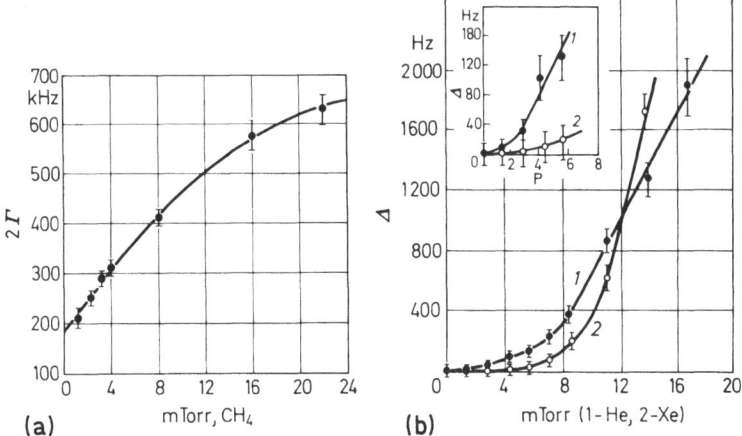

Fig. 4.35a and b. Width and shift of the Lamb dip in CH_4 at 3.39 μm: (a) the width as a function of the pressure of methane; (b) the shift as a function of the pressure of buffer gas of 1. He and 2. Xe, with 1 Torr of CH_4

linear with a slope of 30 MHz/Torr. With an increase in the pressure the slope goes down and at pressures of about 20 mTorr its value is about 10 ± 5 MHz/Torr. At low pressures the shift coefficient is considerably smaller than that measured at higher pressures.

It is also possible to study the dependence of collisional broadening on the relative velocity of colliding atoms (or temperature). By measuring the homogeneous width (or that of the Bennett hole) in different parts of Doppler contour one can measure the broadening by the atoms having a certain fixed velocity component v_z [see (4.21)] and hence the effective temperature [4.58]

$$T_{\text{eff}} = \frac{2}{3} T \left[1 + \left(\frac{v_z}{u} \right)^2 \right]. \tag{4.64}$$

Since the width of the Lamb dip is measured at the centre of the Doppler contour, where $v_z \ll u$, we obtain $T_{\text{eff}} = \frac{2}{3} T$. When there is a saturation resonance observed at the wings of the Doppler contour, the condition $T_{\text{eff}} \gg T$ can be found. The method of counter-running waves with different frequencies is most suitable in observing shifted Lamb dips (Subsect. 4.3.1). A difference in width at various points of Doppler contour was observed by CHEBOTAYEV [4.149] in his experiments on a two-mode He-Ne laser with a Ne absorbing cell. This method has been developed by JAVAN et al. [4.176] in studying the broadening in NH_3. Figure 4.36 shows the typical data obtained for NH_3–Xe collisions. By studying the

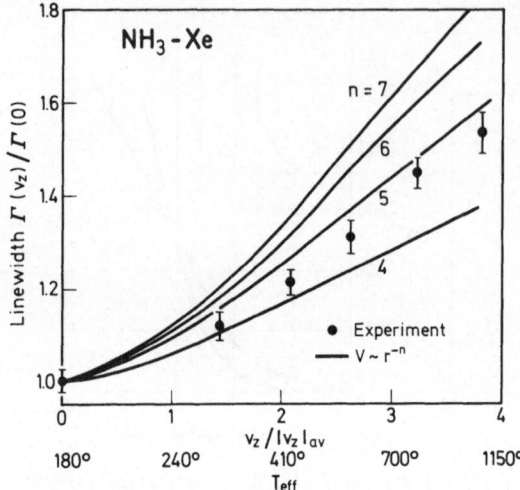

Fig. 4.36. Collisional broadening of the Lamb dip in NH_3 as a function of the molecular velocity component v_z or its effective temperature T_{eff} in a low pressure of NH_3 at room temperature (MATTICK et al. [4.176])

dependence of broadening on the effective temperature, one can define the kind of molecular interaction potential $V \sim r^{-n}$ for colliding particles. This method of saturation spectroscopy holds promise for precision measurements of the temperature dependence of collision cross sections.

2) Inelastic Collisions

The method of optical double resonance is effective in studying inelastic collisions where the quantum state of colliding particles is changed and the velocity of motion varies only slightly. The method makes it possible to obtain information on molecular collisions of the type

$$M(J, K, M, v) + M' \rightarrow M(J', K', M', v) + M' . \qquad (4.65)$$

In the work by BREWER et al. [4.89] collisions of CH_3F molecules with variations in J and M have been studied (see Fig. 4.26). For example, it has been found that for reorienting collisions with $\Delta M = \pm 1 (J, K = 4,3)$ the cross section is 100 Å^2, whereas for the $(J, K = 12,2)$ states it decreases by 100 times. JOHNS et al. investigated collisions of CH_3F, H_2O, and NH_3 molecules and found that reorientating collisions with change in parity of states had a dominant role [4.158]. MEYER and RHODES [4.91] showed that on CO_2–H_2 collisions the change of angular momentum was

$|J - J'| = 2$ or 4. The average change of the velocity component for $J = 20 \leftrightarrow J' = 18$ is about $(3 \pm 2)10^3$ cm/s. The data obtained show that rotational transitions arise mainly due to peripheral collisions that are efficient in angular momentum transfer but alter the linear momentum comparatively little.

4.6.5 Precision Spectroscopy

Firstly, a new length standard (He-Ne/I_2 laser at 6328 Å [4.159–161, 126]) has been practically established on the basis of narrow saturation resonances with its reproducibility of better than 10^{-10}. Using this standard, one can conduct interferometric comparison of laser wavelengths with a precision of no worse than 10^{-10} [4.161]. In combination with the methods of saturation used for exact determination of the centre of the H_α and D_α spectral lines, this has allowed precise determination or the Rydberg constant (HÄNSCH et al. [4.120]). The new Rydberg constant $R_\infty = 109737,3143(10)$ cm^{-1} has an error by one order smaller than the former value.

Secondly, instead of measuring the wavelength, frequency measurements have become possible. Until recently, frequencies in spectroscopy have been determined indirectly and with comparatively low precision. The measured value was the wavelength: and with help of the known value of the speed of light, the frequency could be calculated. The accuracy of such evaluations was always worse than 10^{-7}, since it was limited by the precision of our knowledge of the speed of light. The use of the methods of saturation spectroscopy and the direct measurement of the frequency of light oscillations have improved the precision approximately by 10^2 times. The new value for the speed of light is $c = 299\,792\,458.2(1.2)$ m/s. [4.103, 126, 162]. This subject is considered in detail in the article by EVENSON and PETERSON, volume 2 of this series [4.163]. The new value of the speed of light, in combination with a high precision of the methods of interferometric comparison between laser wavelengths and the He-Ne/I_2 laser wavelength, has made it possible to evaluate frequencies and wavelengths to better than 10^{-9}. The methods of absolute frequency measurement of light [4.102, 163] will open the way for direct measurement of spectral line frequencies with a precision of the international time standard. Although this basically new technique is still in its infancy, it holds great prospects which will show up after simple and efficient nonlinear converters of optical frequency are elaborated.

At present more accessible methods of precision laser spectroscopy are those based on accurate measurement of frequency difference between two light oscillations. By this technique one can measure precisely isotope shifts in atomic and molecular spectra and molecular rotational constants. The frequencies of two lasers were stabilized by narrow fluorescence resonances in saturated absorption of a low-pressure CO_2 cell [4.164, 165] (see (3) in Subsect. 4.3.1). The absolute precision of laser frequency stabilization on the centre of absorption lines of CO_2 molecules was about 2.5 kHz. This allowed the rotational constants to be measured with a relative precision of about 10^{-7}.

4.7 Possibility of Other Applications of Saturation Spectroscopy

In conclusion let us discuss some possibilities of extending the ideas of saturation spectroscopy over the boundaries of traditional application area, i.e., optical spectral lines of atoms and molecules in gases.

4.7.1 γ-Ray Spectroscopy of Nuclear Transitions

The frequency of nuclear γ-transitions is also shifted by the value $k_\gamma v$ due to the Doppler effect. If the distribution of nuclear velocities, that is, of atomic and molecular velocities, is thermal (equilibrium), it gives the Doppler broadening of γ-ray lines. By laser radiation we can excite atoms or molecules with a certain projection of the velocity on the chosen direction (of the laser beam), that is, we can change the velocity distribution of particles at the levels (Fig. 4.37). For example, it is possible to have excited atoms (molecules) with the velocity v_{res} determined by the optical resonance condition

$$k_0 v_{res} = \omega - \omega_0 , \tag{4.66}$$

where k_0 is the laser wave vector, ω is the laser frequency, $\mathscr{E}_i = \hbar\omega_0$ is the atomic (molecular) transition energy. The spectral line of the composite γ-transition, in which atoms (molecules) with a non-equilibrium velocity distribution participate, will have a narrow resonance peak (Fig. 4.37) rather than an ordinary Doppler profile. The frequency of this peak is shifted from the centre of the line $(E_0 + R - \mathscr{E}_i)$, where R is the recoil energy

$$\Omega_\gamma = k_\gamma v_{res} = \frac{\omega - \omega_0}{\omega_0} \omega_\gamma . \tag{4.67}$$

It can be tuned within the whole Doppler contour of the γ-ray line when the laser frequency is tuned along the Doppler-broadened line of the optical transition.

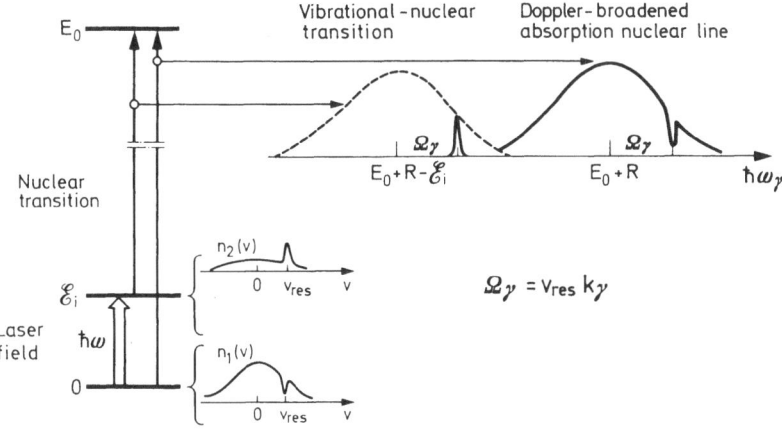

Fig. 4.37. Production of narrow resonances of γ-ray absorption, when an atom or a molecule is excited by a coherent light wave in low-pressure gas

The idea of obtaining narrow tunable γ-resonances of absorption and emission was proposed in Refs. [4.166–168] in 1972. Theoretical treatment of this application of saturation spectroscopy for nuclear spectroscopy is presented in several Refs. (molecules [4.169], atoms [4.170]).

4.7.2 Positronium and Narrow Annihilation Lines

Let us consider briefly the method of producing narrow and tunable lines of positronium annihilation radiation at $E_\gamma = 0.511$ MeV. This method is based on velocity-selective conversion of ortho-positronium (o-Ps) atoms to para-positronium (p-Ps) under the laser radiation and a magnetic field [4.171].

In the ground triplet state the o-Ps atoms are distributed over three magnetic sublevels ($m = 0, \pm 1$). Switching on a stationary magnetic field of several kG causes a mixing of the $m = 0$ sublevels of the states $^1S_0$, p-Ps and $^3S_1$, o-Ps. Due to this, the o-Ps atoms on the $m = 0$ sublevel undergo 2γ-annihilation [4.172]. As a result, only o-Ps atoms on the $m = \pm 1$ states stay in a long-lived triplet state, which must be velocity-selectively converted to the $m = 0$ sublevel.

Assume that the Zeeman σ-components of the line L_α, o-Ps at $\lambda_0 = 2430$ Å interact with a laser wave that induces stimulated transitions from the $1\,^3S_1$, $m = \pm 1$ states to the $2\,^3P_1$, $m = 0$ state. Only those atoms, the Doppler-shifted transition frequency of which coincides with the

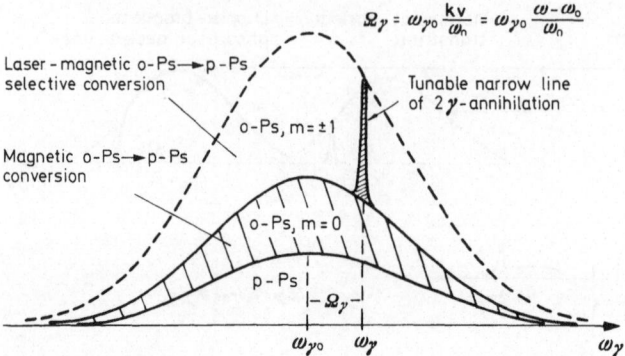

Fig. 4.38. The shape of 2γ-annihilation line which consists of Doppler profile and frequency-tunable narrow peak

laser frequency ω, are excited to the upper level, i.e.,

$$\omega = \omega_0 \frac{\left(1 - \dfrac{v^2}{c^2}\right)^{1/2}}{\left(1 - \dfrac{v}{c}\boldsymbol{n}\right)} \quad \text{or} \quad (\omega - \boldsymbol{k}\boldsymbol{v}) = \omega_0 \sqrt{1 - \frac{v^2}{c^2}}, \tag{4.68}$$

where ω_0 is the L_α-line frequency for fixed o-Ps atoms, $\boldsymbol{k} = \boldsymbol{n}\omega/c$ is the wave vector, $\boldsymbol{v}$ is the velocity of positronium. By the additional microwave field the excited atoms in the $^3P_1$ state from the $m = 0$ sublevel can be transferred to $m = \pm 1$ sublevels. Then all the atoms from the $^3P_1, m = \pm 1$ state due to the selection rule $\Delta m_s = 0$ will return spontaneously with $\tau_{\mathrm{rad}} = 3.10^{-9}$ s to the $^3S_1, m = 0$ state, and thereby full transfer of o-Ps atoms from the $^3S_1, m = \pm 1$ state to $^3S_1, m = 0$ state will occur, producing velocity-selective conversion of the remaining (50%) o-Ps atoms into p-Ps.

The radiation of 2γ-annihilation of the converted o-Ps atoms with the resonant velocity $\boldsymbol{v}$ determined by condition (4.68), that is observed in the laser wave direction, has the frequency (Fig. 4.38)

$$\omega_\gamma = \omega_{\gamma 0}\left(1 - \frac{v^2}{c^2}\right)^{1/2} \Big/ \left(1 - \frac{v}{c}\boldsymbol{n}\right) = \omega_{\gamma 0}\frac{\omega}{\omega_0}, \tag{4.69}$$

where $\hbar\omega_{\gamma 0} = m_0 c^2$. Thus, tuning the laser frequency along the Doppler profile of the ortho-positronium line L_α, we can tune the annihilation radiation frequency along the Doppler profile of the 2γ-annihilation line.

In this method the line shape of 2γ-annihilation radiation should be complex in structure, as shown in Fig. 4.38. A narrow line appears at the frequency of (4.69) against the background of the Doppler-broadened line due to 2γ-annihilation of p-Ps and o-Ps from the $m=0$ state.

It should be noted that the realization of the method discussed for obtaining narrow lines of 2γ-annihilation seems to be quite real in the light of recent brilliant experiments, in which the L_α-radiation of positronium was observed for the first time [4.173] and the hyperfine structure for the first excited positronium state was measured [4.174]. For the laser-nuclear effects mentioned in Subsection 4.7.1 the main experimental difficulty is that it is necessary to apply very intensive sources of γ-ray and high-radioactivity samples. As distinct from them, the basic difficulty for the method of 2γ-annihilation line shape control, mentioned here, consists in developing a cw laser in the region of $\lambda_0 = 2430\,\text{Å}$ with its power from 10^{-2} to 10^{-1} W. Yet one may hope that a rapid progress in tunable dye lasers and in the technique for frequency doubling in non-linear crystals will remove this temporary difficulty.

4.7.3 Elementary Excitations in Solids

Though at first sight it seems unusual, the methods of saturation spectroscopy can be applied to the Doppler-broadened spectral lines in solids. The spectral lines of excitons in a solid are thought to be broadened due not only to their interaction with phonons but also to the Doppler effect [4.175]. If it is so, this Doppler broadening can be detected by the method of saturation spectroscopy. This refers, of course, to the spectral lines of other elementary excitations in solids at low temperatures.

The author wishes to thank Prof. V. P. Chebotayev for cooperation in this area of research as well as for useful remarks made, reading the manuscript of this article.

References

4.1   A. JAVAN, W. R. BENNETT, JR., D. R. HERRIOTT: Phys. Rev. Letters **6**, 106 (1961)

4.2   W. R. BENNETT, JR.: Phys. Rev. **126**, 580 (1962)

4.3   W. E. LAMB, JR.: Phys. Rev. **134A**, 1429 (1964)

4.4   A. SZÖKE, A. JAVAN: Phys. Rev. Letters **10**, 521 (1963)

4.5   R. A. MCFARLANE, W. R. BENNETT, JR., W. E. LAMB, JR.: Appl. Phys. Letters **2**, 189 (1963)

4.6   V. S. LETOKHOV: Pis'ma Zh. Eksper. I. Teor. Fig. **6**, 597 (1967)

4.7   P. H. LEE, M. L. SKOLNICK: Appl. Phys. Letters **10**, 303 (1967)

4.8   V. N. LISITSYN, V. P. CHEBOTAYEV: Zh. Eksper. I. Teor. Fiz. **54**, 419 (1968)

4.9   R. L. BARGER, J. L. HALL: Phys. Rev. Letters **22**, 4 (1969)

4.10 V. S. LETOKHOV, V. P. CHEBOTAYEV: *Principles of Nonlinear Laser Spectroscopy* (Izd. Nauka, Moscow 1975)

166   V. S. LETOKHOV

4.11 V. P. CHEBOTAYEV, V. S. LETOKHOV: in *Progr. Quant. Electr.* ed. by J. H. SANDERS and S. STENHOLM (Pergamon Press, London 1975) vol. 4, p. 1
4.12 I. M. BETEROV, V. P. CHEBOTAYEV: in *Progr. Quant. Electr.* ed. by J. H. SANDERS and S. STENHOLM, vol. 3 (Pergamon Press, London 1974) pp. 1—106
4.13 R. G. BREWER: Science **178**, 247 (1972)
4.14 M. S. FELD, V. S. LETOKHOV: Sci. Am. **229**, 69 (1973)
4.15 V. S. LETOKHOV: Science **190**, 344 (1975)
4.16 S. G. RAUTIAN, I. I. SOBELMAN: Zn. Eksper. I. Teor. Fiz. **44**, 834 (1963)
4.17 S. G. RAUTIAN: Proceedings of Lebedev Physical Institute "Nonlinear Optics" **43** (1968)
4.18 H. GREENSTEIN: Phys. Rev. **175**, 438 (1968)
4.19 S. STENHOLM, W. E. LAMB, JR.: Phys. Rev. **181**, 618 (1969)
4.20 B. J. FELDMAN, M. S. FELD: Phys. Rev. **A1**, 1375 (1970)
4.21 K. UEHARA, K. SHIMODA: Japan. J. Appl. Phys. **10**, 623 (1971)
4.22 K. SHIMODA, K. UEHARA: Japan. J. Appl. Phys. **10**, 460 (1971)
4.23 E. V. BAKLANOV, V. P. CHEBOTAYEV: Zh. Eksper. I. Teor. Fiz. **62**, 541 (1972)
4.24 E. V. BAKLANOV, E. TITOV: Optika I. Spectroscopia. **38**, 169 (1975)
4.25 V. S. LETOKHOV, V. P. CHEBOTAYEV: Pis'ma Zh. Eksper. I. Teor. Fiz. **9**, 364 (1969)
4.26 N. G. BASOV, I. N. KOMPANETZ, O. N. KOMPANETZ, V. S. LETOKHOV, V. V. NIKITIN: Pis'ma Zh. Eksper. I. Teor. Fiz. **9**, 568 (1969)
4.27 YU. A. MATIUGIN, B. I. TROSHIN, V. P. CHEBOTAYEV: Optika I Spectroscopia **31**, 111 (1971); Digest of Physics Gas Laser Symposium, June 1969, Novosibirsk
4.28 E. V. BAKLANOV, V. P. CHEBOTAYEV: Zh. Eksper. I Teor. Fiz. **60**, 552 (1971)
4.29 E. V. BAKLANOV, V. P. CHEBOTAYEV: Zh. Eksper. I Teor. Fiz. **61**, 922 (1971)
4.30 S. HAROCHE, F. HARTMANN: Phys. Rev. **6A**, 1280 (1972)
4.31 M. S. FELD: in *Fundamental and Applied Laser Physics*, ed. by M. S. FELD, A. JAVAN, N. KURNIT (Wiley-Interscience Publ. 1973) pp. 369—420
4.32 H. R. SCHLOSSBERG, A. JAVAN: Phys. Rev. **150**, 267 (1966)
4.33 G. E. NOTKIN, S. G. RAUTIAN, A. A. FEOKTISTOV: Zh. Eksper. I. Teor. Fiz. **52**, 1673 (1967)
4.34 H. K. HOLT: Phys. Rev. Letters. **19**, 1275 (1967)
4.35 M. S. FELD, A. JAVAN: Phys. Rev. Letters. **20**, 12 (1968)
4.36 M. S. FELD, A. JAVAN: Phys. Rev. **177**, 540 (1969)
4.37 T. YA. POPOVA, A. K. POPOV, S. G. RAUTIAN, R. I. SOKOLOVSKII: Zn. Eksper. I. Teor. Fiz. **57**, 850 (1969)
4.38 A. K. POPOV: Zh. Eksper. I. Teor. Fiz. **58**, 1623 (1970)
4.39 T. HÄNSCH, P. TOSCHEK: Z. Physik. **236**, 213 (1970)
4.40 B. J. FELDMAN, M. S. FELD: Phys. Rev. **A5**, 899 (1972)
4.41 K. SHIMODA: Japan. J. Appl. Phys. **11**, 564 (1972)
4.42 M. S. FELD, A. SANCHEZ, A. JAVAN, B. J. FELDMAN: Proceedings of Aussois Symposium *"Methodes de Spectroscopie Sans Largeur Doppler de Niveaux Excited de Systemes Moleculaires Simples"*, May, 1973 (Publ. N 217, CNRS, Paris 1974), pp. 87—104
4.43 I. M. BETEROV, YU. A. MATIUGIN, V. P. CHEBOTAYEV: Pis'ma Zh. Eksper. I. Teor. Fiz. **12**, 174 (1970); Zh. Eksper. I Teor. Fiz. **64**, 1495 (1973)
4.44 V. S. LETOKHOV: Comments on Atomic and Molecular Physics **2**, 181 (1971)
4.45 V. S. LETOKHOV, B. D. PAVLIK: Zh. Eksper. I. Teor. Fiz. **64**, 804 (1973)
4.46 H. GREENSTEIN: J. Appl. Phys. **43**, 1732 (1972)
4.47 YU. V. BRZHAZOVSKY, V. P. CHEBOTAYEV, L. S. VASILENKO: IEEE J. QE-5, 146 (1969); Zh. Eksper. I. Teor. Fiz. **55**, 2096 (1968)
4.48 N. G. BASOV, O. N. KOMPANETZ, V. S. LETOKHOV, V. V. NIKITIN: Zh. Eksper. I. Fiz. **59**, 394 (1970)

4.49 O. N. KOMPANETZ, V. S. LETOKHOV: Zh. Eksper. I. Teor. Fiz. **62**, 1302 (1972)

4.50 C. V. SHANK, S. E. SCHWARZ: Appl. Phys. Lett. **13**, 113 (1968)

4.51 T. W. HÄNSCH, P. TOSCHEK: IEEE J. QE-**4**, 467 (1968)

4.52 F. KEILMANN, R. L. SHEFFIELD, M. S. FELD, A. JAVAN: Appl. Phys. Lett. **23**, 612 (1973)

4.53 N. G. BASOV, V. S. LETOKHOV: Report on URSI Conference Laser Measurements, Sept. 1968, Warsawa, Poland; Electron. Technology **2**, 15 (1969)

4.54 C. FREED, A. JAVAN: Appl. Phys. Lett. **17**, 53 (1970)

4.55 Annual Report of Optical and Infrared Laser Laboratory, MIT, 1973, p. 18; A. JAVAN, Private Communication

4.56 M. S. SOREM, A. L. SCHAWLOW: Opt. Commun. **5**, 148 (1972)

4.57 V. S. LETOKHOV: In *Fundamental and Applied Laser Physics*, ed. by M. S. FELD, A. JAVAN, N. KURNIT (Wiley-Interscience Publ. 1973), pp. 335—367

4.58 A. T. MATTICK, A. SANCHEZ, N. A. KURNIT, A. JAVAN: Appl. Phys. Lett. **23**, 675 (1973)

4.59 W. R. BENNETT, JR., V. P. CHEBOTAYEV, J. W. KNUTSON: 5th Intern. Conf. on Physics of Electron and Atomic Collisions, Abstract of Papers. "Nauka", 1967

4.60 R. H. CORDOVER, P. A. BONCZYK, A. JAVAN: Phys. Rev. Petters **18**, 730 (1967)

4.61 W. G. SCHWEITZER, M. M. BIRKY, J. A. WHITE: **57**, 1226 (1976)

4.62 H. K. HOLT: Phys. Rev. Letters **20**, 410 (1968)

4.63 T. W. DUCAS, M. S. FELD, L. W. RYAN, JR., N. SKRIBANOWITZ, A. JAVAN: Phys. Rev. A**5**, 1036 (1972)

4.64 I. M. BETEROV, V. P. CHEBOTAYEV: Pis'ma Zh. Eksper. I Teor. Fiz. **9**, 216 (1969)

4.65 T. HÄNSCH, R. KEIL, A. SCHABERT, CH. SCHMELZER, P. TOSCHEK: Z. Physik. **226**, 293 (1969)

4.66 R. G. BREWER: Phys. Rev. Letters **25**, 1639 (1970)

4.67 M. TAKAMI, K. S. SHIMODA: Japan. J. Appl. Phys. **11**, 1648 (1972)

4.68 K. SHIMODA: In *Laser Spectroscopy*, ed. by R. BREWER and A. MOORADIAN (Plenum Press, New York 1974) pp. 29—44

4.69 K. SHIMODA, T. SHIMIZU: in *Progress in Quantum Electronics*, ed. by J. H. SANDERS and S. STENHOLM (Pergamon Press, Oxford 1972) vol. 2, pp. 43—139

4.70 H. R. SCHLOSSBERG, A. JAVAN: Phys. Rev. Letters. **17**, 1242 (1966)

4.71 G. W. FLYNN, M. S. FELD, B. J. FELDMAN: Bull. Am. Phys. Soc. **12**, 669 (1967)

4.72 M. DUMONT: Phys. Rev. Letters **28**, 1357 (1972)

4.73 M. DUMONT: In Proceedings of Colloques Internationaux C.N.R.S. N217 *Methodes de Spectroscopie sans Largeur Doppler de Niveaux Excites de Systemes Moleculaires Simples* (Aussois, 23—26 May, 1973) p. 235

4.74 A. C. G. MITCHELL, M. W. ZEMANSKY: *Resonance Radiation and Excited Atoms* (Cambridge University Press, London 1971)

4.75 V. G. POKAZAN'EV, G. V. SKROTSKII: Uspekhi Fiz. Nauk. **107**, 623 (1972)

4.76 M. S. FELD: Ph. D. thesis (MIT, 1967)

4.77 J. S. LEVINE, P. A. BONCZYK, A. JAVAN: Bull. Am. Phys. Soc. **12**, 7 (1967); Phys. Rev. Letters **22**, 267 (1968)

4.78 M. I. DIAKONOV, V. I. PEREL: Zh. Eksper. I. Teor. Fiz. **50**, 448 (1966)

4.79 A. C. LUNTZ, R. G. BREWER, K. L. FOSTER, J. D. SWALEN: Phys. Rev. Letters **23**, 951 (1969)

4.80 E. E. UZGIRIS, J. L. HALL, R. L. BARGER: Phys. Rev. Letters **26**, 289 (1971)

4.81 R. H. CORDOVER, J. PARKS, A. SZÖKE, A. JAVAN: in *Physics of Quantum Electronics*, ed. by P. L. KELLEY, B. LAX and P. E. TANNENWALD (Mc-Graw Hill, New York 1966) p. 591

4.82 B. DECOMPS, M. DUMONT: J. de Phys. **29**, 443 (1968); IEEE J. QE-**4**, 916 (1968)

4.83 T. HÄNSCH, P. TOSCHEK: Phys. Letters **20**, 273 (1966); ibid. **22**, 150 (1966)

4.84 M. TSUKAKOSHI, K. SHIMODA: J. Phys. Soc. Japan **26**, 758 (1969)

4.85 M. DUCLOY: Optics Comm. **3**, 205 (1971)

4.86 M. DUCLOY, J. VIGUE, M. BROYER: in Proceedings of Colloques Internationaeux du C.N.R.S. N 217 *Methodes de Spectroscopie sans Largeur Doppler de Niveaux Excites de Systemes Moleculaires Simples* (Aussois, 23—26, May, 1973) p. 241

4.87 S. N. ATUTOV, A. G. NIKITENKO, S. G. RAUTIAN, E. G. SAPRIKIN: Pis'ma Zh. Eksper. I Teor. Fiz. **13**, 232 (1971)

4.88 J. SCHMIDT, P. R. BERMAN, R. G. BREWER: Phys. Rev. Letters **31**, 1103 (1973)

4.89 R. G. BREWER, R. L. SHOEMAKER, S. STENHOLM: Phys. Rev. Letters **33**, 63 (1974)

4.90 S. M. FREUND, J. W. C. JOHNS, A. R. MCKELLAR, T. OKA: J. Chem. Phys. **59**, 3445 (1973)

4.91 T. W. MEYER, C. K. RHODES: Phys. Rev. Letters **32**, 637 (1974)

4.92 T. OKA: in *Advances in Atomic and Molecular Physics*, ed. by D. R. BATES (Academic Press, New York 1973) vol. 9, p. 127

4.93 J. L. HALL: In *Lectures in Theor. Phys.*, XII, ed. by K. T. MANAUTHEGA and W. E. BOTIN (Gordon Brendi and Co., New York 1970), p. 161

4.94 J. L. HALL: Proceedings of Conference *Methodes de Spectroscopie sans largeur Doppler de niveaux excites de systems moleculaires simples*, May, 1973 (Publ. N 217, CNRS, Paris, 1974) pp. 105—125

4.95 E. V. BAKLANOV, B. YU. DUBETSKII, V. M. SEMIBALAMUT, E. A. TITOV: Soviet J. Quant. Electron. **2**, 2518 (1975)

4.96 J. L. HALL, C. BORDE: Phys. Rev. Letters **30**, 1101 (1973)

4.97 V. S. LETOKHOV: Zh. Eksper. I. Teor. Fiz. **56**, 1748 (1969)

4.98 T. S. JASEJA, A. JAVAN, C. H. TOWNES: Phys. Rev. Letters **10**, 165 (1963)

4.99 W. RADLOFF, E. BELOW: Optics Comm. **13**, 160 (1975)

4.100 J. A. MAGYAR, J. L. HALL: IEEE J. QE-**8**, 567 (1971)

4.101 K. M. EVENSON, J. S. WELLS, F. R. PETERSEN, B. L. DANIELSON, G. W. DAY: Appl. Phys. Lett. **22**, 192 (1973)

4.102 A. JAVAN, A. SANCHEZ: In *Laser Spectroscopy* ed. by R. BREWER and A. MOORADIAN (Plenum Press, N.Y. 1974) pp. 11—28

4.103 K. M. EVENSON, F. R. PETERSEN, J. S. WELLS: in *Laser Spectroscopy*, ed. by R. BREWER, A. MOORADIAN (Plenum Press, N.Y. 1974) pp. 143—156

4.104 A. S. PROVOROV, V. P. CHEBOTAYEV: Dokl. Akad. Nauk USSR **208**, 318 (1973)

4.105 I. M. BETEROV, V. P. CHEBOTAYEV, A. S. PROVOROV: Opt. Commun. **7**, 410 (1973)

4.106 R. L. ABRAMS: Digest of Technical Papers on VIII Intern. Quant. Electr. Conf., June 10—13, 1974, San Francisco, USA, p. 74

4.107 T. W. HÄNSCH, I. S. SHAHIN, A. L. SCHAWLOW: Phys. Rev. Letters **27**, 707 (1971)

4.108 T. W. HÄNSCH, I. S. SHAHIN, A. L. SCHAWLOW: Nature **235**, 63 (1972)

4.109 T. W. HÄNSCH: Appl. Opt. **11**, 895 (1972)

4.110 R. L. BARGER, M. S. SOREM, J. L. HALL: Appl. Phys. Lett. **28**, 573 (1973)

4.111 R. E. GROVE, F. Y. WU, L. A. HACKEL, R. G. YOUMANS, S. EZEKIEL: Appl. Phys. Lett. **23**, 442 (1973)

4.112 W. HARTIG, H. WALTHER: Appl. Phys. **1**, 171 (1973)

4.113 V. S. LETOKHOV, B. D. PAVLIK: Kvantovai Elektronika **3**, N 1 (1976)

4.114 S. R. BRUECK, A. MOORADIAN: IEEE J. QE-**10**, 634 (1974)

4.115 C. K. N. PATEL: Appl. Phys. Lett. **25**, 112 (1974)

4.116 J. L. HALL: Report on IV Vavilov Nonlinear Optics Conference, June 1975, Novosibirsk, USSR

4.117 V. S. LETOKHOV: Phys. Letters **41A**, 333 (1972)

4.118 V. S. LETOKHOV: Phys. Letters **53A**, 275 (1975)

4.119 PH. CAHUZAC, R. VETTER: Phys. Rev. Letters **34**, 1070 (1975)

4.120 T. W. HÄNSCH, M. H. NAYFEH, S. A. LEE, S. M. CURRY, I. S. SHAHIN: Phys. Rev. Letters **32**, 1336 (1974)

4.121 T. W. HÄNSCH, M. D. LEVENSON, A. L. SCHAWLOW: Phys. Rev. Letters **26**, 946 (1971)

4.122 M. D. LEVENSON, A. L. SCHAWLOW: Phys. Rev. **A6**, 10 (1972)

4.123 G. R. HANES, J. LAPIERRE, P. R. BUNKER, K. C. SHOTTON: J. Molec. Spectros. **39**, 506 (1971)

4.124 T. W. MEYER, J. F. BRILANDO, C. K. RHODES: Chem. Phys. Lett. **18**, 382 (1973)

4.125 O. N. KOMPANETZ, A. R. KOOKOODJANOV, V. S. LETOKHOV, V. G. MINOGIN, E. L. MIKHAILOV: Zh. Eksper. I. Teor. Fiz. **69**, 32 (1975)

4.126 Comite Consultatif pour la Definition du Metre, 5th session, Rapport (Bureau International des Poids et Mesures, Serves, France 1973)

4.127 A. JAVAN: in *Fundamental and Applied Laser Physics*, ed. by M. S. FELD, A. JAVAN, N. KURNIT (Wiley-Interscience Publ. 1973) pp. 295—334

4.128 J. DATCHARY, M. DUCLOY: Compt. Rend. Acad. Sc. **274B**, 337 (1972)

4.129 M. P. GORZA, B. DECOMPS, M. DUCLOY: Optics Comm. **8**, 323 (1973)

4.130 M. DUMONT: J. Phys. **33**, 971 (1972)

4.131 B. DECOMPS, M. DUMONT, M. DUCLOY: Linear and Nonlinear Phenomena in Laser Optical Pumping; in *Topics in Applied Physics*, Vol. 2: Laser Spectroscopy of Atoms and Molecules, ed. by H. WALTHER (Springer, Berlin, Heidelberg, New York 1976)

4.132 R. G. BREWER, M. J. KELLY, A. JAVAN: Phys. Rev. Letters **23**, 559 (1969)

4.133 R. G. BREWER, J. D. SWALEN: J. Chem. Phys. **52**, 2774 (1970)

4.134 M. J. KELLY, R. E. FRANCKE, M. S. FELD: J. Chem. Phys. **53**, 2979 (1970)

4.135 A. C. LUNTZ, J. D. SWALEN, R. G. BREWER: Chem. Phys. Lett. **14**, 512 (1972)

4.136 S. M. FREUND, G. DUXBURY, M. RÖMHELD, J. T. TIEDJE, T. OKA: J. Molec. Spectr. **52**, 38 (1974)

4.137 A. C. LUNTZ, R. G. BREWER: J. Chem. Phys. **54**, 3641 (1971)

4.138 K. UEHARA: J. Phys. Soc. Japan, **34**, 777 (1973)

4.139 A. C. LUNTZ, R. G. BREWER: J. Chem. Phys. **53**, 3380 (1970)

4.140 A. JAVAN: Report on 2nd Intern. Laser Spectroscopy Conf., 23—27 June 1975, Megeve, France

4.141 P. W. SMITH: J. Appl. Phys. **37**, 2089 (1966)

4.142 G. A. MIKHENKO, E. D. PROTZENKO: Optica and Spectr. **30**, 668 (1969)

4.143 W. DIETEL: Phys. Lett. **29A**, 268 (1969)

4.144 A. SZOKE, A. JAVAN: Phys. Rev. **145**, 137 (1966)

4.145 S. N. BAGAEV, YU. D. KOLOMNIKOV, V. N. LISITSYN, V. P. CHEBOTAYEV: IEEE J. QE-4, 868 (1968)

4.146 T. R. SOSNOWSKI, W. B. JOHNSON: IEEE J. QE-4, 56 (1968)

4.147 K. A. BIKMUHAMETOV, B. M. KLEMENT'EV, V. P. CHEBOTAYEV: Soviet Quant. Electr. **3**, 74 (1972)

4.148 L. S. VASILENKO, M. N. SKVORTZOV, V. P. CHEBOTAYEV, G. I. SHERSHNEVA, A. V. SHISHAEV: Optika I Spektroskopia. **32**, 1123 (1972)

4.149 YU. A. MATIUGIN, A. S. PROVOROV, V. P. CHEBOTAYEV: Zh. Eksper. I Teor. Fiz. **63**, 2043 (1972)

4.150 I. M. BETEROV, YU. A. MATUGIN, S. H. RAUTIAN, V. P. CHEBOTAYEV: Zh. Eksper. i Teor. Fiz. **58**, 1243 (1970)

4.151 IM THEK-DE, A. P. KAZANTZEV, S. G. RAUTIAN, E. G. SAPRIKIN, A. M. SHALAGIN: Soviet Quant. Electr. **1**. 416 (1974)

4.152 O. N. KOMPANETZ, A. R. KOOKOODJANOV, V. S. LETOKHOV, E. L. MIKHAILOV: Kvantovai Elektronika, N 16, 28 (1973)

4.153 L. S. VASILENKO: Thesis, Novosibirsk, Physics Semiconductor Institute, Siberian Branch of Academy of Sciences USSR (1971)

4.154 S. N. BAGAEV, E. V. BAKLANOV, V. P. CHEBOTAYEV: Pis'ma Zh. Eksper. I. Teor. Fiz. **16**, 15 (1972)

4.155 S. N. BAGAEV, E. V. BAKLANOV, V. P. CHEBOTAYEV: Pis'ma Zh. Eksper. I. Teor. Fiz. **16**, 344 (1972)

4.156 V. A. ALEKSEEV, T. L. ANDREEVA, I. I. SOBEL'MAN: Zh. Eksper. i Teor. Fiz. **64**, 813 (1973)
4.157 S. N. BAGAEV, A. K. DMITRIEV, V. P. CHEBOTAYEV: Pis'ma Zh. Eksper. i Teor. Fiz. **15**, 91 (1972)
4.158 J. W. C. JOHNS, A. R. W. MCKELLAR, T. OKA, M. RÖMHELD: J. Chem. Phys. **62**, 1488 (1975)
4.159 G. R. HANES, K. M. BAIRD, J. DEREMIGIS: Appl. Opt. **12**, 1600 (1973)
4.160 A. J. WALLARD: J. Phys. E: Sci. Instr. **5**, 926 (1972)
4.161 W. G. SCHWEITZER, JR., E. G. KESSLER, JR., R. D. DESLATTES, H. P. LAYER, J. R. WETSTONE: Appl. Opt. **12**, 2927 (1973)
4.162 K. M. EVENSON, J. S. WELLS, F. R. PETERSEN, B. L. DANIELSON, G. W. DAY, R. L. BARGER, J. L. HALL: Phys. Rev. Lett. **29**, 1346 (1972)
4.163 K. M. EVENSON, F. R. PETERSEN: Laser Frequency Measurements, the Speed of Light, and the Meter; in *Topics in Applied Physics*, Vol. 2: Laser Spectroscopy of Atoms and Molecules, ed. by H. WALTHER (Springer, Berlin, Heidelberg, New York 1976)
4.164 F. R. PETERSEN, D. G. MCDONALD, J. D. CUPP, B. L. DANIELSON: Phys. Rev. Lett. **31**, 573 (1973)
4.165 F. R. PETERSEN, D. G. MCDONALD, J. D. CUPP, B. L. DANIELSON: in *Laser Spectroscopy*; ed. by R. BREWER and A. MOORADIAN (Plenum Press, New York 1973) pp. 555—569
4.166 V. S. LETOKHOV: Pis'ma Zh. Eksper. i Teor. Fiz. **16**, 428 (1972)
4.167 V. S. LETOKHOV: Phys. Rev. Letters **30**, 729 (1973)
4.168 V. S. LETOKHOV: Phys. Letters **43A**, 179 (1973)
4.169 V. S. LETOKHOV: Phys. Rev. A **12**, 1954 (1975)
4.170 L. N. IVANOV, V. S. LETOKHOV: Zh. Eksper. i Teor. Fiz. **68**, 1748 (1975)
4.171 V. S. LETOKHOV: Phys. Letters **49A**, 275 (1974)
4.172 V. I. GOL'DANSKII: *Physical Chemistry of Positron and Positronium* (Publ. "Nauka", Moscow 1968)
4.173 K. F. CANTER, A. P. MILLS, S. BERKO: Phys. Rev. Letters **34**, 277 (1975)
4.174 A. P. MILLS, S. BERKO, K. F. CANTER: Phys. Rev. Letters **34**, 1541 (1975)
4.175 E. F. GROSS, S. A. PERMAGOROV, B. S. RAZBIRIN: Z. Phys. Ch. Sol. **27**, 1647 (1966); Sov. J. of Solid State Phys. **8**, 1483 (1966)
4.176 A. T. MATTICK, N. A. KURNIT, A. JAVAN: Lecture Note 1 by JAVAN in Les-Houches Summer School for Theoretical Physics, July 1975, Les-Houches, France
4.177 V. P. CHEBOTAYEV, T. M. BETEROV, V. N. LISITSYN: IEEE J. QE-4, 788 (1968)
4.178 V. N. LISITSYN, V. P. CHEBOTAYEV: Zh. Eksper. I. Teor. Fiz. **54**, 419 (1968)
4.179 V. M. TATARENKOV, A. N. TITOV, A. V. USPENSKY: Optika I Spectroscopia **28**, 572 (1970)
4.180 V. M. TATARENKOV, A. N. TITOV: Optika I Spectroscopia **30**, 803 (1971)
4.181 G. R. HANES, K. M. BAIRD: Metrologia. **5**, 32 (1969)
4.182 G. R. HANES, C. E. DAHLSTROM: Appl. Phys. Letters **14**, 362 (1969)
4.183 J. D. KNOX, YOH-HAN PAO: Appl. Phys. Letters **16**, 129 (1970)
4.184 J. T. LATOURETTE, R. S. ENG: IEEE J. QE-8, 561 (1972)
4.185 W. G. SCHWEITZER, JR.: Appl. Phys. Letters **13**, 367 (1968)
4.186 V. N. LISITSYN, V. P. CHEBOTAYEV: Optika I Spectroscopia **26**, 856 (1969)
4.187 N. G. BASOV, M. V. DANILEIKO, V. V. NIKITIN: Zh. Prikl. Spectroscopii **11**, 543 (1969)
4.188 S. N. BAGAEV, V. P. CHEBOTAYEV: Pis'ma Zh. Eksper. I Teor. Fiz. **15**, 91 (1972)
4.189 R. L. BARGER, J. L. HALL: Appl. Phys. Letters **22**, 196 (1973)
4.190 S. N. BAGAEV, E. V. BAKLANOV, E. TITOV, V. P. CHEBOTAYEV: Pis'ma Zh. Eksper. I. Teor. Fiz. **20**, 292 (1974)

4.191 Yu. M. Malushev, V. M. Tatarenkov, A. N. Titov: Pis'ma Zh. Eksper. I Teor. Fiz. **13**, 592 (1971)

4.192 L. S. Vasilenko, V. P. Chebotayev, G. I. Shershneva: Optika I Spectroscopia **19**, 204 (1970)

4.193 T. Kan, G. J. Volga: IEEE J. QE-7, 141 (1971)

4.194 P. W. Smith, T. Hansch: Phys. Rev. Letters **26**, 740 (1971)

4.195 C. Borde: C. R. Acad. Sci. Paris **271**, 371 (1970)

4.196 P. Rabinowitz, R. Keller, J. T. La Tourette: Appl. Phys. Letters **14**, 376 (1969)

4.197 F. Shimizu: Appl. Phys. Lett. **14**, 378 (1969)

4.198 E. R. Petersen, B. L. Danielsen: Bull. Amer. Phys. Soc. **15**, N11 (1970)

4.199 I. M. Beterov, L. S. Vasilenko, B. Gangardt, V. P. Chebotayev: Kvantovai Elektronika **1**, 970 (1974)

4.200 Yu. A. Gorokhov, O. N. Kompanetz, V. S. Letokhov, G. A. Gerasimov, Yu. J. Posudin: Opt. Commun. **7**, 320 (1973)

4.201 C. Costain: Canad. J. Physics **47**, 2431 (1969)

4.202 R. S. Winton, W. Gordy: Phys. Letters **32**A, 219 (1970)

5. High Resolution Saturated Absorption Studies of Methane and Some Methyl-Halides

J. L. HALL and J. A. MAGYAR

With 6 Figures

It is clear that lasers have opened new vistas in spectroscopy. Indeed the high laser intensity and monochromaticity are leading to a virtual explosion in the field of molecular spectroscopy. For example in the visible region, selective laser excitation [5.1] of a single upper level gives simple spectral patterns in fluorescence that contain information only about the accessible lower states. By contrast, in absorption spectroscopy, one such pattern is contributed by each lower state in a vibration-rotation manifold of lower states. Without laser selective excitation, the apparent chaotic overlapping of lines requires prodigious effort to unravel, even for diatomic molecules such as I_2.

In the infrared however, the pre-laser situation was more tenable. The basic point is that infrared transitions are between vibration-rotation levels of the same electronic state and therefore the transition can produce at most a very small change in the average internuclear distance. This fact in turn leads to the approximate selection rule that the vibrational and rotational quantum numbers can change by only 0 and ± 1. Thus the spectrum is collapsed from a virtual continuum as in the case of the visible spectrum of I_2, to a clear and unambiguous infrared line spectrum. For molecules as complex as the methyl-halides, however, infrared spectrographs of the appropriate resolution ($\sim 100,000$) have been available only quite recently with the development of large high-resolution diffraction grating instruments by HENRY et al. [5.2] and by PLYLER et al. [5.3], and with the techniques for Fourier-transform spectroscopy of CONNES [5.4] and his colleagues. Thus, for example, it is only very recently that a definitive analysis of the methyl bromide v_1 spectrum has been available [5.5].

In counterpoint to the above-discussed developments in "classical" spectroscopy has been the development of several techniques that can provide resolution far superior to the usual Doppler limit of the classical methods. Although we will only here make use of saturation spectroscopy, several additional sub-Doppler techniques besides saturation are considered in other parts of this volume. Nor is there a shortage of molecules and lasers that can interact with them. For example, the 3.39 μm He–Ne laser is just in the midst of the CH_3-stretching band: we will list approximately 40 wavelength/frequency coincidences observed within the $\pm 0.15\,\text{cm}^{-1}$ tuning range of our saturation spectroscopy apparatus. The lifetimes of these transitions are milliseconds and longer, so spectral resolving power approaching 10^{12} is possible in principle. The laser technology and optical components are well developed for this spectral region so that the uninitiated might be led to assume that attention can mainly be focussed on the physical parameters of the observed quantum systems. Indeed—after one has paid the technology "initiation fee" in time, labor, and apparatus development—it begins to be possible to determine "true molecular properties."

Essentially, then, this article deals with the sub-Doppler spectroscopy of CH_3Cl, CH_3Br, CH_3I, and CH_4 near 3.39 µm. We identify a number of lines and measure their transition frequencies. We display the hyperfine spectra for a few selected transitions. We then deduce hyperfine coupling constants such as eqQ for the ground state and (for the first time) the vibrationally excited state. We report a partly successful attempt to assign (J, K) for a transition based on the observed spectrum combined with a least-squares search procedure using the known J-, K-dependence of the electric quadrupole interaction. Finally, we turn briefly to the question of the ultimate spectral resolution limits and—in the sense of a progress report—present a super-high resolution spectrum of CH_4 that clearly shows the spectral shift/doubling associated with the recoil effect.

5.1 Outline of Sub-Doppler Spectroscopy Near 3.39 µm

Of course a laser/absorber wavelength coincidence is the first requirement for high resolution spectroscopy using saturated absorption methods. A number of coincidences between the Zeeman-tuned helium neon laser and a number of molecules had been established by the early work of GERRITSEN [5.6] and of SHIMODA [5.7], and it was thus natural to look first in these molecules for the saturation absorption resonances. A set of three Zeeman-tuned helium neon lasers was constructed. Two of these devices used powerful Alnico magnets to reach a tuning range of ± 30 ppm, or about $0.1\,\text{cm}^{-1}$. A third laser was built utilizing the transverse field of a laboratory electromagnet to provide a tuning range $1\frac{1}{2}$ times larger and vastly more convenient to adjust. This latter device had a 50 cm intracavity absorption cell for the molecules to be investigated. As indicated in Table 5.1 we found more than 40 coincidences within the magnetic tuning range of this laser. Many of these lines, if not most, showed some evidence of structure under the highest resolution conditions. We will return to this subject later.

Table 5.1. Saturated absorption resonances observed

Identification[a]	B (kG)	Δv (MHz)	$\tilde{v}_B{}^{b}$	$\tilde{v}_{\text{Heterodyne}}{}^{b}$
$^{12}CH_4$ $P(7)$ $F_2^{(2)}$	—	0.0 (Ref.)	—	7.912103 (Ref.)
$^{12}CH_4$ $P(7)$ E	-1.72	-3032.56	7.809	7.810948
$^{12}CH_4$ $P(7)$ A_2	3.47		8.113	
$^{13}CH_4$ $P(6)$ $F_2^{(1)}$	-0.48	-936.53	7.881	7.88086
CH_3OH	0.23	336.85	7.923	7.92334
CH_3F	-2.76		7.748	
CH_3F	-2.55		7.760	
CH_3F	-1.16	-2134.17	7.842	7.84091
CH_3F	-0.35	-696.76	7.889	7.88886
CH_3F	0.87		7.961	
CH_3F $^{Q}P_6(10)^c$	1.27		7.984	
CH_3I $^{Q}P_3(45)$	3.45		8.112	
CH_3I $^{Q}P_0(46)$	2.54		8.058	
CH_3I $^{Q}P_1(46)$	1.63		8.005	
CH_3I $^{Q}P_2(46)$	-0.96	-1731.79	7.854	7.85434
CH_3I $^{Q}P_5(44)$	-1.21	-2181.7	7.839	7.83933
CH_3I $^{Q}P_6(43)$	-1.74	-3105.8	7.808	7.80851
CH_3I $^{Q}P_4(45)$	-2.06		7.790	
CH_3CH_3	-0.45	-826.21	7.884	7.88455
$CH_3{}^{37}Cl$ $^{Q}P_0(23)$	-3.14		7.726	
$CH_3{}^{35}Cl$ $^{Q}P_3(22)$	-1.52	-2732.17	7.821	7.82097
$CH_3{}^{35}Cl$ $^{Q}P_5(21)$	-1.32	-2408.98	7.832	7.83175
$CH_3{}^{37}Cl$ $^{Q}P_3(22)^d$	3.09		8.091	
$CH_3{}^{35}Cl$ $^{Q}P_2(22)$	3.20		8.097	
$CH_3{}^{35}Cl$ $^{Q}P_6(20)$	3.30		8.103	
CH_3Br	-2.84		7.743	
CH_3Br	-2.30		7.775	
$CH_3{}^{81}Br$ $^{Q}P_1^-(37)\rbrace$	-0.80	-1476.8	7.863	7.86287
$CH_3{}^{81}Br$ $^{Q}P_5^+(41)\rbrace$	-0.74	-1394.62	7.866	7.86558
$CH_3{}^{79}Br$ $^{Q}P_1^-(37)$	-0.28	-575.96	7.894	7.89289
$CH_3{}^{81}Br$ $^{Q}P_6^+(39)$	-0.06	-121.37	7.906	7.90806
$CH_3{}^{79}Br$ $^{Q}P_3^-(35)\rbrace^f$	0.75		7.954	
$CH_3{}^{79}Br$ $^{Q}P_4^+(44)\rbrace$	0.93		7.964	
$CH_3{}^{79}Br$ $^{Q}P_6^+(39)$	1.68		8.008	
$CH_3{}^{81}Br$ $^{Q}P_0^-(37)$	1.76		8.013	
$CH_3{}^{79}Br$ $^{Q}P_0^-(37)\rbrace_e$	2.05		8.029	
$CH_3{}^{81}Br$ $^{Q}P_2^-(36)\rbrace$	2.07		8.030	
CH_3Br	2.96		8.083	
CH_3Br	3.57		8.119	

[a] Note: Identified methyl-halide lines are $v_1(a_1)$ band transitions.
[b] Note: Add 2940.000000 cm^{-1} to all $\tilde{v}$ values.
[c] Luntz et al., Ref. [5.12].
[d] $^{Q}P_5(21)$ for $CH_3{}^{37}Cl$ is calculated to have the same frequency.
[e] Identification possibly reversed.
[f] $^{Q}P_9(34)$ of $CH_3{}^{81}Br$ also is calculated to lie nearby.

CH_3I $\lbrace$ Identification: Russell and Overend, Ref. [5.10]
$\quad\quad$ $\lbrace$ Fourier transform spectrum: Amiot, Ref. [5.11]
CH_3Br identification: Morillon and Betrencourt, Ref. [5.5].

There seem to be three interesting types of physical information to be obtained for these several resonances—the absolute optical frequency or equivalently, the frequency intervals from a well-known reference frequency as provided by the methane line [5.8], the hyperfine structure and parameters which enter into it, and the spectral assignment of the transition.

Perhaps the assignments are less interesting in relation to the topic of this volume, and we will first discuss this subject briefly before returning to the other two. Basically, we used the magnetic field values at which the resonance occurred, along with the known value of $g_J \simeq 1.66$ for the neon transitions, to establish a magnetic wavenumber scale called $\tilde{v}_b$. These wavenumbers are used to identify the transitions listed in Table 5.1. Unfortunately, many data in the literature have been taken with practical infrared wavelength standards which suffer some possible spectral offsets. Thus in comparing infrared spectra from several sources, one may find calibration discrepancies of up to $\pm 0.02 \text{ cm}^{-1}$. We have found it useful in analyzing the methyl-halide spectra to plot the frequencies at which we observe resonances alongside the frequencies at which resonances were observed in classical spectroscopy. With some expertise in pattern recognition, and knowledge of the transition strengths, it proves reasonably simple to generate unambiguous identifications. Fortunately CH_3Cl (both Cl isotopes) has been carefully studied by MORILLON-CHAPEY and GRANER [5.9]. Their instrumental resolution of 0.031 cm^{-1} resolved only three "lines" near 2947.9 cm^{-1} although they calculated eight lines to fall in the interval of interest $(\pm 0.2 \text{ cm}^{-1})$. Six lines were observed in our survey work with three of them falling into an interval spanning little more than two Doppler linewidths near 2948.1 cm^{-1}. Sub-megahertz saturation spectroscopy of course resolves them absolutely. The line pair at 2947.821 and 0.831 cm^{-1} are also almost Doppler unresolved. The relative intensities of the saturated absorption signals were useful to further discriminate between potential assignments. Finally, the $^Q P_3(22)$ assignments for the 2947.821 cm^{-1} line is confirmed by the quadrupolar hyperfine analysis to be presented later,

In the case of CH_3Br there are strong interactions with another band $(v_3 + v_5 + v_6)$ occurring throughout this spectral region and it is only quite recently that these transitions have been assigned by BETRENCOURT et al. [5.5]. MORILLON and BETRENCOURT have kindly provided the CH_3Br assignments listed in Table 5.1. The $^Q P_6(39)$ assignment of the line at 2947.908 cm^{-1} has been confirmed by the quadrupole analysis to be presented.

The identification of transitions in CH_3I has been greatly facilitated by some unpublished results made available to us by RUSSELL of Michigan

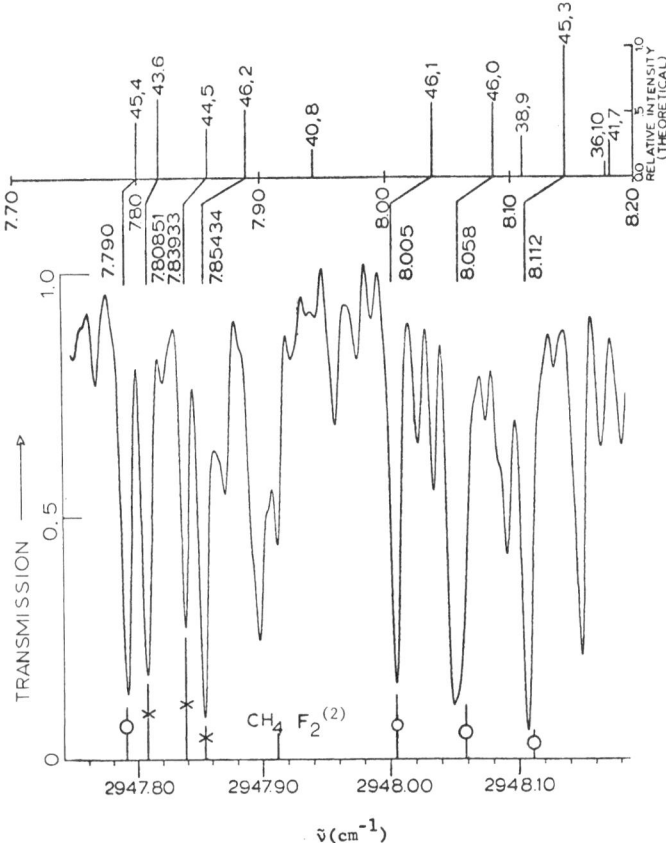

Fig. 5.1. Fourier transform spectrum of CH_3I near 2947.9 cm^{-1} (courtesy C. Amiot, Laboratoire Aime Cotton, Orsay). Absorption path 8 m, pressure 5 Torr, resolution $\Delta\sigma = 5 \times 10^{-3}$ cm^{-1}. The seven bars indicate the frequencies measured; the "X" denotes heterodyne measurements, "0" denotes magnetic field measurement. Upper lines represent theoretical positions and intensities of spectral lines. See text for formula and explanation

State University [5.10] and OVEREND of Minnesota. Additionally a Fourier transform spectrum covering this region was generously made available to us by AMIOT from the Laboratoire Aimé Cotton [5.11]. Essentially we have extended the analysis of RUSSELL and OVEREND beyond the spectral limits they explored, but have been able to check these calculations against the Fourier transform spectrum. The formula $\tilde{v}_K(J) = 2971.225 - 0.503 J - 0.0494 K^2$ reproduces the P-branch experimental frequencies to within ± 0.02 cm^{-1} in most cases. Since the Fourier transform wavenumbers are known in absolute terms we can readily make a correspondence between our lines and those of the Fourier transform spectrum. The success in making unambiguous identification can be traced to the opportunity to compare the spectrum, not only in the region of direct interest to us, but also in a region where the usual energy level expansion and identification techniques are most powerful. Thus we have synthesized the results of these three methods. As an illustration of this pattern recognition procedure, Fig. 5.1 shows the Fourier transform spectra [5.11] of CH_3I and the line positions calculated from the above formula. This figure also shows our seven observed resonances. Three of these resonances (marked by an "X") were measured with the frequency offset methods to be described later. The positions of the other four resonances were measured only with the magnetic field technique, but are probably accurate to about $0.002\,cm^{-1}$. Again we have been able to confirm one assignment, that of the $^QP_2(46)$ line, by analysis of the hyperfine spectrum.

Unfortunately, the identification of CH_3F was not pursued as vigorously, primarily because the spin 1/2 fluorine nucleus cannot give rise to the electric quadrupole interactions of the type we wish to study. LUNTZ et al. identified one CH_3F line using Stark-tuned sub-Doppler laser spectroscopy [5.12]. The residual magnetic effects should be comparable with those in methane, that is, of the order of 30 kHz. We will return to this question of spectral assignments when the hyperfine spectra are analysed.

5.2 Frequency Measurement and Control

Let us turn now to the question of frequency measurements. Basically it is clear that the very high resolution techniques discussed in this book will not be interesting in practice unless methods can be developed for the systematic use of this resolution. Since it is not reasonable to try to build a laser oscillator with the free-running stability needed for this investigation, one is obliged to be interested in servo-controlling the laser frequency. For example we may stabilize two lasers, each independently stabilized on its own saturation peak. The beat frequency therefore gives interesting physical information about the spectral intervals between different transitions. (Lineshape information can be obtained by locking one laser to its peak and locking the second laser at a controlled offset frequency from it—see later.) Of course, the frequency intervals can quickly become too large for convenient frequency measuring methods and it is attractive to synthesize a network of well-known laser frequencies such as is illustrated in Fig. 5.2.

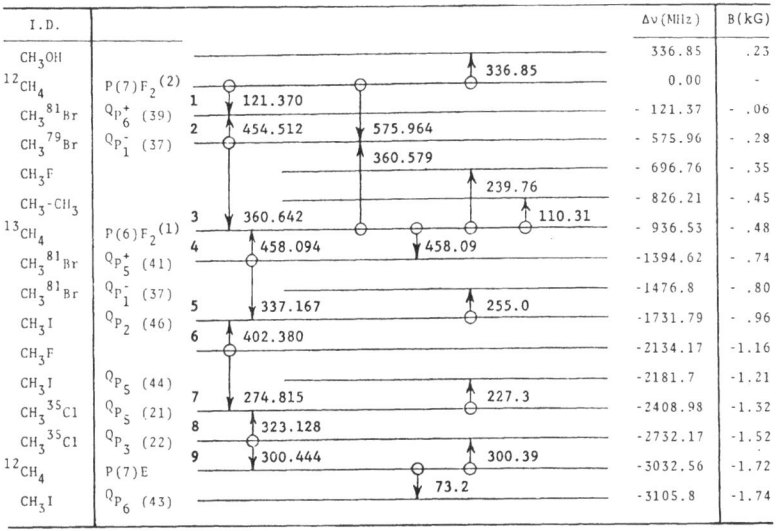

Fig. 5.2. Optical heterodyne ladder. The frequency reference is the usual $^{12}CH_4$ $P(7)$ $F_2^{(2)}$ line, $\nu_{ref}(CH_4) = 88,376,181.627(50)$ MHz, from [5.8]. Circles represent strong lines used as transfer points. See text

Although a number of the lines in Table 5.1 are, in fact, relatively weak, we were lucky to find strong lines nearby so that a possible offset from ideal line center of the lock on one of the weak lines did not affect the whole network. For example, in Fig. 5.2 it may be seen that the CH_3Br line at -121 MHz was used only as a transfer standard, with the electromagnet-tuned laser being stabilized first on methane to give the 121.37 MHz beat, and then stabilized on another CH_3Br transition (a strong one) to give the 454.512 MHz beat. The frequency values reported in this table use as transfer points the locks indicated by the circles. That is, arrows connecting those junctions represent sequential experiments typically separated by hours to days, whereas the measurements to a transition where the arrowheads meet were made with time separations of just minutes. This procedure minimizes the errors introduced into the network by possible offsets occurring on any one of the intermediate frequency locks.

The lasers used in these heterodyne studies gave spectral widths of about 0.75 MHz for the saturation peaks. Considering that reasonable care was taken to operate with laser parameters such that the resonances were viewed on a flat power background, these heterodyne offset frequencies probably are reliable to ± 10 kHz in a typical case. For the very weak lines or the one or two accidental cases in which it was difficult to obtain the desired background profile, the errors could be as high as

± 50 kHz. Recently, the frequency interval between the usual CH_4 F-line and the CH_4 Coriolis E-component has been measured in a one-step heterodyne experiment by BRILLET et al. [5.13]. Their value, $\Delta f = -3032.560$ MHz, is in very satisfactory agreement with the result indicated in Table 5.2.

It is perhaps worth noting here that in separate experiments using microwave signals applied directly to the reverse-biased indium arsenide photodiode, it was found that very good photo-beat signals could be obtained well into the gigahertz region. For example, the usual 3.39 µm neon line $(3s_2 \rightarrow 3p_4)$ has a companion line $(3s_2 \rightarrow 3p_2)$ about 0.88 cm^{-1} to the red. This frequency difference of 26.5 GHz was directly measured [5.14] at JILA in 1972 by WRIGHT, using an InAs diode in the reverse-bias "varactor" mode. However in present experiments measuring the hyperfine spectrum of $^{13}CH_4$, we have returned to the usual "single-mixer" mode since the diode's rf efficiency at the 936 MHz beat frequency is sufficient to ensure effective optical phase-lock of the controlled laser. In the work reported here the beat frequency up to 500 MHz was converted into the 5 MHz range of the frequency lock system using a balanced mixer driven by a variable but stable laboratory oscillator.

The measurements presented up to now have used saturated absorption techniques to provide frequency markers to better perform the tasks usually associated with "classical infrared rotation-vibration spectroscopy". Let us now turn to application of high resolution techniques to problems that are not accessible without these methods, that is, to the study of the internal structure of these "lines". Thus, we now will be discussing the measurement of magnetic and electric quadrupolar hyperfine structures of rotation-vibration transitions studied directly with high resolution optical spectroscopy. To emphasize the change in frequency scale we note that—although the experiments reported above have high precision—the resolution of about 0.75 MHz employed corresponds "only" to ~ 300 resolution elements within the Doppler profile. Now we will discuss data with spectral resolution of about 10 kHz, a resolution improvement factor of nearly 100-fold. This higher spectral resolution level corresponds to resolution of a Doppler velocity distribution into bins of about 5 cm/s width.

This working resolution range near 10 kHz (compared with the optical frequency of 10^{14} Hz) is attractive for these molecular spectroscopic investigations. It is sufficient to nicely separate most of the spectral features we encounter but it does not entail the large price in phase-space (signal-to-noise) experienced when one tries to reach the ultimate resolution as we will discuss at the end of this chapter.

Certainly in the resolution domain of 10^{10} and higher it is necessary to use automatic frequency control methods. Since these techniques have been described before [5.15] we summarize them only briefly here. Basically, the high signal-to-noise ratio of a laser with an internal absorption cell can be used to frequency stabilize that laser very effectively as we have seen. However, having the absorption cell inside the laser resonator makes for complications which (at best) do not facilitate clear understanding of the observed resonances. Also to display the resonance line shapes it is desirable to have a very precise frequency scan that is difficult to accomplish without frequency offset techniques. Thus we prefer to build the laser saturated absorption spectrometer using a laser power oscillator to interrogate molecules contained within an external absorption cell. The absorber will operate in a low intensity, low pressure, regime to minimize the line broadening. Thus we will be prepared to trade signal-to-noise ratio for a perturbation-free environment for our molecules and for a transparent relation between the observable signal and the molecular absorption coefficients. If necessary, the price will be paid by longer data accumulation times. This power oscillator is frequency-controlled, -modulated, and -scanned by heterodyne reference to a local "optical frequency standard" system. This frequency reference system can be optimized to minimize its effective spectral width and long-term frequency drift. On the other hand, it is not necessary that its performance and frequency offsets be completely calculable from first principles; it will suffice if the reference system is spectrally narrow and frequency stable. Short-term stability can best be provided by use of a laser that is stable by virtue of its thoughtful construction. For times longer than ~ 10–100 ms, its drifts will be important and one finds that a laser stabilized to an internal cell saturated-absorption peak gives better performance in this long time domain. Thus an optimized laboratory laser frequency reference system consists of two lasers. The system's frequency output will be taken from the "solid laser", the one optimized for low spectral width and good short-term stability. The slow drifts of this laser will be corrected out by the saturated absorption signal from the other laser. To optimize the saturated absorption reference laser, the short-term stability of the "solid laser" will be transferred also to the reference laser. This "crossed" control topology thus uses each laser in an optimal way. No control system difficulties are encountered in view of the strong difference in servo attack times (100 μs control of the methane-stabilized laser and 10–30 ms updating rate of the solid laser).

In the latest system the power laser of the saturated absorption spectrometer is phase-locked to the solid laser's output with a 30 μs settling time. The free running stability of this laser at short times is sufficient that the phase errors do not exceed ~ 100 mrad peak. Such small phase excursions correspond to very small frequency modulation sideband intensities, well separated from the carrier. Thus the narrow spectral width (~ 200 Hz) of the solid laser as well as the good long-term stability ($\lesssim 10^{-13}$) of the methane-stabilized reference laser is transferred to the power laser. The desired frequency modulation "dither" is developed as a phase modulation in the phase-lock loop. The laser frequency scan function is implemented by mixing the laser rf beat signal with a digitally programmed frequency synthesizer within the phase loop. Correspondence between offset frequency and channel number of the signal averager is thus established in a natural and rigorous way.

Before presentation and discussion of our saturation spectroscopy results on the methyl-halides, it may be in order to sketch the context in which this work was undertaken. A general objective of the present work has been to pursue and develop high resolution methods that facilitate the clear separation between the properties of the apparatus and its interactions with the subject molecules, and the properties of those molecules themselves. We view this as an appropriate and necessary basis for consideration and development of optical frequency standards.

We have already noted that it is attractive to place the absorber gas outside the laser cavity even at the expense of signal-to-noise ratio. The resulting extension of the signal accumulation time then puts a premium on the long-term frequency stability of the system, but these problems have been overcome with appropriate servosystem concepts and technology as described above.

An additional factor is important in the low pressure regime where one is naturally led in the search for precise, reliable high resolution data: this is the transit effect of the molecules through the laser light beam. In this free-flight regime, the unperturbed molecular velocity ensures an intimate relation between the spatial phase fronts of the laser wave and the Fourier spectrum of the exciting pulse experienced by the molecules. Thus the fact is that the observed spectral lineshapes—position and symmetry as well as width—depend essentially on the diameter and optical quality of the light beams. One reluctantly comes to view as inappropriate most of the sensitivity-enhancing optical and modulation tricks that are effective when broader lines are investigated or when less secure spectral information is sufficient.

Consideration of the geometrical aspects of the studied saturation absorption process leads one to appreciate that the signals contain collision information mainly about those collisions occurring at large values of impact parameter. A useful way to appreciate one aspect of the collision effects in saturated absorption spectroscopy is for each intermolecular separation to associate a radial force with the gradient of the intermolecular potential at that position. Integration of the force over the collision history yields directly the transferred impulse and therefore the trajectory deviation angle. Thus ultimately we have a new first-order Doppler shift corresponding to the new axial components of velocity. Physically it seems reasonable to distinguish those collisions as "hard" that importantly deviate the velocity vector and thus carry the frequency of the interacting dipole away from its driving frequency by more than one homogeneous linewidth γ_2. Collisions occurring at larger values of the impact parameter apparently have no essential effect. Since the linewidth γ_2 is itself primarily associated with the transit effect, we come to expect a dependence of the saturated absorption resonance's pressure broadening coefficient upon the laser beam's diameter and angular content. Although this idea has been considered [5.16] and analyzed [5.17] before, up to now there have been no really satisfactory data illustrating or denying the effect. Our unpublished results are instructive and show the expected trend, but probably are not quantitatively definitive.

Thus a channel of investigation that seemed interesting in this collisional connection was to change the value and long-range behavior of the intermolecular forces. For example, a spherical molecule such as CH_4 interacting with a rare gas may be expected to have a Van der Waals $V = -C_6 r^{-6}$ asymptotic behavior at long distances. A molecule with a strong permanent electric dipole moment (such as a methyl-halide, for example) interacting with a rare gas may have $V(r) = -C_4 r^{-4}$ and, finally one expects a component like $V(r) = -C_3 r^{-3}$ for resonant interaction with other molecules of the same type. Thus it seemed interesting to develop a repetoire of usable molecules of differing dipole moment, i.e., usable in precision saturated absorption spectroscopy experiments. Of course some new tuning and detection techniques would be necessary to allow unambiguous identification of the expected saturation resonances, to facilitate their careful study, and finally to allow the extraction of the desired collisional and spectroscopic information. This joint experimental/analytical challenge formed the thesis topic of one of the authors (JAM): the methyl-halide spectroscopic results to be presented here arise essentially from this work. Before the collision effects can be well addressed, it was clearly necessary for us to appreciate the spectroscopy of these molecules in some detail, and the remainder of the chapter deals with this subject.

5.3 Observed Spectra of Methyl-Halides and Their Analysis

Spectra of the heavier methyl-halides might at first be expected to show rather dramatic frequency splittings since the nuclear electric quadrupole interaction term is of the order of 100 MHz. However, we must not forget that lines observed here are rotation-vibration transitions within the ground electronic state with high J-values and $\Delta K = 0$. As noted in the introduction, the vibrationally excited state has essentially the same nuclear electric quadrupolar level structure. Thus we will have all the main transition components between quadrupolar-labeled states falling at basically the same transition frequency. Those small differences that do exist between ground and excited state quadrupole interaction constants will be manifested as splitting of the spectral line, providing us a kind of "vernier" measurement technique. Magnetic interactions are still smaller by another two or three orders of magnitude.

5.3.1 CH$_3$Cl: Observed Spectrum and Assignment

As an example of the type of spectra obtained we show in Fig. 5.3 a derivative spectrum obtained from $CH_3{}^{35}Cl$ in the vicinity of 2947.82 cm^{-1}. The measured spectrum, comprised of the points, displays four (partly overlapping) derivative lines corresponding to the four possible vector additions of the chlorine nuclear spin ($|I| = 3/2$) with the rotational angular momentum J. The question of the correspondence between F-labels and the observed transitions can in principle be decided by using the relative intensity information from the spectrum along with the known quantum-mechanical intensity factors. Because of the rather strong overlapping between lines it is profitable to use an iterative least-squares line profile-fitting program to obtain line center information as well as the correct intensity ordering of the components. The "best-fitting" sum of Lorentzians of equal widths is displayed as the solid curve of Fig. 5.3. It can be seen that this rather simple lineshape function gives surprisingly effective representation of the observed data. The output parameters for this fit are listed in Table 5.2. The linewidth 16.7 kHz, contains about a 10% correction due to the finite fm dither used for phase-sensitive signal recovery [5.18].

It had been hoped that the mass increase between CH_4 and CH_3Cl (and later, CH_3I) would lead to a considerable narrowing of the resonance linewidth; the larger mass would lower the mean thermal velocity $u = (2kT/m)^{1/2}$ and thus reduce the time-of-flight contribution to the spectral width. Unfortunately the data were taken using as reference laser the "Alnico-magnet" Zeeman-tuned laser stabilized to the E Coriolis component of CH_4. At that time these heavy magnets contributed an unwelcome but generous frequency sensitivity to vibrations. Thus it is felt that the operating resolution was degraded at least by a factor of 2 by such technical problems. Ultimately these problems disappear as more labor and money are invested, but it *is* attractive to try some spectroscopic experiments along the way....

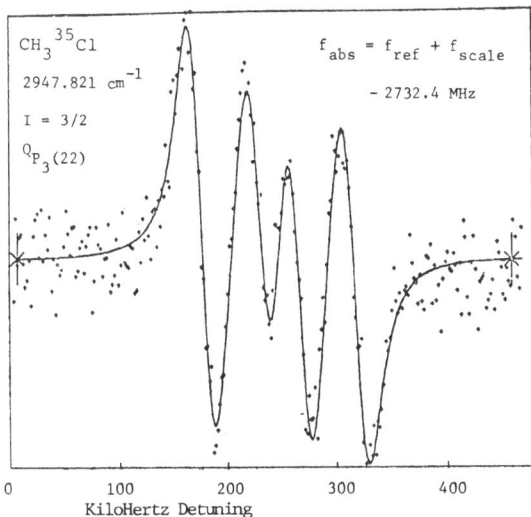

Fig. 5.3. Saturated absorption (derivative) spectrum of $CH_3{}^{35}Cl$ at 2947.821 cm^{-1}. The solid curve is a least-squares fit of 4 modulated Lorentzians to the measured points. The common linewidth was 16.7 kHz. The four lines correspond to the 4 values of $F_1 = I + J$ for $I = 3/2$

Although the signal-to-noise ratio displayed by Fig. 5.3 is only modest, the least-squares output relative intensities of Table 5.2 lead to an essentially unique ordering according to intensity. Inspection of the intensity tables [5.19] for $I = 3/2$ shows that for our observed signals J is certainly greater than 10. This is consistent with our identification in Table 5.1 of this transition as $^QP_3(22)$ of $CH_3{}^{35}Cl$. Some skepticism

Table 5.2. Fitted parameters and theoretical intensities for the CH_3Cl line at 2947.821 cm^{-1}

ID $(F_1 - J)$	Center frequency $f_c{}^a$ (kHz)	$\mathscr{I}_{calc}{}^b$ (%)	$\mathscr{I}_{expt}{}^c$ (%)	$\mathscr{I}^2_{calc}{}^b$ (%)
$-1/2$	140.5 ± 0.4	24.4	23.9 ± 0.8	23.7
$1/2$	89.7 ± 0.4	25.5	25.5 ± 0.9	26.0
$-3/2$	53.5 ± 0.4	23.3	23.0 ± 0.9	21.6
$3/2$	0.0 ± 0.4	26.8	27.6 ± 0.8	28.6
		range = 3.5%	range = 3.6%	range = 7.0%

A common linewidth was assumed in the fitting. Its fitted value is $\Delta v/2 = (16.7 \pm 0.4)$ kHz.

[a] These are relative frequencies. The absolute frequencies are given by $f_{abs} = f_r + f_c - 2732.3$ MHz, where f_r is the $^{12}CH_4$ F-line reference frequency.

[b] $\mathscr{I}$ is the relative intensity for linear absorption calculated for $J = 22$.

[c] The mean relative intensity is $100\%/4 = 25\%$.

may be in order here, but for the present purposes we will state the ordering, from left to right in Fig. 5.3, to be $F_1 = J - 1/2$, $J + 1/2$, $J - 3/2$, $J + 3/2$. Verification of these assignments and, indeed, of $J = 22$ will develop as we go along. We may begin with an attempt to use the intensity information in a quantitative way.

A proper lineshape theory of saturated absorption in the free-flight regime is only slowly evolving [5.20], mainly due to complications associated with the anomalous contribution of the very slow molecules. However, such effects are not essential in this work due to the conditions of our experiment—notably the low laser intensity and rather gross spectral widths, taken with the appreciable pressure broadening even at our 1/2 millitorr pressure. Thus we may confidently expect that at low intensities the size of the saturation peaks grows with saturation parameter

$$ S = \left(\frac{d^2 E^2}{\hbar^2 \gamma_2 T_1^{-1}} \right) $$

at a rate proportional to S^2, softening towards a linear dependence on S at very high intensities. The transition dipole moment d will depend on the hyperfine quantum numbers in the usual way [5.21]. For a fixed laser power then we expect the relative intensities of hyperfine components to be bounded by the square of the linear intensity factor $\mathscr{I}^2 \propto (d^2)^2$ and its first power $\mathscr{I} \propto d^2$ in the same two limits. Inspection of Table 5.2 shows that the experimental relative intensity values do indeed fall between these two limits, calculated for $J = 22$. Using the experimental intensity range of 3.6% as the measure, we may calculate that acceptable J assignments are bounded by the values $J = 22$ from the linear limit and $J = 42$ from the quadratic, S^2, dependence.

5.3.2 Theoretical Analysis

We now turn to analyze the spectra. Of course electric quadrupolar hyperfine energies have been well studied using microwave-induced rotational transitions. We can take these results [5.22] over intact for each state of our rotation-vibration transitions. The electric quadrupole interaction associates rather large energy changes with the orientation of the nuclear spin of halide nucleus. One thus comes to a coupling scheme where the molecular rotational angular momentum J couples to I to form states labeled by $F_1 = J + I$. Projections of F_1 on a laboratory axis are labeled by M_F. Finally the remaining three proton spins can be coupled to F_1 to form F_2, but at the present level of resolution, only the projection of the three-proton spins on F_1 is of interest. From perturbation theory one has the following first- and second-order values for the total quadrupolar energy:

$$E_Q = E_Q^{(1)} + E_Q^{(2)}$$
$$= \langle IJKFM_F | H_Q | IJKFM_F \rangle$$
$$+ \sum_{J'K'} \frac{|\langle IJKFM_F | H_Q | IJ'K'FM_F \rangle|^2}{W_{JK} - W_{J'K'}}, \tag{5.1}$$

where H_Q is the quadrupolar interaction Hamiltonian [5.22]. The second-order matrix elements in the sum of (5.1) are zero unless $K' = K$ and $J' = J \pm 1$ or $J' = J \pm 2$. Due to their rather complicated form the reader is referred to Eqs. (6–10) of Ref. [5.22]. In view of the very small contribution of the second-order terms, we can take the rotational energy in their denominators to be

$$W_{JK} = BJ(J+1) + (A-B)K^2 ,$$

where A and B are the rotational constants.

For our case in which the nuclear quadrupole-bearing atom lies on the molecular symmetry axis, the first-order quadrupolar energy has the value

$$E_Q^{(1)} = eqQ \, \frac{(3/4)\,C(C+1) - I(I+1)\,J(J+1)}{2I(2I-1)\,(2J-1)\,(2J+3)} \left[\frac{3K^2}{J(J+1)} - 1 \right] \tag{5.2}$$

where $C \equiv F_1(F_1+1) - I(I+1) - J(J+1)$. Here eqQ is the molecular electric quadrupole coupling energy associated with a non-rotating molecule's electric field gradient q at the axial site of the nucleus of quadrupole moment Q. The last factor in brackets gives the modification of q due to rotational averaging as the molecular axis rapidly precesses around $\mathbf{J}$. The second term in braces is the quantum analog of the classical dependence, $(3\cos^2\theta_{Z,I} - 1)$, where the axis $\mathbf{Z}$ is to be identified with $\mathbf{J}$.

Corresponding to the very small shift of the interatomic distances when the molecule is vibrationally or rotationally excited, there can be a (very) small possible change of the electric field gradient tensor at the halide site. It is an interesting situation that the resulting small shifts in electric quadrupole coupling can perhaps be better studied with the optical techniques described here than with the microwave rotational transitions, even though we deal here with far higher frequencies and so must achieve *much* higher fractional resolution. Basically vibrationally excited states, and very high J values (where the effects of centrifugal distortion of the molecule increase), are just not available without lasers due to the unfavorable Boltzmann factors.

Finally there are three small magnetic interactions to consider: 1) the on-axis nuclear spin interacting with the (magnetic field induced by the) rotation ($\propto \boldsymbol{I} \cdot \boldsymbol{J}$); 2) the three off-axis protons interacting with the rotation $\left(\propto \boldsymbol{J} \cdot \sum_{i=1}^{3} \boldsymbol{I}_i \right)$; 3) the dipolar interaction of the three protons with the on-axis nuclear spin $\boldsymbol{I}$. It may be shown [5.23] that at high J ($\gtrsim 20$) and for the case of interest here $\Delta K = 0$, $\Delta J = \Delta F_1 = -1$, we may write the magnetic energy in the approximate form

$$\Delta v^{P}_{\text{Mag}} \simeq C_I (J - F_1). \tag{5.3}$$

Thus an interesting task now is to verify that the quadrupole and magnetic interactions expressed by (5.1) and (5.3) can give a good account of the observations. In the process, we will deduce the electric quadrupole coupling constant for lower and vibrationally excited states separately. Finally we can investigate use of the "quality" of the least-squares fit to decide the spectral assignments.

The observed infrared transition frequencies all contain a common vibration-rotation term which may be conveniently subtracted off to yield the hyperfine spectrum

$$\Delta v_Q(J', F'_1, J'', F''_1, K, I)$$
$$= h^{-1} [E'_Q(I, J', K, F'_1) - E''_Q(I, J'', K, F''_1)]. \tag{5.4}$$

It is useful to separate the explicit dependence on eqQ from the factor $C(I, J, K, F_1)$ containing the dependence on quantum numbers. The form of the first-order part of the function $C(I, J, K, F_1)$ may be seen from (5.2).

We then have

$$\Delta v_Q(J', F'_1, J'' \; F''_1, K, I)$$
$$= Q'_v C(I, J', K, F'_1) - Q''_v C(I, J'', K, F''_1) \tag{5.5}$$

where Q'_v and Q''_v represent, in frequency units, eqQ in the upper and lower state, respectively. From the earlier discussion we know that $Q'_v \simeq Q''_v$, so it is convenient to write

$$Q'_v = Q''_v + \Delta Q_v. \tag{5.6}$$

Also, since all the transitions observed in this work are $\Delta K = 0$, $\Delta J = \Delta F = -1$, we may specialize to this case.

Making the several substitutions from (5.1), (5.3), (5.5), and (5.6) gives the spectrum

$$\Delta v^{\mathrm{P}}(I_N, J, K, F_1) = Q_v''[C(I, J-1, K, F_1 -1) - C(I, J, K, F_1)]$$
$$+ \Delta Q_v\, C(I, J-1, K, F_1 -1)$$
$$+ C_I(J - F_1)$$
$$+ E_Q^{(2)'} - E_Q^{(2)''}.$$

(5.7)

Here the first term is associated with the change in angle between J and the molecular symmetry axis due to the J change of one unit in the transition. This term arises from the rotational averaging of q. The second term is specifically due to the difference in the quadrupole coupling constants in the two states; we will find fractional differences of less than 1 %. The third term adequately represents the magnetic energy at our level of resolution, and the fourth term schematically indicates the difference between two rather cumbersome expressions for the second-order quadrupolar energies. Equation (5.7) gives the frequency shift, measured from the unperturbed rotation-vibration transition frequency, of the $2I+1$ major hyperfine components due to both nuclear electric quadrupolar and internal magnetic interactions.

One may regard Q_v'', ΔQ_v, and C_I appearing in (5.7) as unknown molecular parameters to be recovered from the experimental data by some kind of curve-fitting analysis procedure.

So we come immediately to the question of reasonable initial values to be used in the iterative least-squares fitting. Values of $C_I \approx 20\,\mathrm{kHz}$ may be reasonably expected. Our analysis [5.24] strongly suggests that Q_v'' will hardly depart at $J \sim 40$ from its microwave value (corresponding to $J \lesssim 10$). On the other hand this analysis does suggest possible influence due to the vibrational excitation, with $\Delta Q_v/Q_v$ expected to be up to $\approx 1\%$. Due to the relatively large value of J, the magnitudes of these two first terms in (5.7) may be comparable.

It was necessary in our least-squares fitting procedure to allow for the possibility that the ordering of the transitions changes as the values of the parameters are changed. This problem was handled by first computing the parameter values at which the hyperfine components cross. No iteration was allowed to jump these crossing values without relabeling the transitions and testing the fit again. In practice only those solutions were satisfactory that began and converged to final values within one such band of label ordering.

Although the intensity ordering of the experimental lines in principle gives the assignment $F_1 - J$, it seemed desirable to be able to obtain all sets of Q_v'', ΔQ_v, and C_I that could match the observed spectrum

Table 5.3. Experimentally determined hyperfine constants for the $CH_3{}^{35}Cl$ lines at 2947.821 cm^{-1}

Fitted for the assignment $^QP_3(22)$, using $Q_v^0 = -74.74$ MHz

SSR (MHz)^2	ΔQ_v (MHz)		Q_v (MHz)		C_I (kHz)	
	ΔQ_v	σ	Q_v	σ	C_I	σ
4.6×10^{-8}	-0.0282	0.001	-74.74	F	1.49	0.10
1.1×10^{-5}	-0.0290	0.010	-74.74	F	0.00	F
b	b		b		0.00	F
a	-0.0316	a	-75.39	a	1.29	a

F means this parameter held fixed at the indicated value. $\Delta Q_v(J) \simeq 0.0045$ MHz.

^{a} No information obtainable since this is a fit of four free parameters to four linear equations.

^{b} No result obtained within the specified limits for ΔQ_v, Q_v, and the SSR.

within an assigned sum-square-residuals (SSR) tolerance value. Usually a few accidental "fits" turned up when we attempted to find fits for an assumed range of values for J and K.

Unfortunately, $I = 3/2$ leads only to three measurable frequency intervals while the effective hyperfine Hamiltonian contains three parameters. The system is therefore not overdetermined. We can still obtain some appreciation of the tolerance range of the fitted parameters by fixing one or more of them. Using the spectral constants from Table 5.2, we calculated the best fitting molecular constants under several different assumptions. These results are presented in Table 5.3.

Since no effort has been taken to orthogonalize the fitting parameters we may assume that the essential stability of $\Delta Q_v (\simeq 4 \times 10^{-4} Q_v)$ in the three cases is evidence that this parameter is significantly different from zero. From Table 5.2 we may estimate the expected sum-square-residuals due to random noise as

$$\sigma^2 \approx 4 \times (0.4 \text{ kHz})^2 \simeq 6 \times 10^{-7} \text{ MHz}^2 .$$

Looking at the first entry in Table 5.3 we see that a SSR value smaller than this expected random noise level is obtained with Q_v fixed at its microwave (ground state) value. This is consistent with our estimate that $\Delta Q_v(J) \simeq 0.0045$ kHz, vastly smaller than the difference indicated between lines 1 and 4. This spurious change in line 4 of Q_v is a typical feature of least-squares fitting of data sets that are not strongly overdetermined. In contrast, fixing the magnetic interaction term C_I at zero gives rise to a relatively large value of SSR—some factor of 18 increase over the value estimated from the random noise. Thus we must conclude that a nonzero value for C_I is required by the data. We feel that the parameters listed in the first line are to be preferred, but at this stage we can offer no realistic guide to systematic error effects.

Certainly, one of the interesting future opportunities in this work will be to study a few more spectra of the CH_3Cl transitions listed in Table 5.1. The J values cover only a 10% range. Thus probably nothing secure about $\Delta Q_v(J)$ will be available. However, the vibrationally induced change, ΔQ_v, and the magnetic interaction could probably be put on a very secure basis by such work.

5.3.3 CH_3Br and CH_3I

It should also be interesting to study methyl-halide quadrupolar spectra where larger values of the halide nuclear spin give rise to more transitions and hence to definite values and error estimates for the quadrupolar and magnetic interaction parameters. At present we have two other spectra of reasonably high resolution: $CH_3{}^{81}Br$ at 2947.908 cm^{-1} and CH_3I at 2947.854 cm^{-1}. Unfortunately, the ^{81}Br nuclear spin is also only 3/2, and again only three spectral intervals are determined. Also from the work of MORILLON-CHAPEY and BETRENCOURT [5.5] it is known that the $v_1(a_1)$ band is in strong Fermi interaction in this spectral region with another band, probably the $v_3 + v_5 + v_6(A_1 + A_2)$. Thus CH_3Br may not be an ideal case for our analysis but we will present our results before passing on to the case of CH_3I.

In Figs. 5.4a, and b we present the observed derivative spectrum of CH_3Br. It was recorded in two parts due to sweep-range limitations in the digital frequency synthesis electronics. No effort was made to make the two intensity scales comparable; thus we can obtain only the frequency information. The two pairs of lines were separately fitted with two Lorentzians of the same width. The influence of the other pair of resonances was minimized by projecting them out using their least-squares representation. After iteration, the values displayed in Table 5.4 were obtained.

From some preliminary analog records of the entire spectrum one could appreciate that the four transitions were very nearly equal in intensity, as would be expected for the rather high J value, $J=39$. Considering that the intensity ordering information was not available, a number (~ 20) of assignments J, K were able to essentially reproduce the observed line positions. In the range $35 \leq J \leq 50$ and $0 \leq K \leq 10$, nine spurious fits in $CH_3{}^{81}Br$ were found of SSR less than or equal to that of the correct assignment, and six such accidental fits were found in $CH_3{}^{79}Br$. Only four spurious fits in ^{81}Br and another four in ^{79}Br have the same ordering of intensities as the correct assignment. An

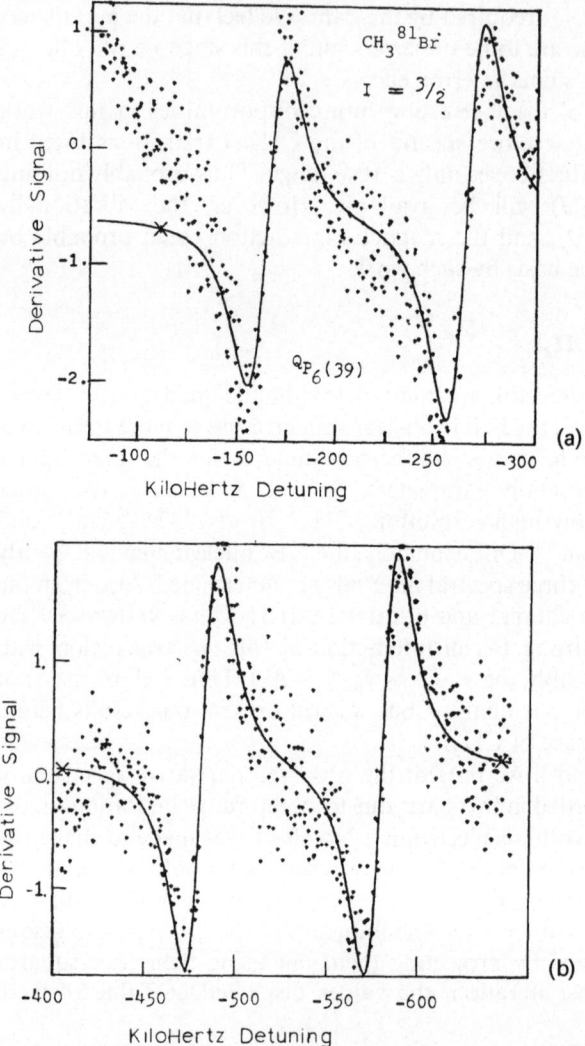

Fig. 5.4a and b. Saturated absorption spectrum of $CH_3{}^{81}Br$ at 2947.908 cm^{-1}. $f_{abs} = f_{ref} -$ 121.0 MHz $+ f_{scale}$. Absorber pressure about 0.4 mTorr. Severe baseline tilt visible on the left side of Fig. 5.4a is instrumental. In lieu of a calculated correction, a lower weight was assigned to these components. See Table 5.4

additional eight spurious fits in each isotope had SSR within a factor 2 of that of the correct assignment. Often unphysical values were required for ΔQ_v, C_I, or both, but is clear that at least some intensity information would help to provide a unique assignment. It is interesting to note that the resolution level presently achieved (which will be discussed

Table 5.4. Fitted parameters for the hyperfine components of the $CH_3{}^{81}Br$ line at 2947.908 cm^{-1}

Linewidth (kHz)		Center frequency[a] f_c (kHz)		Frequency interval (kHz)	
$\Gamma_L/2$	σ	f_c	σ	f_c	σ
15.0[b]	0.4	411.7	1.4	100.4	1.6
15.0[b]	0.4	311.3	0.7	212.6	0.8
14.7[c]	0.4	98.7	0.3	98.7	0.4
14.7[c]	0.4	0.0	0.3		

[a] These are relative frequencies. The absolute frequencies are given by $f_{abs} \simeq f_r + f_c - 121.579$ MHz, where f_r is the $^{12}CH_4\ F_2^{(2)}$ reference frequency.
[b] These components belong to the upper frequency line pair; Fig. 5.4a.
[c] These components belong to the lower frequency line pair; Fig. 5.4b.

Table 5.5. Experimentally determined hyperfine constants for the $CH_3{}^{81}Br$ line at 2947.908 cm^{-1}

Fitted for the assignment $^QP_6(39)$, using $Q_v^0 = 482.16$ MHz

SSR (MHz)^2	ΔQ_v (MHz)		Q_v (MHz)		C_I (kHz)	
	ΔQ_v	σ	Q_v	σ	C_I	σ
0.30×10^{-6}	0.547	0.001	482.16	F	-0.72	0.20
3.2×10^{-6}	0.546	0.004	482.16	F	0	F
2.2×10^{-6}	0.574	0.015	489.6	3.8	0	F
0[a]	0.532	[a]	478.1	[a]	1.0	[a]

F means parameter held fixed at indicated value.
[a] No information available since this is a fit of 4 free parameters to 4 linear equations.

momentarily) yields line center positions with a factor > 10 better accuracy. At this level very likely the quadrupole pattern alone would fix the assignment uniquely, although the magnetic interactions will have to be treated more fully.

Using the identification $^QP_6(39)$ from Table 5.1, we have the results summarized in Table 5.5. The value of SSR expected from random noise is $\simeq 2.6 \times 10^{-6}$ MHz^2. It may be seen that again ΔQ_v is stable, this time at $+1.1 \times 10^{-3} Q_v$, indeed with rather smaller apparent variations than in the CH_3Cl case.

Unfortunately the smaller apparent value of C_I and the somewhat larger uncertainty almost obscure its value. (One might have expected that a $2.75 \times$ increase in the ^{81}Br nuclear magnetic moment compared

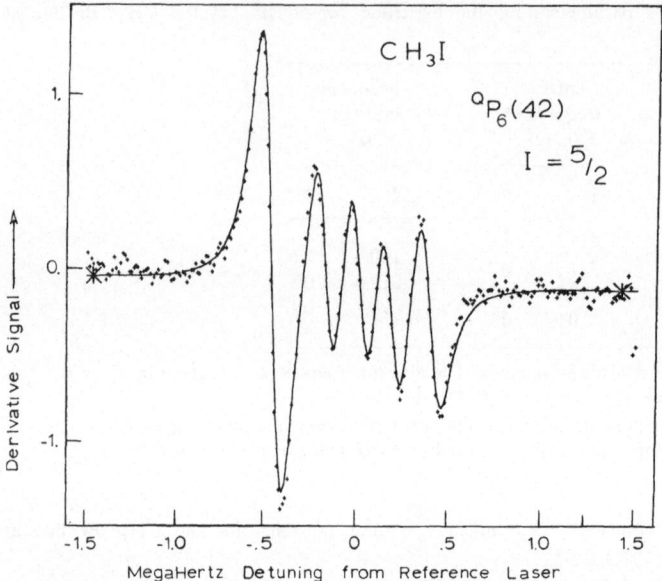

Fig. 5.5. Intracavity saturated absorption spectrum of CH_3I at 2947.854 cm^{-1}. Laser mode radius 0.7 mm with 1 mW one way intracavity power. Pressure less than 1 mTorr, $f_{abs} \simeq f_{ref} - 1731.7 + f_{scale}$

with Cl would have compensated the somewhat smaller B_0 value, yielding a C_I value similar to that of CH_3Cl). Table 5.1 shows that a rich selection of CH_3Br lines is available for future investigations of these questions.

Finally, in Fig. 5.5 we come to present our data for CH_3I. Unfortunately these data exhibit the rather "low" resolution of $\simeq 100$ kHz HWHM, having been taken in an early intracavity experiment. Inspection and numerical experiment show that the overlapping is not completely hopeless, however, thanks to the large CH_3I quadrupole coupling constant of $-1934,0$ MHz. Counting the large low frequency peak as 2, we see the $2I_N + 1 = 6$ (derivative) transitions expected for the case $I_N = 5/2$. Table 5.6 presents the parameters obtained by fitting six (partly overlapping) Lorentzians of the same width. Again, the rather small modulation broadening (0.195 linewidths) was treated following WAHLQUIST [5.18]. The centers of all peaks including the double one are well determined, but the expected correlation between splitting and width of the doublet is observed. The heights of all but two of the components lie within the S and S^2 intensity limits described previously, even though the intensity range covers only 1.9%. Thus, although these CH_3I data are inferior to the other two examples, they are interesting because the

Table 5.6. Fitted parameters for the hyperfine components of the CH_3I line at 2947.854 cm^{-1}

Component number	ID $(F_1 - J)$	Center frequency f_c^a (kHz)		$\mathscr{I}$ (%) Calculated[b]	$\mathscr{I}_{exp}$ (%)	$\mathscr{I}^2$ (%) Calculated[b]
		f_c	σ			
1	$-5/2$	900.2	1.2	15.7	15.1	14.7
2	$-3/2$	681.7	1.2	16.1	15.7	15.6
3	$5/2$	509.6	1.5	17.5	17.8	18.5
4	$-1/2$	318.7	1.3	16.4	17.0	16.3
5	$1/2$	31.9	1.9	16.8	17.2[c]	17.0
6	$3/2$	0.0	1.9	17.2	17.2[c]	17.9

All components are fitted using a common linewidth. Its value of $\Gamma_L/2 = 98.2 \pm 1.1$ kHz was obtained from a fit to components 1 through 4. This was subsequently applied as a fixed parameter fitting components 1 through 6.

[a] These are relative frequencies. The absolute frequencies are given by $f_{abs} \simeq f_r + f_c - 1732.4$ MHz, where f_r is the $^{12}CH_4$ F-line reference frequency.

[b] $\mathscr{I}$ is the relative intensity for linear absorption calculated for $J = 45$. The mean relative intensity is $100\%/6 = 16.7\%$.

[c] These relative intensities were constrained to be equal in the fitting calculations.

Table 5.7. Experimentally determined hyperfine constants for the CH_3I line at 2947.854 cm^{-1}

Fitted for the assignment: $^Q P_2(46)$, using $Q_v^0 = -1934.0$ MHz

SSR (MHz)^2	ΔQ_v (MHz)		Q_v (MHz)		C_I (kHz)	
	ΔQ_v	σ	Q_v	σ	C_I	σ
1.86×10^{-4}	-2.92	0.03	-1934.0	F	6.37	1.64
8.61×10^{-4}	-2.92	0.06	-1934.0	F	0.00	F
[a]	[a]		[a]		0.00	F
1.69×10^{-4}	-2.93	0.04	-1956.6	54.8	4.84	3.23

Fitted for the assignment: $^Q P_4(45)$, using $Q_v^0 = -1934.0$ MHz

SSR (MHz)^2	ΔQ_v (MHz)		Q_v (MHz)		C_I (kHz)	
	ΔQ_v	σ	Q_v	σ	C_I	σ
1.86×10^{-4}	-4.58	0.03	-1934.0	F	3.32	1.46
3.55×10^{-4}	-4.59	0.04	-1934.0	F	0.00	F
2.75×10^{-4}	-4.63	0.05	-1966.5	31.3	0.00	F
1.69×10^{-4}	-4.55	0.07	-1902.6	52.2	4.76	3.32

F means this parameter held fixed at the indicated value. $\Delta Q_v(J) \simeq 0.2$ MHz.

[a] No result obtained within the specified fit limits for ΔQ_v, Q_v, and the SSR.

five spectral intervals should allow us to obtain a unique and over-determined set of hyperfine coupling parameters.

We present in Table 5.7 these fitted quadrupolar and magnetic interaction constants. Again one observes that ΔQ_v is remarkably stable, this time at the value $+0.15\% Q_v$. The random-noise generated SSR should be

$$\sigma^2 \sim 6 \times (1.5\,\text{kHz})^2 \sim 1.4 \times 10^{-5}\,\text{MHz}^2\,.$$

This value is not really approached by the fits listed in the upper block of Table 5.7, probably because of inadequate fitting of the doublet. As before, it may be seen that when Q_v is treated as a free parameter, a large shift in Q_v occurs, of some 250 times the value $\Delta Q_v(J)$ estimated by expansion of Q_v [5.24]. One must however also observe that the shift amounts to less than $(1/2)\sigma Q_v$. Thus the only conclusion possible at present is that we have no evidence here for a possible J-dependence of the quadrupole interaction within the vibrational ground state.

A further caveat may be appreciated from inspection of the lower block of Table 5.7. A comparative satisfactory fit, including the correct component ordering, can be produced by a totally spurious assignment when the line center positions are inadequately known. Evidently with line center accuracies of $\geq 1\,\text{kHz}$ it is simply essential to know the assignment by other techniques. At $\sim 0.3\,\text{kHz}$ we appear to be near some kind of a threshold. One may estimate that assignments might be unambiguously made from hyperfine patterns alone if the line centers can be known to $\lesssim 0.1\,\text{kHz}$. For reference, the present setting precision in CH_4 is about 30 times better than this latter value, and > 100 times better than the data analyzed here. It is interesting to pursue this assignment idea because the determination of an assignment from measurement in a narrow spectral window (of width $\Delta v/v \lesssim 10^{-8}$) would be amusing in view of the 1–10% fractional bandwidth otherwise necessary for this task.

5.4 Ultra-High Resolution Spectroscopy of CH_4

As a final brief topic it may be of interest to present in Fig. 5.6 a "typical" spectrum obtained under conditions of ultra-high resolution. It happens to be of the usual Coriolis F line in $^{12}CH_4$ although the equally high resolution of $1.0\,\text{kHz}$ has recently been obtained for a $^{13}CH_4$ line [5.25].

In Fig. 5.6 one sees the three main magnetic hyperfine components studied previously with appreciably poorer resolution [5.26]. Each hyperfine component is doubled by the recoil [5.27]. As a molecule is put into

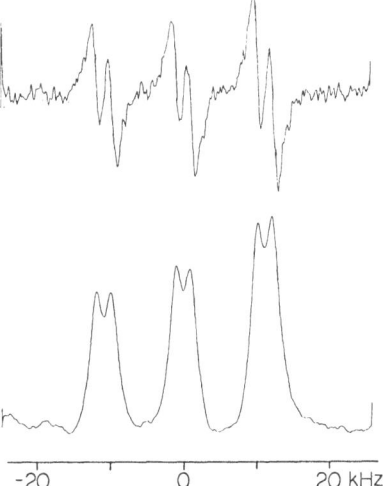

Fig. 5.6. "Ultrahigh-resolution" spectrum of CH_4. The integral form of the spectrum is displayed below the actually recorded derivative spectrum. The three peaks arise from magnetic hyperfine interactions discussed in [5.26]. The splitting of each hfs component into a doublet is due to recoil. See text

the excited state it absorbs the photon's momentum as well as energy. The return beam of the saturated absorption spectrometer thus sees two velocities (frequencies) of anomalously high transmission. One peak occurs in the usual way by interrogation of a population "hole" in the absorbing, lower state. The other peak corresponds to spectrally narrow amplification of the return beam by the velocity-selected and -displaced molecules placed into the excited state by interaction with the other laser beam [5.28, 29].

These new high resolution results have been obtained with a new large diameter (38 cm) absorption cell of 13 m length. Careful attention to compensation of optical imaging errors allows us to produce wavefronts of better than $\lambda/5$ absolute flatness at $\lambda = 3.39 \, \mu m$. The mode waist size has been chosen to be 11 cm which leads to an expected time-of-flight linewidth of ~ 650 Hz. Pressure broadening at the $15 \, \mu Torr$ operating pressure contributes about 60 Hz. The laser spectral width was measured to be $\simeq 200$ Hz. Considering that the cancellation of the earth's field may be imperfect and that there is some remnant power and modulation broadening, it appears that one should not be disappointed with the obtained resolution of $\lesssim 1.0$ kHz HWHM. The corresponding spectral resolving power is $\sim 4.5 \times 10^{10}$ and is evidently enough to provide incipient resolution of the spectral doubling due to recoil as mentioned above. Computer fits to data such as these give the expected kinematic splitting

$$\Delta v = v \frac{hv}{Mc^2} = 2.163 \text{ kHz}$$

and allow one to study the influence of collisions on the intensity partition between the two components. Since the low (high) frequency peak corresponds to a nonlinear interaction in excited (ground) states, this kind of study allows us to separate the collision effects on the coherent lifetimes in the two states. A preliminary report on this topic already has been presented [5.30] and a more detailed discussion is being prepared [5.31].

The obtained resolution level of $\geq 4.5 \times 10^{10}$ corresponds to dividing the Doppler profile into 3×10^5 velocity bins of 3 mm/s width and is believed to be unprecedented resolution in coherent spectroscopy.

One might hope—and perhaps expect—that molecular spectroscopy under this new resolution level will yield some interesting new effects and new ideas, as well as evidently providing the opportunity for learning a few spectral quantities more precisely. Experiments with the heavier methyl-halides under "super resolution" conditions may lead to very interesting frequency standards results, assuming the collisional effects are not too serious.

Acknowledgements. The authors are grateful to their colleagues for stimulating discussions, especially P. L. BENDER and A. C. GALLAGHER. In the methane work, collaboration with C. J. BORDÉ and K. UEHARA has been enjoyed by one of the authors (JLH). We especially appreciate the generosity of J. W. RUSSELL in making his unpublished CH_3Br and CH_3I analysis available to us. The Fourier transform spectrum of CH_3I was essential in our work and we are especially grateful to C. AMIOT and his colleagues for providing it to us. We also are extremely grateful to M. MORILLON-CHAPEY and M. BETRENCOURT for their help in clarifying the CH_3Br spectrum and assigning the transitions of interest. This research has been supported over a number of years by the National Bureau of Standards as part of its research program on improved precision measurement techniques for applications to basic standards.

References

5.1   W. DEMTRÖDER, M. MCCLINTOCK, R. N. ZARE: J. Chem. Phys. **51**, 5495 (1969) and references therein

5.2   L. HENRY, N. HUSSON, R. ANDIA, A. VALENTIN: J. Mol. Spectrosc. **36**, 511 (1970)

5.3   W. L. BARNES, J. SUSSKIND, R. H. HUNT, E. K. PLYLER: J. Chem. Phys. **56**, 5160 (1972)

5.4   P. CONNES: Thesis, Paris (1957); Rev. Opt. **38**, 157–201 (1959); **38**, 416–446 (1959); **39**, 402–436 (1960)

5.5   M. BETRENCOURT, M. MORILLON-CHAPEY, C. AMIOT, G. GUELACHVILI: J. Mol. Spectrosc. **15**, 402–415 (1975); and private communication May 3, 1974

5.6 H. J. GERRITSEN, M. E. HELLER: Appl. Opt. Suppl. no. 2, 73 (1965); H. J. GERRITSEN: In *Physics of Quantum Electronics*, ed. by P. L. KELLY, B. LAX, P. E. TANNENWALD (McGraw-Hill, New York 1966)

5.7 K. SAKURAI, K. SHIMODA: Japan J. Appl. Phys. **5**, 744 (1966)

5.8 K. M. EVENSON, J. S. WELLS, F. R. PETERSEN, B. L. DANIELSON, G. W. DAY: Appl. Phys. Lett. **22**, 192 (1973)

5.9 M. MORILLON-CHAPEY, G. GRANER: J. Mol. Spectrosc. **31**, 155 (1969)

5.10 J. W. RUSSELL: Private communication, April 17, 1968

5.11 C. AMIOT: Private communication, June 12, 1973

5.12 A. C. LUNTZ, J. D. SWALEN, R. G. BREWER: Chem. Phys. Lett. **14**, 512 (1972)

5.13 A. BRILLET, P. CEREZ, S. HAJDUKOVIC, F. HARTMANN: Paper presented at the 5th Intern. Conf. on Atomic Masses and Fundamental Constants, Paris, June 2–6, 1975, ed. by J. SANDARS and C. AUDOIN (Plenum Press, New York)

5.14 This heterodyne work was in connection with a proposed determination of the speed of light using a differential interferometric method. An early report may be found in J. L. HALL, R. L. BARGER, P. L. BENDER, H. S. BOYNE, J. E. FALLER, J. WARD: Electron. Technol. (Warsaw) **2**, 53 (1969)

5.15 See, for example, R. L. BARGER and J. L. HALL: Phys. Rev. Lett. **22**, 4 (1969); and J. L. HALL: in *Méthodes de Spectroscopie sans Largeur Doppler de Niveaux Excités de Systèmes Moléculaires Simples*, ed. by J. C. LEHMAN, J. C. PEBAY-PEYROULA (CNRS Press, Paris 1974) pp. 105–125

5.16 J. L. HALL: In *VI Conference on the Physics of Electronic and Atomic Collisions: Abstracts of Papers* (MIT Press, Cambridge, Mass. 1969) p. 994

5.17 J. L. HALL: In *Lectures in Theoretical Physics*, ed. by K. T. MAHANTHAPPA, W. E. BRITTIN (Gordon and Breach, New York 1973), p. 161 ff.; V. A. ALEKSEEV, T. L. ANDREEV, I. I. SOBEL'MAN: Zh. Eksp. Teor. Fiz. **62**, 614 (1972); Sov. Phys.-JETP **35**, 325 (1972) 1; S. N. BAGAEV, E. V. BAKLANOV, V. P. CHEBOTAYEV: JETP Lett. **16**, 9 (1972)

5.18 H. WAHLQUIST: J. Chem. Phys. **35**, 1708 (1961)

5.19 C. H. TOWNES, A. L. SCHAWLOW: *Microwave Spectroscopy* (McGraw-Hill, New York 1955) p. 502

5.20 See, for example, C. J. BORDÉ, J. L. HALL, C. V. KUNASZ, D. G. HUMMER: "Saturated Absorption Line Shape. I. Calculation of the Transit Time Broadening by a Perturbation Approach", Phys. Rev. A **13** (1976) (in press)

5.21 C. H. TOWNES, A. L. SCHAWLOW: *Microwave Spectroscopy* (McGraw-Hill, New York 1955) p. 152

5.22 C. H. TOWNES, A. L. SCHAWLOW: *Microwave Spectroscopy* (McGraw-Hill, New York 1955) Chap. 6, especially pp. 156, 157

5.23 J. A. MAGYAR: PhD Thesis, University of Colorado (1974), Appendix I

5.24 See [5.23], p. 41 ff. Also see H. J. ZEIGER, D. J. BOLEF: Phys. Rev. **85**, 788 (1957), and I. N. LEVINE: *Molecular Spectroscopy* (Allyn & Bacon, Boston 1970) pp. 145–154

5.25 K. UEHARA, J. L. HALL: To be published

5.26 J. L. HALL, C. J. BORDÉ: Phys. Rev. Lett. **30**, 1101 (1973)

5.27 A. P. KOL'CHENKO, S. G. RAUTIAN, R. I. SOKOLOVSKII: Zh. Eksp. Teor. Fiz. **55**, 1864 (1968) [Sov. Phys.-JETP **28**, 986 (1969)]

5.28 C. J. BORDÉ, J. L. HALL: In *Laser Spectroscopy*, ed. by R. G. BREWER, A. MOORADIAN (Plenum Press, New York 1974) p. 125

5.29. S. STENHOLM: J. Phys. B **7**, 1235 (1974)

5.30 J. L. HALL, C. J. BORDÉ, K. UEHARA: In Proceedings, 2nd Symp. on Gas Laser Physics, Novosibirsk (June 1975)

5.31 J. L. HALL, C. J. BORDÉ, K. UEHARA: To be published

6. Three-Level Laser Spectroscopy

V. P. CHEBOTAYEV

With 19 Figures

Resonances arise from the nonlinear interaction of optical fields with a gas of atoms or molecules. Their use provides nonlinear laser spectroscopy of super-high resolution. Three-level laser spectroscopy (TLS) is the whole complex of methods based on the use of absorption or emission resonances arising from the nonlinear interaction of two or several radiations with a gas of three-level atoms or molecules. The schemes of levels and transitions used in TLS are shown in Fig. 6.1.

The progress in obtaining and using resonances on Doppler-broadened transitions is achieved mainly by using the methods of nonlinear absorption. Using this method, resonances in an absorption line arise from the nonlinear interaction of several waves whose frequencies coincide with the transition frequencies within the Doppler linewidth. The physical principles of the methods and their various applications to spectroscopy are described in Ref. [6.1].

The resonances in three-level systems possess new features which are of interest for applications. The present paper is dedicated to studying the properties of resonances in three-level systems and to using them in solving spectroscopic problems.

The TLS methods are not so widely used as, for example, the methods of nonlinear absorption. It is not due to some principal limitations of the TLS methods but to the absence of tunable sources of monochromatic radiation. At present this problem is being solved in the infrared and optical ranges of wavelengths.

The range of problems connected with three-level systems is rather extensive. It comprises well-known and developed methods of microwave and optical double resonances, optical pumping, etc. These problems are now being given much consideration (see [6.2, 6.3]). We shall not therefore consider phenomena directly related to the above-mentioned methods. We shall only consider resonance phenomena in the three-level schemes connected with the investigations of resonant stimulated Raman scattering in the optical region. We shall assume that the following conditions are met:

1) The two frequencies are within the Doppler linewidths of transitions, that is, $\Omega \lesssim k\bar{v}$ and $\Omega' \lesssim k'\bar{v}$ where $\bar{v}$ is an average thermal velocity,

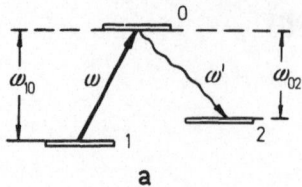

a

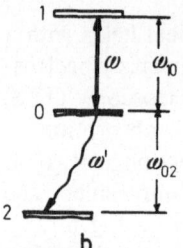

b

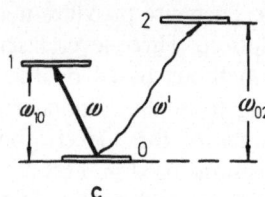

c

Fig. 6.1. The scheme of transitions in three-level systems

k and k' are wavenumbers of the $1 \to 0$ and $2 \to 0$ transitions, $\Omega = \omega - \omega_{01}$ and $\Omega' = \omega' - \omega_{02}$ are the frequency detunings about the transition frequencies.

2) The homogeneous widths $2\Gamma_{01}$ and $2\Gamma_{02}$ of the $0 \to 1$ and $0 \to 2$ transitions are much smaller than the Doppler widths $k\bar{v}$ and $k'\bar{v}$.

3) The energy of the interaction of atoms with the optical fields $p_{01}E/\hbar$, $p_{02}E'/\hbar$ (in units of $\hbar$) is much less than the Doppler width, p_{01} and p_{02} are matrix element of dipole moments of the $0 \to 1$ and $0 \to 2$ transitions, respectively, $2E$ and $2E'$ are amplitudes of the fields with the frequencies ω and ω'.

4) Decay constants of levels are different.

The conditions 1, 2, and 3 imply an inhomogeneous character of saturation of a gas of two-level atoms on the coupled transitions. The inhomogeneity in saturation results in a well-known effect of formation of Bennett "holes" in the velocity distribution of atoms [6.4]. A monochromatic light effectively interacts with atoms whose velocities satisfy the condition of resonance: the frequencies seen by a moving atom (frequency in the c-system) are equal to the transition frequencies $\omega \to \omega - kv$ and $\omega' \to \omega' - k'v$.

The velocity interval of the atoms which effectively interact with the optical field in determined by the homogeneous widths $2\Gamma_{01}$ and $2\Gamma_{02}$. When $\Gamma_{01} \ll k\bar{v}$ and $\Gamma_{02} \ll k'\bar{v}$, the change of level populations takes place in a narrow velocity interval. Interaction of a standing wave with a two-level system gives rise to a dip with a homogeneous width at

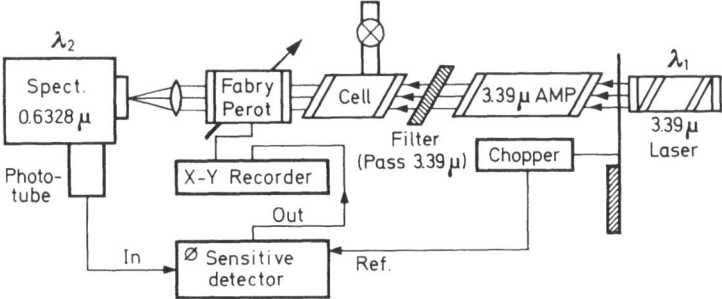

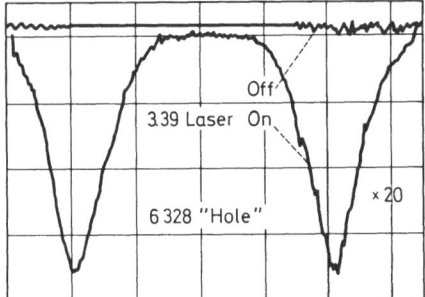

Fig. 6.2. Experimental arrangement for the observation of emission at $\lambda = 0.63\,\mu m$ in the presence of the laser radiation at $\lambda = 3.39\,\mu m$ (upper). Change of the line shape at $\lambda = 0.63\,\mu m$ under the influence of radiation at $\lambda = 3.39\,\mu m$ (lower)

the centre of the line [6.5]. The processes taking place in a gas of three-level atoms are closely connected with those in two-level systems. We should note that one of the first investigations of a three-level system in the optical region [6.6] was aimed at detection of the non-equilibrium velocity distribution which arose under the action of a strong laser field. Indeed, a dip or a peak in the velocity distribution of excited particles may be recorded by a change in the line shape of spontaneous emission on the coupled transitions. The change of emission spectrum of the Ne line at $\lambda = 0.63\,\mu m$ $(3s_2 - 2p_4$ transition) under the action of the laser field at $\lambda = 3.39\,\mu m$ $(3s_2 - 3p_4$ transition) has been observed [6.6] (see Fig. 6.2). The induced transitions from the $3s_2$ level to the $3p_4$ level under the irradiation at $\lambda = 3.39\,\mu m$ reduce the population of the $3s_2$ level; this is observed in the change of spontaneous radiation at $\lambda = 0.63\,\mu m$. The observation of the narrower line as compared with the Doppler linewidth permits us to study the influence of collisions upon the homogeneous width of the $3s_2 - 2p_4$ transition. In the other experiments the change of the spontaneous emission was used for measuring isotopic shifts [6.7] and resonance broadening [6.8]. The technique of observation of the line shape of the spontaneous radiation does not possess adequate sensitivity. Most of the subsequent experiments were therefore carried out in observation of the stimulated emission on an adjacent transition. It is the case we shall analyse in this chapter.

The above-mentioned features of the absorption (emission) line shape are explained by the change of distribution of atoms of a common level under the action of the laser field. Its influence upon the line shape on the adjacent transition is only discussed as due to the saturation effects. Let the laser field E be resonant with the Doppler-broadened $0 \rightarrow 1$ transition and propagate along the z axis. As a result of the interaction of the laser field E with atoms of the level 1, a peak arises on the level 0 in the distribution of the velocity projections of excited atoms onto the z axis (see Fig. 2.5). The laser field E' which is resonant with the $0 \rightarrow 2$ transition and the atoms on the level 0 produce an induced dipole moment on the $0 \rightarrow 2$ transition which is responsible for the absorption (emission) of energy at frequency ω'. We must emphasize that the energy of absorption (emission) at the frequency ω' is due to the population of the level 0. Since the distribution of atoms on the level 0 is narrower as compared with the Maxwell one we shall observe a narrow line on the $0 \rightarrow 2$ transition.

It is easy to obtain the line shape on the adjacent transition arising from the saturation effects only. An individual atom with the velocity projection v onto the z axis emits in the direction of the z axis with the spectrum

$$G(v) = G_0 \frac{\Gamma_{02}^2}{\Gamma_{02}^2 + (\Omega' - k'v)^2}, \tag{6.1}$$

where G_0 is the cross section of the radiative $0 \rightarrow 2$ transition. The line shape on the $0 \rightarrow 2$ transition is given by the expression

$$\alpha(\omega') = \int G(v)[N_0(v) - N_2(v)] \, dv \tag{6.2}$$

where $\alpha(\omega')$ is a linear absorption coefficient at the frequency ω', $N_0(v)$ and $N_2(v)$ are the velocity distributions of atoms on the level 0 and 2. The velocity distribution of the population difference is given by the expression

$$N_0(v) - N_2(v)$$

$$= \left\{ N_0^{(0)}(v) - N_2^{(0)}(v) + [N_1^{(0)}(v) - N_0^{(0)}(v)] \frac{\gamma_{1/2} \Gamma_{01} \kappa}{(\Omega - kv)^2 + \Gamma_{01}^2 (1 + \kappa)} \right\} \tag{6.3}$$

where $N_i^{(0)}$ is the population of the level i in the absence of the fields, γ_i is the rate of decay of the level i, κ is the saturation parameter on the

$0 \rightarrow 1$ transition

$$\kappa = \left(\frac{p_{01} E}{\hbar} \right)^2 \frac{4}{\gamma_0 \gamma_1} .$$

Substituting (6.3) into (6.2), we obtain for the Maxwell distribution $N_i^{(0)}(v)$

$$\alpha(\omega') = \alpha_0 \exp[-(\Omega'/k'v)^2]$$

$$\cdot \left[1 + \frac{N_1^{(0)} - N_0^{(0)}}{N_0^{(0)} - N_2^{(0)}} \frac{k'}{2k} \frac{\varkappa}{\sqrt{1+\varkappa}} \frac{\Gamma_+ \dot{\gamma}_1}{\left(\Omega' \mp \frac{k'}{k} \Omega \right)^2 + \Gamma_+^2} \right], \qquad (6.4)$$

where $\Gamma_+ = \Gamma_{02} + (k'/k) \Gamma_{01} \sqrt{1+\kappa}$, the signs $(+)$ and $(-)$ correspond to the cases of opposite and unidirectional propagation of the waves E and E'. It is easy to see that the resonances at the frequencies ω' correspond to the case when both waves interact with the same atoms. The line shape is the convolution of two contours: the contour of the line with the width $2\Gamma_{02}$ and the velocity distribution of atoms with the width $2(k'/k)\Gamma_{01}\sqrt{1+\kappa}$.

In the above consideration we ignore coherent processes which can occur in the nonlinear interaction of waves with the three-level system. We shall give a qualitative explanation of coherent processes. A three-level system can be presented as a nonlinear oscillator with resonance frequencies ω_{01}, ω_{02}, and ω_{12}. The dipole oscillations arise at the corresponding frequencies in such a system under the action of the fields E and E'. Due to nonlinearity, the oscillations at the frequency ω and the field E' at the frequency ω' lead to dipole oscillations at the frequencies $\omega' - \omega$ and $\omega' + \omega$ (the oscillation amplitude depends to a great extent how far away the frequency $|\omega' - \omega|$ or $\omega' + \omega$ is from the frequency ω_{12}). Similarly the nonlinear interaction of the oscillations at the frequency $|\omega' - \omega|$ or $\omega' + \omega$ and the field E leads to an appearance of the dipole moment at the frequency ω'. We should note that the dipole moment at the frequency ω', which we are interested in, arises due to only the nonlinear features of the system and the coherent features of the optical fields E and E'.

The theoretical analysis of the influence of an external signal upon the line shape and the experiments have shown that the coherent processes can also strongly influence the line shape on the coupled transition [6.9–11]. When the fields are weak, these processes can be considered as the two-quantum ones, that is, of the type of Raman scattering. Indeed, the three-level scheme in Fig. 6.1a corresponds to the classical scheme of the Raman scattering if the frequency of the incident photon ω is far away from the resonance frequency ω_{01}. The scattering of a photon is interpreted by the simultaneous processes of absorption of the initial photon and emission of the other photon. The scattering effect (or two-quantum absorption) can appear only in the second order of the perturbation theory [6.12]. The part of the matrix element for the process under consideration is performed by the sum

$$V_{21} = \sum_n{}' \left(\frac{V'_{2n} V_{n1}}{\mathscr{E}_1 - \mathscr{E}'_n} + \frac{V_{2n} V'_{n1}}{\mathscr{E}_1 - \mathscr{E}''_n} \right)$$

where $\mathscr{E}_1$ is the initial energy of the system atom + photons, $\mathscr{E}_1 = E_1 + \hbar\omega$, $\mathscr{E}'_n = E_n$, and $\mathscr{E}''_n = E_n + \hbar\omega + \hbar\omega'$ are energies of intermediate states, V and V' are matrix elements of photon absorption and emission. Far from resonance the scattering cross section is small and its observation in gases is possible at very high pressures only. As the frequency approaches the resonance frequency ω_{01}, both the scattering and the resonance fluorescence cross sections increase. Near resonance the cross section increases so that it can be observed at moderate intensities of incident light in low-pressure gases on the excited levels. The interpretation of the processes of radiation becomes complicated in the conditions of resonance. Indeed, far from resonance the scattering line is far away from the resonance transition at the frequency ω_{02}. The emission at the transition frequency ω_{02} also arises in the second order of the perturbation theory and is a consequence of excitation (appearance of population) of the real level 0. The matrix element of the transition from the level 0 to the level 2 at the frequency ω_{02} will also comprise the product of matrix elements of absorption and emission, as in the case of scattering. The excitation of a real level leads to a finite probability of emission at the resonance frequency. The second process can be considered as step-by-step one-quantum transitions. The contribution of the second process to scattering at the frequency which is far away from resonance may be neglected. In this case the interpretation of the processes is trivial. Thus, in the conditions of resonance both the above-mentioned processes contribute to the line which cannot be unambiguously interpreted as a line of Raman scattering. The interference of both processes determines the resulting line shape of resonant scattering.

In analysing the line shape of scattering near resonance, we have shown that the part of these two processes in a scattering line depends on the ratio of relaxation constants of the levels [6.13]. When the rate of decay of the common level is much faster than the rate of decay of the initial level, the resonant scattering is connected with two-quantum transitions and, consequently, can be considered as a "pure" resonant Raman scattering. When $\gamma_0 \ll \gamma_1$, on the other hand, the main contribution to the scattering is provided by step-by-step one-quantum transitions. For equal relaxation constants, the observed line shape is the result of composition of probability amplitudes of two processes[1]. Thus, the difference between the relaxation constants of levels results in the interesting features of scattering in a gas which are characteristic of the optical range.

6.1 Fundamental Equations of Three-Level Spectroscopy

We shall be interested in the line shape of absorption of a probe wave at the frequency ω' close to ω_{02}. The absorbed power can be found from the transition probability of an atom from the level 1 to the level 2 or polarization of the medium at the frequency ω'. The second approach is more general and permits us to consider different relaxation processes. It is convenient to use equations for a density matrix. The transition probability of an atom can be also found by using Schrödinger equations of probability amplitudes. Such an approach enables us to analyse elementary microscopic processes of interaction. We shall consider both approaches. It is necessary to do this because the same elementary processes of radiation exhibit macroscopic properties of the medium in different ways depending on the directions of wave propagation at the frequencies ω and ω', the ratio of wavenumbers k and k', the scheme of levels and so on. A great variety of different variants has led the authors of many works to concentrate their attention on different aspects of properties of a three-level gas system, using the different terminology and interpretation of phenomena predicted by the theory (see surveys [6.14, 15]).

NOTKIN et al. [6.10] interpreted changes in the line shape of spontaneous emission as nonlinear interference effects which arose in mixing stationary states of an isolated atom by an external field. HOLT [6.11] associated them with the frequency correlation in the two-photon transition. SCHLOSSBERG and JAVAN [6.9], FELD and JAVAN [6.16], FELDMAN and FELD [6.25] associated the line narrowing effects under the action of a laser field with the two-quantum transitions as well as with the change of the one-quantum transition probabilities in the presence of the laser field on the adjacent transition. In Ref. [6.16, 25] the transition from the level 1 to the level 2 is considered to be two-quantum, where no difference between the two-quantum proper and step-by-step transitions is discussed.

---

[1] We should note that in some cases the effect of the second order of perturbation arises as a result of the composition (interference) of amplitudes of the processes of the first and third orders of perturbation. It is important in the initial excitation of the common level.

Reference [6.18] introduced the following classification of effects which lead to the change of an emission and absorption spectrum of a gas placed in an external monochromatic light field which is resonant with an adjacent transition. First is the production of non-equilibrium velocity distribution of atoms; the second effect is the splitting of atomic levels; and the third is the nonlinear interference effect reflecting the coherence which is brought into an atom by the strong laser field. Hänsch and Toschek [6.14], restricting their considerations to polarization of the third order, analysed two effects: the effect of frequency correlation and the effect of level splitting.

At last in Ref. [6.13] the phenomena observed within the framework of the perturbation theory were interpreted by two-quantum and one-quantum processes and by their interference. The part of the processes in resonance largely depends on the ratio between relaxation constants of the levels. Using an explanation on the basis of two-quantum and one-quantum step-by-step processes one can find a connection with the classical works on scattering. It is evident that the physical processes which occur in different configurations of three-level schemes (see Fig. 6.1) are similar. In order to be more concrete we shall therefore consider and analyse the three-level system corresponding to the scheme of Raman scattering which is acted upon by an electric field of two frequencies

$$\mathscr{E}(t) = E(t) + E'(t) = E e^{-i\omega t + ikr} + E' e^{-i\omega' t + ik' r} + \text{c.c.} \tag{6.5}$$

If an atom is on the level 1 at the initial moment, it can be also in the levels 0 and 2 under the action of the field. The wave function of an atom describing its state is of the form

$$\Psi = a_1(t) \Psi_1 e^{-\frac{iE_1 t}{\hbar}} + a_0(t) \Psi_0 e^{-\frac{iE_0 t}{\hbar}} + a_2(t) \Psi_2 e^{-\frac{iE_2 t}{\hbar}} \tag{6.6}$$

where Ψ_i and E_i are, respectively, the wave function and the energy of the stationary state, a_i are the probability amplitudes of the state i.

6.1.1 Transition Probabilities

The magnitude $|a_i|^2$ defines the probability of finding a particle on the level i. Then $\gamma_i |a_i|^2 dt$ is the probability of decay of the level i during the time dt. Since $|a_2|^2 = 0$ in the absence of field E', the total probability of decay of the level 2 during an infinitely long period of time is equal to the transition probability from the level 1 to the level 2 under the action of the fields E and E'. Thus the transition probability from the level 1 to the level 2 is

$$W_{1 \to 2} = \gamma_2 \int_0^\infty |a_2|^2 dt . \tag{6.7}$$

The energy emitted by an atom under the action of the field at the frequency ω' is

$$\Delta E = \hbar \omega' \cdot W_{1 \to 2} . \tag{6.8}$$

Sometimes the energy absorbed (emitted) by an atom at the frequency ω'_+ is determined through a dipole moment by an optical electron in the atom. The power absorbed (emitted) by the atom at the frequency ω' is

$$P = \langle j \rangle E' = e \frac{\partial \langle r \rangle}{\partial t} E'(t) , \tag{6.9}$$

where j is a current operator, r is the totality of coordinates of electrons, e is an electron charge. The mean value of r is

$$\langle r(t) \rangle = \int \Psi^* r \Psi \, dq .$$

We shall be interested in the mean value of $\langle r \rangle$ at the frequency of the $0 \to 2$ transition:

$$\langle r(t) \rangle = \langle r_{02} \rangle \, e^{i\omega_{02} t} a_0^* a_2 + \text{c.c.}$$

where

$$\langle r_{02} \rangle = \int \Psi_0^* r \Psi_2 \, dq .$$

The energy absorbed (emitted) by the atom at the frequency ω' is equal to

$$\mathscr{E}' = \int_0^\infty P \, dt .$$

Finally we have

$$\mathscr{E}' = e \langle r_{02} \rangle E' 2 \operatorname{Re}(i\omega_{02} \int_0^\infty a_0^* a_2 \, e^{-i\Omega' t} dt) \tag{6.10}$$

where $p = e \langle r_{02} \rangle$ is the matrix element of the $0 \to 2$ dipole transition.

The equations of the probability amplitudes, allowing that the frequencies ω and ω' are close to the resonance frequencies ω_{01} and ω_{02}, are of the form [2]

$$\begin{aligned}
&1) \quad \dot{a}_0 + (\gamma_0/2) a_0 = G e^{-i\Omega t} a_1 + G' e^{-i\Omega' t} a_2 \\
&2) \quad \dot{a}_1 + (\gamma_1/2) a_1 = -G^* e^{i\Omega t} a_0 \\
&3) \quad \dot{a}_2 + (\gamma_2/2) a_2 = -G'^* e^{i\Omega' t} a_0
\end{aligned} \tag{6.11}$$

[2] For the scheme of cascade transitions the system of equations for the probability amplitude is obtained from (6.11) by replacing Ω' with $-\Omega'$, G with $-G'^*$. Therefore all the results obtained for the scheme may be automatically extended to the cascade scheme by making corresponding replacement in the final formulae.

where γ_i is the rate of decay of the level i, $\Omega = \omega - \omega_{01}$, $\Omega' = \omega' - \omega_{02}$,
$$G = \frac{ip_{01}E}{\hbar}, \quad G' = \frac{ip_{02}E'}{\hbar}.$$

We should note some properties of the system of equations (6.11). Let us write down two evident equalities

$$\dot{a}_2 a_2^* + \frac{\gamma_2}{2}|a_2|^2 = -G'^* e^{i\Omega't} a_0 a_2^*$$

$$a_2 \dot{a}_2^* + \frac{\gamma_2}{2}|a_2|^2 = -G' e^{-i\Omega't} a_0^* a_2.$$

Adding them we have

$$\frac{d}{dt}|a_2|^2 + \gamma_2|a_2|^2 = -2\,\mathrm{Re}(G' e^{-i\Omega't} a_0^* a_2). \tag{6.12}$$

Integration of (6.12) becomes

$$|a_2|^2\,|_0^\infty + \gamma_2 \int_0^\infty |a_2|^2\,dt = -2\,\mathrm{Re}(G' \int_0^\infty e^{-i\Omega't} a_0^* a_2\,dt), \tag{6.13}$$

since $a_2 = 0$ at $t = 0$ and $t = \infty$. Then

$$W_{1\to2} = \gamma_2 \int_0^\infty |a_2|^2\,dt = -2\,\mathrm{Re}(G' \int_0^\infty e^{-i\Omega't} a_0^* a_2\,dt). \tag{6.14}$$

Comparing (6.8), (6.10) and (6.14), we are convinced of their identity. Writing down and adding the equalities similar to (6.12) for the first two equations, we obtain

$$\gamma_1 \int_0^\infty |a_1|^2\,dt + \gamma_2 \int_0^\infty |a_2|^2\,dt + \gamma_0 \int_0^\infty |a_0|^2\,dt = 1. \tag{6.15}$$

Solution of (6.11). We give the solution of (6.11) in the perturbation theory for weak fields.

In zero perturbation (in the absence of fields) we have

$$a_1 = \exp(-\gamma_1 t/2), \quad a_0 = 0 \quad \text{and} \quad a_2 = 0. \tag{6.16}$$

Substituting $a_1(t) = \exp(-\gamma_1 t/2)$ into (6.11), we obtain

$$\dot{a}_0 + \frac{\gamma_0}{2}a_0 = G\,e^{-i\Omega t - \frac{\gamma_1 t}{2}}. \tag{6.17}$$

With the initial conditions $a_2 = 0$ at $t = 0$ we have

$$a_0(t) = G \frac{e^{-i\Omega t - \frac{\gamma_1 t}{2}} - e^{-\frac{\gamma_0 t}{2}}}{(\gamma_0 - \gamma_1)/2 - i\Omega}. \tag{6.18}$$

The amplitude $a_0(t)$ contains two exponential terms. The first one carries information about the frequency of incident radiation and the lifetime of the initial level 1. The second one decays with the lifetime of the level 0.

The behavior of the probability amplitude $a_0(t)$ as a function of time considerably differs from the case of incoherent excitation, for example, by an electron impact. The dipole moment of an atom contains two frequencies: the transition frequency and the frequency of the incident wave; each component decays with the rates γ_1 and γ_0, respectively.

At great frequency detuning Ω ($\Omega \gg \gamma_0, \gamma_1$) the power absorbed by an atom is related to the induced dipole moment only. In the resonance, however, both dipole moments contribute to absorption. (Due to decay of the dipole moment at the frequency ω_{02} there is always a Fourier component at the frequency ω). At exact resonance and equal relaxation constants $\gamma_0 = \gamma_1$, the dipole moments coincide and, accordingly, their contributions turn out to be the same.

The transition probability $W_{1 \to 0}$ is equal to

$$W_{1 \to 0} = \frac{2\Gamma}{\gamma_1} \frac{1}{\Omega^2 + \Gamma^2} \frac{(p_{10} E)^2}{\hbar^2}. \tag{6.19}$$

The amplitude a_2 is found by integrating the third equation of (6.11) after substituting (6.18).

$$\dot{a}_2 + \frac{\gamma_2}{2} a_2 = -G'^* G \frac{e^{i(\Omega' - \Omega)t - \frac{\gamma_1 t}{2}} - e^{i\Omega' t - \frac{\gamma_0 t}{2}}}{\frac{\gamma_0 - \gamma_1}{2} - i\Omega}. \tag{6.20}$$

Hence

$$a_2 = -G'^* G \frac{1}{\frac{\gamma_0 - \gamma_1}{2} - i\Omega} \left[\frac{e^{i(\Omega' - \Omega)t - \frac{\gamma_1 t}{2}} - e^{-\frac{\gamma_2 t}{2}}}{\frac{\gamma_2 - \gamma_1}{2} + i(\Omega' - \Omega)} \right.$$
$$\left. - \frac{e^{i\Omega' t - \frac{\gamma_0 t}{2}} - e^{-\frac{\gamma_2 t}{2}}}{(\gamma_2 - \gamma_0)/2 + i\Omega'} \right]. \tag{6.21}$$

The amplitude $a_2(t)$ contains some oscillating and damping terms with different constants. The oscillating term with the frequency $\Omega' - \Omega$ is responsible for the dipole moment arising on the $1 \rightarrow 2$ transition with the frequency $\omega' - \omega[\langle r_{12}(t) \rangle = \langle r_{12} \rangle a_2 a_1^* \exp(i\omega_2, t)]$. The dipole moment on the $0 \rightarrow 2$ transition contains the frequencies ω_{02} and ω' as well as in the two-level scheme. In the presence of the field on the adjacent transition, an induced dipole moment arises at the frequencies $\omega_{02} + \Omega$, $\omega_{02} + (\Omega' + \Omega)$ and $\omega_{02} + (\Omega' - \Omega)$ with different damping coefficients. Fourier components of the dipole moment at the frequency ω_{02} determine the features of absorption at this frequency.

Substituting (6.18) and (6.21) into (6.14), integrating and transforming the expression to a convenient form, we have (6.22). After integration and algebraic transformations we can write the probability in a form which is more convenient for analysis [6.19]

$$
W_{1 \rightarrow 2} = \frac{2|G|^2 |G'|^2}{\left(\dfrac{\gamma_1 + \gamma_0}{2}\right)^2 + \Omega^2} \operatorname{Re} \left\{ \frac{1}{\gamma_0} \frac{1}{\left(\dfrac{\gamma_0 + \gamma_2}{2}\right) + i\Omega'} \right.
$$

$$
+ \frac{1}{\gamma_1} \frac{1}{\dfrac{\gamma_1 + \gamma_2}{2} + i(\Omega' - \Omega)}
$$

$$
\left. + \frac{1}{\left[\dfrac{\gamma_0 + \gamma_2}{2} + i\Omega'\right]\left[\dfrac{\gamma_1 + \gamma_2}{2} + i(\Omega' - \Omega)\right]} \right\}. \tag{6.22}
$$

The transition probability contains three terms which describe the characteristic features of the line shape. The first term describes the characteristics of step-by-step one-quantum processes. The maximum is at the transition frequency. And the line intensity depends on the frequency detuning from the transition frequency.

The line shape described by the second term corresponds to the two-quantum transition. It has a width of the $1 \rightarrow 2$ forbidden transition, and the intensity maximum at the detuning where $\Omega' - \Omega = 0$. The third term may be associated with the interference of these processes. It is noticeable in the conditions of resonance under certain circumstances (see Fig. 6.3).

At large detuning of the frequency Ω from resonance $\Omega \gg \gamma_1 + \gamma_0$, two maxima are observed in the stimulated emission line at frequencies $\Omega' = 0$ and $\Omega' = \Omega$ with the widths, respectively, $\gamma_1 + \gamma_0$ and $\gamma_1 + \gamma_2$. Their relative intensities depend on the ratio of relaxation constants of the levels 1 and 0. A line at the frequency $\Omega' = 0$ corresponds to that of the step-by-step transition. The part of the field at the frequency ω

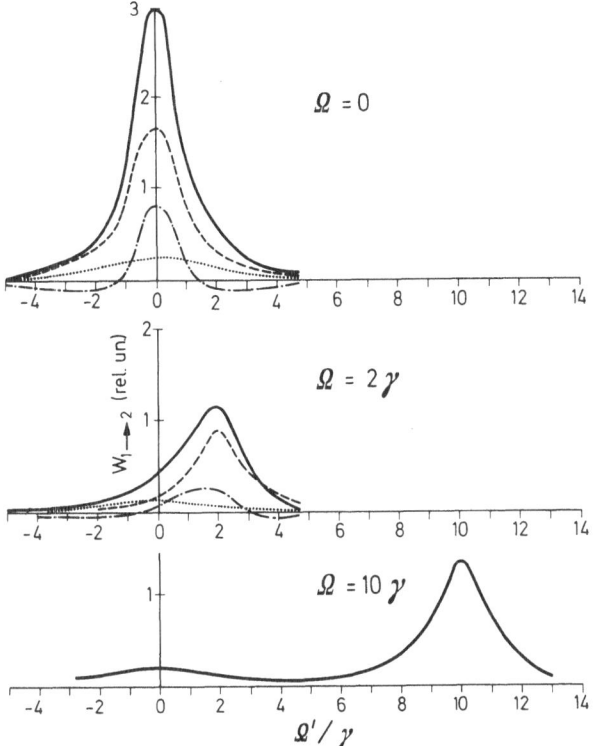

Fig. 6.3. Transition probability $W_{1\rightarrow2}$ as a function of the frequency of detuning Ω' at $\gamma_1 = \gamma_2 = \gamma$, and $\gamma_0 = 3\gamma$ at $\Omega = 0$. $\Omega = 2\gamma$, and $\Omega = 10\gamma$ respectively. $\cdots\cdots$ corresponds to the first term of (6.22), $---$ corresponds to the second term of (6.22), $-\cdot-\cdot-$ corresponds the the third term of (6.22)

is ascribed to excitation to the level 0 and subsequent radiation at the transition frequency.

For $\gamma_0 \gg \gamma_1$, the first and the last terms in (6.22) may be neglected and the scattering line can be considered as resonant SRS ($\Omega \sim \gamma_0$):

$$W_{1\rightarrow2} = \frac{2|G|\,|G'|^2}{\Omega^2 + \left(\dfrac{\gamma_0}{2}\right)^2}\,\frac{\gamma_1 + \gamma_2}{2\gamma_1}\,\frac{1}{(\Omega'-\Omega)^2 + \left(\dfrac{\gamma_1+\gamma_2}{2}\right)^2}\,. \tag{6.23}$$

For $\gamma_0 \ll \gamma_1$, the transition probability takes the form

$$W_{1\rightarrow2} = \frac{2|G|^2\,|G'|^2}{\Omega^2 + (\gamma_1/2)^2}\,\frac{\gamma_0 + \gamma_2}{2\gamma_0}\,\frac{1}{\left(\dfrac{\gamma_0+\gamma_2}{2}\right)^2 + \Omega'^2}\,. \tag{6.24}$$

The formula (6.24) corresponds to the expression for the probability of the step-by-step transition.

Before turning to the line shape of Doppler-broadened transitions, we give the equations for a density matrix and their solution.

6.1.2 Equations for Density Matrix Elements

In addition, we should note that the density matrix equations permit us to simplify the procedure of calculations, and to include phenomenological relaxation constants due to collisions. The interpretation of phenomena on the basis of the density matrix equations becomes simpler too.

We shall give the equations averaged over atoms excited at random moments. In the same notation the equations take the form

$$\left(\frac{d}{dt}+\gamma_0\right)\varrho_{00}=G(r,t)\varrho_{10}+G^*(r,t)\varrho_{01}+G'(r,t)\varrho_{20}$$
$$+G'^*(r,t)\varrho_{02}$$

$$\left(\frac{d}{dt}+\gamma_1\right)\varrho_{11}=-[G(r,t)\varrho_{10}+G^*(r,t)\varrho_{01}]+\gamma_1\varrho_{11}{}^{(0)}$$

$$\left(\frac{d}{dt}+\gamma_2\right)\varrho_{22}=-[G'(r,t)\varrho_{20}+G^*(r,t)\varrho_{02}]$$

$$\left(\frac{d}{dt}+i\omega_{02}+\Gamma_{02}\right)\varrho_{02}=-G'(r,t)(\varrho_{00}-\varrho_{22})+G(r,t)\varrho_{12}$$

$$\left(\frac{d}{dt}+i\omega_{01}+\Gamma_{01}\right)\varrho_{01}=-G(r,t)(\varrho_{00}-\varrho_{11})+G'(r,t)\varrho_{21}$$

$$\left(\frac{d}{dt}+i\omega_{12}+\Gamma_{12}\right)\varrho_{12}=-G^*(r,t)\varrho_{02}-G'(r,t)\varrho_{10}$$

(6.25)

$$\varrho_{ik}=\varrho_{ki}^*, \quad G(r,t)=\frac{id_{01}E(r,t)}{\hbar}, \quad G'(r,t)=\frac{id_{02}E'(r,t)}{\hbar},$$

$$\frac{d}{dt}=\frac{\partial}{\partial t}+v\frac{\partial}{\partial r},$$

$\Gamma_{ik}=(\gamma_i+\gamma_k)/2+v_{ik}$, v_{ik} is the collision frequency.

Here we assume the conditions used earlier in solving (6.11): namely, pumping is made up to the level 1. The atom interacts with two fields $E(r, t) = E \exp(-i\omega t + ikr) + \text{c.c.}$ and $E'(r, t) = E' \exp(-i\omega' t + ik'r) + \text{c.c.}$ The frequencies ω and ω' are close to the frequencies ω_{01} and ω_{02}, respectively. The solution of the system of Eqs. (6.25) enables us to find powers of emission and absorption and the absorption coefficient at the frequency ω' of the $0 \to 2$ transition.

The linear coefficient of absorption of the field at the frequency ω' is defined by

$$\alpha' = -4\pi k' \, \text{Im} \, \chi' \tag{6.26}$$

where χ' is the polarizability at the frequency ω' which is connected with polarization of the medium by

$$P(z, t) = \chi' E' \, e^{-i\omega' t + ik'r} . \tag{6.27}$$

The polarization of the medium is defined by the non-diagonal elements

$$P(r, t) = (P_{02}\varrho_{02} + P_{02}\varrho_{20}) N_0 . \tag{6.28}$$

The solution of a set of Eqs. (6.25) can be obtained in the general case of arbitrary fields. However the expressions prove to be rather unwieldy. The case when the field E', which we shall call a probe, is weak and the field E is arbitrary is of great importance for spectroscopy.

In the stationary case we take

$$\varrho_{01} = G \, e^{-i\omega t + ikr} r_{01} , \qquad \varrho_{02} = G' \, e^{-i\omega' t + ik'r} r_{02}$$

$$\varrho_{12} = G' G^* \, e^{i(\omega - \omega')t - i(k - k')r} r_{12} . \tag{6.29}$$

Then (6.25) are reduced to algebraic equations

$$\gamma_0 \varrho_{00} = |G|^2 (r_{01} + r_{10}) + |G'|^2 (r_{02} + r_{20})$$

$$\gamma_1 \varrho_{11} = -|G|^2 (r_{01} + r_{10}) + \gamma_1 \varrho_{11}^{(0)}$$

$$\gamma_2 \varrho_{22} = -|G'|^2 (r_{02} + r_{20})$$

$$[\Gamma_{01} - i\Omega] r_{01} = (\varrho_{11} - \varrho_{00}) + |G'|^2 r_{21}$$

$$[\Gamma_{02} - i\Omega'] r_{02} = (\varrho_{22} - \varrho_{00}) + |G|^2 r_{12}$$

$$[\Gamma_{12} - i(\Omega' - \Omega)] r_{12} = -(r_{02} + r_{10}) . \tag{6.30}$$

By omitting the term $|G'|^2$ in the three equations, we find

$$\varrho_{11} - \varrho_{00} = \frac{\Omega^2 + \Gamma_{01}^2}{\Omega^2 + \Gamma_{01}^2(1+\kappa)} \varrho_{11}^{(0)}$$

$$\varrho_{00} = \frac{|G|^2}{\gamma_0} \frac{2\Gamma_{01}}{\Omega^2 + \Gamma_{01}^2(1+\kappa)} \varrho_{11}^{(0)} \tag{6.31}$$

$$r_{01} = \frac{\Gamma_{01} + i\Omega}{\Omega^2 + \Gamma_{01}^2(1+\kappa)} \varrho_{11}^{(0)}$$

where $\kappa = 2|G|^2(\gamma_0^{-1} + \gamma_1^{-1})/\Gamma_{01}$ is the saturation parameter on the $0 \to 1$ transition by the field E. Substituting ϱ_{00} and r_{01} into the fourth and fifth equations of (6.30) we obtain two equations to find r_{02}, r_{12} and

$$r_{02} = -\frac{|G|^2\{2\Gamma_{01}[\Gamma_{12} - i(\Omega' - \Omega)] + \gamma_0(\Gamma_{01} - i\Omega)\}\varrho_{11}^{(0)}}{\gamma_0\{(\Gamma_{02} - i\Omega')[\Gamma_{12} - i(\Omega' - \Omega)] + |G|^2\}[\Omega^2 + \Gamma_{01}^2(1+\kappa)]}. \tag{6.32}$$

When $\kappa \ll 1$, we have

$$r_{02} = -|G^2| \left\{ \frac{2\Gamma_{01}}{\gamma_0(\Gamma_{02} - i\Omega')(\Omega^2 + \Gamma_{01}^2)} \right.$$

$$\left. + \frac{1}{(\Gamma_{01} + i\Omega)[\Gamma_{12} - i(\Omega' - \Omega)]\Gamma_{02} - i\Omega')} \right\} \varrho_{11}^{(0)}. \tag{6.33}$$

With the help of (6.33) we can obtain expressions for the transition probability, the absorption coefficient and so on.

In accordance with (6.25) the dipole moment on the $0 \to 2$ transition at the frequency ω' arises due to population difference between the levels 0 and 2 as well as due to the induced dipole moment p_{12}. The main features of the line shape are related to the latter. The appearance of the dipole moment at the frequency ω' may be schematically expressed in the following way:

a) $\underset{n_1}{\circ} \xrightarrow{E} p_{01} \xrightarrow{E} \underset{n_0}{\circ} \xrightarrow{E'} p_{02} \xrightarrow{E'} \circ$

b) $\underset{n_1}{\circ} \xrightarrow{E} p_{01} \xrightarrow{E'} p_{12} \xrightarrow{E} p_{02} \xrightarrow{E'} \underset{n_2}{\circ}.$ $\tag{6.34}$

The first process corresponds to an ordinary step-by-step transition. The field E polarizes the $0 \to 1$ transition. The interaction of the dipole moment induced by the field E with the field itself leads to population of the level 0. Then a similar process occurs on the $0\rightarrow2$ transition and with the field E'.

Considering the interaction of fields with excited levels, one must take into account processes associated with excitation of the other levels. It results in new processes in addition to the above processes a) and b) (see [6.19]). Within the framework of the perturbation theory all the principal processes at the population of all the levels are as follows:

a) $\quad\circ\xrightarrow{\ E\ } p_{01}\xrightarrow{\ E\ }\!\!\circ\xrightarrow{\ E'\ } p_{02}\xrightarrow{\ E'\ }\!\!\circ$
$\qquad n_1-n_0 \qquad\qquad \Delta n_0 \qquad\qquad \Delta n_2$

b) $\quad\circ\xrightarrow{\ E\ } p_{01}\xrightarrow{\ E'\ } p_{12}\xrightarrow{\ E\ } p_{02}\xrightarrow{\ E'\ }\!\!\circ$
$\qquad n_1-n_0 \qquad\qquad\qquad\qquad\qquad \Delta n_2$

c) $\quad\circ\xrightarrow{\ E'\ } p_{02}\xrightarrow{\ E'\ }\!\!\circ \qquad\qquad\qquad\qquad$ (6.35)
$\qquad n_0-n_2 \qquad\qquad \Delta n_2$

d) $\quad\circ\xrightarrow{\ E'\ } p_{02}\xrightarrow{\ E\ } p_{12}\xrightarrow{\ E'\ } p_{02}\xrightarrow{\ E'\ }\!\!\circ$.
$\qquad n_0-n_2 \qquad\qquad\qquad\qquad\qquad \Delta n_2$

The process c) gives the linear absorption coefficient on the $0\rightarrow2$ transition. The process a) is associated with the change of the population of the level 0 due to the population difference between the levels 1 and 0. The coherent processes b) and d) which give rise to the dipole moment d_{02} are not associated with the change of the population of the levels 0 and 2. If the interpretation of the process b) corresponds to the $1\rightarrow2$ two-quantum transition within the framework of the second order of the perturbation theory, the coherence effects on the population of the level 0 arise as a result of interference of the amplitude $a_2^{(1)}$ in the first order and the amplitude $a_2^{(3)}$ in the third order of the perturbation theory.

6.2 The Line Shape on the Adjacent Transition in a Gas

The expressions (6.22) and (6.32) may be used for finding the absorption line shape in a gas. In the system of the centre of inertia a moving atom sees the waves at frequencies: $\omega\rightarrow\omega-\mathbf{k v}$ and $\omega'\rightarrow\omega'\mp\mathbf{k'v}$. Then the corresponding detunings Ω and Ω' must be replaced with $\Omega-\mathbf{k v}$ and $\Omega'\mp\mathbf{k'v}$, respectively. The case of collinear propagation of the waves E and E' is of interest for spectroscopy. Let us choose the z axis in the direction of propagation of the wave E. Then the corresponding Doppler additions to the frequency are equal to $-\mathbf{k v}$ and $\mp\mathbf{k'v}$.

The above sign $(-)$ corresponds to the propagation of the waves E and E' in the same direction, the sign $(+)$ in the opposite direction. Later on for brevity we shall distinguish between these cases by calling the absorption (emission) line shape of the wave E' in the presence of the wave E of the same direction α_- the line shape of forward scattering; correspondingly, α_+ is the line shape of backward scattering.

The linear absorption coefficient of the wave E' is equal to the ratio of the absorbed energy in a unit volume to the energy flux density of incident radiation:

$$\alpha(\omega') = h\omega' Q\langle W_{1\to2}\rangle(8\pi/cE'^2),\tag{6.36}$$

rate of particle excitation for 1st level. The probability of the transition $W_{1\to2}$ is of the form

$$W_{1\to2}(v) = \frac{2|G|^2|G'|^2}{(\Omega-kv)^2+\left(\frac{\gamma_0+\gamma_1}{2}\right)^2}\,\mathrm{Re}\left\{\frac{1}{\gamma_1}\frac{1}{\left[\left(\frac{\gamma_1+\gamma_2}{2}\right)-i(\Omega'-\Omega\mp k'v+kv)\right]}\right.$$

$$+\frac{1}{\gamma_0}\frac{1}{\left(\frac{\gamma_0+\gamma_2}{2}\right)-i(\Omega'\mp k'v)}$$

$$+\left.\frac{1}{\left[\left(\frac{\gamma_1+\gamma_2}{2}\right)-i(\Omega'-\Omega\mp k'v+kv)\right]\left[\frac{\gamma_0+\gamma_2}{2}-i(\Omega'\mp k'v)\right]}\right\}.$$

$$\tag{6.37}$$

Let us consider two cases depending on the magnitude of detuning Ω as compared with the Doppler width.

1) $\Omega \gg k\bar{v}$.

This case corresponds to the classical stimulated Raman scattering. The first and the third terms may be neglected near the detuning frequencies $\Omega' \sim \Omega$. In spite of the velocity, all the atoms emit the same line shape but with Doppler shift depending on the velocity. The line shape is given by the expression

$$W_{1\to2}^{(\pm)} = \frac{2|G|^2|G'|^2}{\Omega^2}\,\mathrm{Re}\left[\frac{1}{\gamma_1}\frac{1}{\sqrt{\pi}\bar{v}}\int_{-\infty}^{\infty}\frac{e^{-\left(\frac{v}{\bar{v}}\right)^2}dv}{\left(\frac{\gamma_1+\gamma_2}{2}\right)-i(\Omega'-\Omega\mp k'v+kv)}\right]$$

$$\tag{6.38}$$

As usual we assume the Maxwell velocity distribution.

The integration in (6.38) can be made in an analytical form for two cases:

a) $|k' - k|\bar{v} \ll \dfrac{\gamma_1 + \gamma_2}{2}$,   b) $|k' - k|\bar{v} \gg \dfrac{\gamma_1 + \gamma_2}{2}$.

In the case a) the Doppler broadening of the forward Raman scattering line may be neglected. All the atoms emit the line shape of a dispersive form at the frequency $\omega' = \omega_{02} + \Omega$. Averaging over velocities is equivalent to summation over a number of atoms. In the case b) the lines of forward and backward Raman scattering have a Gaussian profile with a characteristic width $|k' - k|\bar{v}$ and $(k' + k)\bar{v}$, respectively.

A two-photon resonance arises in the centre of an absorption line in the cascade scheme at $\Omega \gg k\bar{v}$ [6.20]. The compensation of Doppler shifts occurs at the interaction of oppositely moving waves in the cascade schemes in contrast to the scheme of Raman scattering. As we have already noted, the probability of the $1 \to 2$ transition is obtained by replacing Ω' with $-\Omega'$ in the scheme of the cascade transitions. The transition probability of two-photon absorption is of the form

$$W_{1 \to 2}(v) = \frac{8|G|^2|G'|^2}{\Omega^2} \frac{1}{\gamma_1} \mathrm{Re} \left[\frac{1}{\sqrt{\pi}\bar{v}} \int_{-\infty}^{\infty} \frac{e^{-\left(\frac{v}{\bar{v}}\right)^2} dv}{\left(\dfrac{\gamma_1 + \gamma_2}{2}\right) - i(\Omega' + \Omega + k'v - kv)} \right].$$

(6.39)

For the oppositely moving waves of the same frequency near resonance we have $2\omega = \omega_{12}$ and $k' = +k$. We should note that all the atoms have the same line shape of two-quantum absorption irrespective of their velocities, and the integration over velocities means summation over a number of atoms. The two-photon resonance[3]

$$W_{1 \to 2} = \frac{8|G|^2|G'|^2}{\Omega^2} \frac{1}{\gamma_1} \mathrm{Re} \left[\frac{1}{\gamma_1 + \gamma_2 - i(2\omega - \omega_{12})} \right]$$

(6.40)

possesses the features which are of interest for spectroscopy and applications (see [6.21]). A detailed treatment of these features may be found in Chapter 8 and Ref. [6.22].

2) $\Omega \lesssim kv$, $\Omega' \lesssim k'v$.

---

[3] An additional factor of "4" is related to the fact that an atom "does not distinguish" between the pumping and the probe waves.

In this case, both fields are resonant with the Doppler-broadened transition. It is the case which is of great interest for us now. The condition of resonance with the Doppler-broadened transition means that there are always atoms which are in exact resonance with the fields E and E'. A number of authors [6.9–11, 14, 16–18, 23, 24] have made detailed theoretical and experimental studies of phenomena in three-level schemes in the conditions of resonance. The line shape of gaseous atoms proves to be dependent on the ratio of transition frequencies, the direction of wave propagation, level populations and so on.

The principal theoretical results obtained in various works permit us to consider any situation which can be found in spectroscopy of three-level gas systems. On the basis of the investigations of resonant SRS in neon made in the Institute of Semiconductor Physics, Siberian Branch, USSR Academy of Sciences, we shall illustrate the use of TLS methods in solving concrete spectroscopic problems. We give the principal results for some cases.

6.2.1 The Weak Field Case ($\kappa \ll 1$)

a) $k' > k$.

$$\alpha_+ = \alpha_0 \, e^{-(\Omega'/k'\bar{v})^2} \left[1 + |G|^2 \frac{N_1^{(0)} - N_0^{(0)}}{N_0^{(0)} - N_2^{(0)}} \left(\frac{k'2\Gamma_+}{k\gamma_0} \right) \frac{1}{\left(\Omega' + \frac{k'}{k}\Omega \right)^2 + \Gamma_+^2} \right]$$

(6.41)

$$\alpha_- = \alpha_0 \, e^{-(\Omega'/k'\bar{v})^2} \left[1 + |G|^2 \frac{N_1^{(0)} - N_0^{(0)}}{N_0^{(0)} - N_2^{(0)}} \left(\frac{2\Gamma_- k'}{\gamma_0 k} \right) \frac{1}{\left(\Omega' - \frac{k'}{k}\Omega \right)^2 + \Gamma_-^2} \right]$$

α_0 is the linear absorption coefficient on the $0 \to 2$ transitions; N_1, N_0, and N_2 are populations of the levels 1, 0, and 2, respectively.

$$2\Gamma_+ = \gamma_2 + \gamma_0 + \frac{k'}{k}(\gamma_1 + \gamma_0)$$

(6.42)

$$2\Gamma_- = \gamma_1 + \gamma_2 + \left(\frac{k'}{k} - 1 \right)(\gamma_1 + \gamma_0).$$

Γ_+ and Γ_- are HWHM of the lines of forward and backward scattering.

b) $k' < k$.

$$\alpha_+ = \alpha_0 \exp[-(\Omega'/k'\bar{v})^2]$$

$$\cdot \left[1 + \frac{N_1^{(0)} - N_0^{(0)}}{N_0^{(0)} - N_2^{(0)}} \left(\frac{k'}{k}\right) \frac{2\Gamma_+}{\gamma_0} \frac{1}{\left(\Omega' + \frac{k'}{k}\Omega\right)^2 + \Gamma_+^2} |G|^2 \right]$$

$$\alpha_- = \alpha_0 \exp[-(\Omega'/k'\bar{v})^2] \cdot \left\{ 1 + \frac{2k'}{k}(1 - k'/k) \frac{\Gamma_-^2 - \Omega_-^2}{(\Gamma_-^2 + \Omega_-^2)^2} |G|^2 \right.$$

$$\left. + \left(\frac{k'}{k}\right) \frac{N_1^{(0)} - N_0^{(0)}}{N_0^{(0)} - N_2^{(0)}} |G|^2 \left[\frac{2\Gamma_-}{\gamma_0} \frac{1}{(\Omega_-^2 + \Gamma_-^2)} \right] \right\},$$

$$(6.43)$$

where

$$\Omega_- = \Omega' - \frac{k'}{k}\Omega, \qquad 2\Gamma_- = \frac{k'}{k}(\gamma_1 + \gamma_2) + \left(1 - \frac{k'}{k}\right)(\gamma_2 + \gamma_0),$$

$$2\Gamma_+ = \gamma_2 + \gamma_0 + \frac{k'}{k}(\gamma_1 + \gamma_0).$$

The main feature of stimulated emission line in a gas is a distinction between α_+ and α_-. When the width of the common level γ_0 is large as compared with γ_1 and γ_2, the distinction is notable. For $\gamma_0 \gg \gamma_1, \gamma_2$, the line width of forward emission is determined by the width of the $1 \to 2$ two-quantum transition which is Doppler broadened due to the distinction between k' and k. The backward emission line has a shape characteristic of a step-by-step transition. Its width is due to the width of the $0 \to 2$ transition and to the velocity distribution of atoms on the level 0.

One must pay attention to the sharp difference in the emission line shape of an ensemble of atoms and of an isolated atom. An ensemble of atoms emits a pure line of resonant SRS irrespective of the ratio of relaxation constants of individual levels. We should remember that individual atoms emit a line which cannot be unequivocally assigned to the two-quantum process. The whole thing is that by averaging over the velocities with uniform distribution, the contribution of the processes corresponding to the step-by-step transition [the first term in (6.36)] and the interference of the step-by-step and two-quantum processes [the third term in (6.36)] are compensated (or give a line-width of the two-quantum process). Thus, a gas of atoms permits us to observe two-quantum processes in a pure form. The appearance of a backward emission line which coincides with a step-by-step transition line is explained in the same way. For $k' < k$ and with some population of the common level, the forward line shape is not Lorentzian. As has been noted, the additional changes in the line shape are due to the interference process in which amplitudes $a_2^{(1)}$ and $a_2^{(3)}$ are added. The interference processes are essential when oscillating terms in the probability amplitude $\exp(-i\Omega't)$ and $\exp(i(\Omega'-\Omega)t)$ have close frequencies. Turning to Doppler-shifted frequencies, we shall obtain the condition at which interference will give maximum contribution to the line

$$k'v=(k-k')v. \tag{6.44}$$

Hence $k'=k/2$. In this case the total compensation of Doppler frequency shift occurs and, as we have already seen, an emission line of the ensemble coincides with emission of an individual atom. Thus, the case $k'=k/2$ at forward scattering in the ensemble of atoms permits us to observe directly quantum-mechanical effects of interference which occur at transitions in an isolated atom. The changes of the line shape on the $0\rightarrow2$ transition occur when $N_1=N_0$. This case has been studied in Refs. [6.23, 24].

Broadening of the SRS Line. At greatly different k and k' the line-width of forward and backward scattering of the ensemble of atoms proves to be much larger than the width of radiation of an isolated atom. It yields inhomogeneous broadening of a resonant SRS line which is due to motion of atoms. The degree of inhomogeneous broadening depends on the direction of observation and the ratio of relaxation constants. In the forward resonant SRS, the degree of inhomogeneous broadening is determined by the ratio k'/k. In the backward scattering, the resonant SRS line is always inhomogeneously broadened. The character of inhomogeneous broadening can affect consideration of the saturation effect in SRS of a gas laser. We shall dwell on these phenomena later.

In accordance with the character of broadening, the non-equilibrium velocity distribution in the three-level system has its own features too. The expression (6.37) can be used to determine the velocity distribution of atoms on the level 2

$$N_2(v)=QW_{1\rightarrow2}(v)\cdot\frac{1}{\gamma_2}. \tag{6.45}$$

It is seen from (6.22) that, when $\gamma_0\gg\gamma_1$, the peaks have the widths $\sim\Gamma_{12}/(k'-k)$ and $\Gamma_{12}/(k'+k)$. These widths can be much narrower than those in the two-level $0\rightarrow2$ transition under the action of the field E' with the velocity distribution. When $k'/k\gg1$ and $\gamma_0\gg\gamma_1,\gamma_2$, one can easily see that atoms in a narrow velocity interval take part in the two-quantum process of a region, whilst each field E and E' interacts independently with atoms in a greater velocity interval.

6.2.2 The Strong Field Case

a) $k' > k$.

The emission line shape on the $0 \rightarrow 2$ transition in the presence of arbitrary field E has been considered for this case in Ref. [6.17, 19]

$$\alpha_+ = \alpha_0 \, e^{-(\Omega'/k'\bar{v})^2} \left[1 + \frac{N_1^{(0)} - N_0^{(0)}}{N_0^{(0)} - N_2^{(0)}} \cdot \frac{Q^2 - 1}{2Q} \cdot \frac{k'}{k} \cdot \frac{\gamma_1 \Gamma_+}{\Gamma_+^2 + \left(\Omega' + \frac{k'}{k}\Omega\right)^2} \right]$$

$$\alpha_- = \alpha_0 \, e^{-(\Omega'/k'\bar{v})^2} \left[1 + \frac{N_1^{(0)} - N_0^{(0)}}{N_0^{(0)} - N_2^{(0)}} \cdot \frac{Q^2 - 1}{2Q} \cdot \frac{k'}{k} \cdot \frac{\gamma_1 \Gamma_-}{\left(\Omega' - \frac{k'}{k}\Omega\right)^2 + \Gamma_-^2} \right]$$

$$(6.46)$$

$$2\Gamma_- = \gamma_2 + \gamma_1 Q + \left(\frac{k'}{k} - 1\right)(\gamma_1 + \gamma_0)Q$$

$$2\Gamma_+ = \gamma_2 + \gamma_0 Q + \frac{k'}{k}(\gamma_0 + \gamma_1)Q, \quad Q = \sqrt{1 + \kappa}.$$

b) $k' < k$.

This case has been considered in Refs. [6.25, 26]. The line shape strongly differs from the case of $k' > k$ for waves of the same direction. The changes in the line shape can be of interest for spectroscopy.

The analysis made in Refs. [6.25, 26] has shown that at strong saturation the factor of the line shape is approximated by

$$\alpha' \sim \frac{1}{\sqrt{2S\Omega_-}} \left[\frac{1}{(\Omega_- - S)^2 + \gamma^2} - \frac{1}{(\Omega_- + S)^2 + \gamma^2} \right]^{1/2},$$

$$2\gamma = \gamma_2 + \gamma_0 + \frac{k'}{k}(\gamma_1 - \gamma_0). \qquad (6.47)$$

The line shape is described by two resonances with the widths $\sqrt{3}\gamma$ which are situated on the frequencies S and $-S$, respectively (see Fig. 6.4). Since $S \gg 2\gamma$, the distance between the peaks is large as compared with their widths. For large enough $|G|$ the distance between maxima is given with a good accuracy by

$$\Delta = 2S = 4|G|[k'(k - k')]^{1/2}/k. \qquad (6.48)$$

The authors of Refs. [6.17, 26] associate the sharp structure which arises in strong fields with a high-frequency Stark effect [6.27] which is

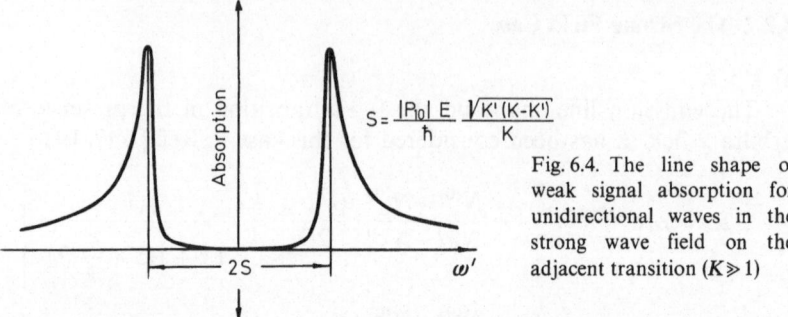

$$S = \frac{|P_0|\,E}{\hbar} \frac{\sqrt{K'\,(K-K')}}{K}$$

Fig. 6.4. The line shape of weak signal absorption for unidirectional waves in the strong wave field on the adjacent transition ($K \gg 1$)

well known in the microwave region where Doppler broadening is negligible.

It is known that, in a strong wave at the frequency of the $0 \rightarrow 1$ transition, the probability of finding an atom on the level 0 oscillates with a low frequency depending on the magnitude of the field. This oscillation leads to modulation of the dipole moment on the $0 \rightarrow 2$ transition. As a result, additional resonances which can be interpreted as splitting of the level 0 arise in the line shape on the $0 \rightarrow 2$ transition.

In the strong field on the $0 \rightarrow 1$ transition the expression for $a_1(t)$ with the initial conditions $a_1 = 1$, $a_0 = 0$ is of the form

$$a_0(t) = \frac{|G|}{\alpha_1 - \alpha_2} (e^{-i\alpha_1 t} - e^{-i\alpha_2 t}) \tag{6.49}$$

where

$$\alpha_{1,2} = \frac{-i\Gamma + \Omega}{2} \pm \sqrt{|G|^2 + \left(\frac{\Omega - i\gamma^2}{2}\right)}$$

$$\Gamma = \frac{\gamma_1 + \gamma_0}{2}, \qquad \gamma = \frac{\gamma_1 - \gamma_0}{2}.$$

A wave function of an atom

$$\Psi = a_1 \Psi_1 \exp(-iE_1 t/\hbar) + a_0 \Psi_0 \exp(-iE_0 t/\hbar)$$

will have oscillating terms at the frequencies $\mathscr{E}$ and $\mathscr{E}_0$ which correspond to new quasi-stationary states of the "atom + field" system.

The condition under which one can use a model of splitting is $|G| \gg \Gamma$. When level damping is neglected, the following energies correspond to new states of the atom:

$$E_0^{(1)} = E_0 + \{(\Omega/2) + [(\Omega/2)^2 + |G|^2]^{1/2}\} \cdot \hbar$$
$$E_0^{(2)} = E_0 + \{(\Omega/2) - [(\Omega/2)^2 + |G|^2]^{1/2}\} \cdot \hbar \tag{6.50}$$

The resonance frequencies for the $0 \rightarrow 2$ transition are

$$\omega_{02}^{(1,2)} = \omega_{02} + \frac{\Omega}{2} \pm \sqrt{\left(\frac{\Omega}{2}\right)^2 + |G|^2}. \tag{6.51}$$

Formula (6.48) can be obtained for qualitative reasons in the model of level splitting in a strong electromagnetic field. The resonance frequencies on the $0 \rightarrow 2$ transition in the presence of a strong field on the $1 \rightarrow 0$ transition allowing for Doppler shift (for unidirectional waves)

$$\omega' - k'v = \omega_{02}^{(1,2)} = \omega_{02} + k'v - \frac{kv}{2} \mp \sqrt{\left(\frac{kv}{2}\right)^2 + |G|^2}. \tag{6.52}$$

For the sake of simplicity we assume $\Omega = 0$. Then the velocities of the atoms resonantly interacting with the fields can be easily found from (6.52):

$$v_p = \frac{\Omega'(k' - k/2) \pm \sqrt{(k\Omega'/2)^2 + (k'^2 - k'k)|G|^2}}{k'^2 - k'k}. \tag{6.53}$$

Doing the similar calculations, we have for oppositely moving waves:

$$v_p = \frac{-\Omega'(k' + k/2) \pm \sqrt{(k'\Omega'/2)^2 + (k'^2 + k'k)^2 |G|^2}}{k'^2 + k'k}. \tag{6.54}$$

Analysing (6.53), we notice that for the case of unidirectional waves there is detuning Ω' for which there are no atoms resonantly interacting with the probe wave.

The magnitude of detuning Ω' at which there are resonantly interacting atoms is determined from the condition under which the expression under the root sign in (6.53) is positive

$$|\Omega'| \geqq 2|G|(kk' - k'^2)^{1/2}/k = S.$$

The absence of resonantly interacting atoms means zero absorption of the probe wave in a wide range of frequencies. For oppositely moving waves there are always resonant atoms and no similar peculiarities arise. The result obtained coincides with the exact calculation. It enables us to notice that manifestation of the dynamic Stark effect on Doppler-broadened transitions consists not in line splitting but in formation of a single range of frequencies where absorption is small. We note that the dynamic Stark effect shows up in the two-level system in a similar way [6.28]. The line shape in ranges $\Omega > S$ can be obtained from the following reasons. The absorption of probe wave is proportional to the number of atoms having resonant velocities. The velocity distribution on the common level 0 under the action of the strong wave on the $0 \rightarrow 1$ transition is

$$\Delta n_b(v) = n_1(v) \frac{\Gamma_{01}}{\gamma_0} \frac{|G|^2}{(kv)^2 + \Gamma_{01}^2 \dfrac{|G|^2}{\gamma_1 \gamma_0}} . \tag{6.55}$$

For the sake of simplicity let us consider the case $2k' = k$. Then we obtain from (6.53)

$$v_p = \pm \frac{\sqrt{\Omega'^2 - 2|G|^2}}{k} . \tag{6.56}$$

Substituting (6.56) into (6.55), we obtain for the absorption of the probe wave

$$\alpha' \sim n_1(v) \frac{\Gamma_0}{\gamma_0} \frac{|G|^2}{\Omega'^2 - 2|G|^2 + 4\Gamma_0^2 \dfrac{|G|^2}{\gamma_1 \gamma_0}} , \quad |\Omega'| > \sqrt{2}|G|. \tag{6.57}$$

The formula describes the correct dependence of the absorption on the frequency qualitatively.

6.3 Investigations of Resonant SRS Line in Neon

The simplest investigations of peculiarities of the spectrum in the three-level system were made on observation of the line shape of spontaneous emission [6.6–8]. The investigations of the line shape of spontaneous emission permitted us to detect not only effects of the change of level populations but also coherence effects in the emission spectrum [6.29]. However, a small intensity of optical signals and rather low accuracy which is due to the instrumental resolution of interferometers limit possibilities of these methods. Therefore, methods of investigation of the line shape of laser-stimulated radiation are now widely used [6.13, 14, 23, 30]. The investigations were first made in neon where generation was obtained on coupled transitions. The detailed spectroscopic investigations of resonant SRS were made on the $2s_2 - 2p_1 (\lambda = 1.52 \ \mu m)$ and $2s_2 - 2p_4 (\lambda = 1.15 \ \mu m)$ transitions in neon.

6.3.1 Choice of Transitions

This system of transitions has been chosen for the following reasons: the common level $2s_2$ is resonant, that is, connected with the ground state by a strong optical transition. A lifetime of the $2s_2$ level can be much less than that of the $2p_1$ and $2p_4$ levels. The wavelengths of the $2s_2 - 2p_1$ and $2s_2 - 2p_4$ transitions are so close that one can expect that compensation of Doppler shifts in the two-quantum transition will be effective enough and the contribution of Doppler broadening will be small. As a result of numerous investigations, considerable data on times of relaxation of the $2p_1$ and $2s$ levels of neon have been collected. For the scheme under investigation an interference effect of the one- and two-quantum processes must not appear in the approximation of weak field at the finite population of the $2s_2$ level. All this enables us to consider an experiment as direct observation of the line shape of stimulated resonant Raman scattering.

The lifetimes of the $2p$ levels have been measured in Ref. [6.31]. These measurements give the values $\gamma(2p_1) = 6.95 \times 10^7$ rad/s, $\gamma(2p_4) = 5.24 \times 10^7$ rad/s. The rate of decay of the $2s_2$ level has the value $\gamma(2s_2) = 1.6 \times 10^8$ rad/s by the data of Ref. [6.32]. A width of the forbidden $2p_1 \rightarrow 2p_4$ transition turns out to be 19 MHz. A peak width in the velocity distribution on the $2s_2$ level is 36.8 MHz. As has been pointed out, the forward scattering line width is equal to that of two-quantum scattering which consists of widths of the forbidden $2p_1 \rightarrow 2p_4$ transition and a Doppler part $(k'/k - 1)[\gamma(2p_1) + \gamma(2s_2)]$. Thus, for $2\Gamma_-$ we have 30.8 MHz. The width of backward scattering line is equal to that of the step-by-step transition. It is due to contribution of the non-equilibrium velocity distribution of atoms equal to $(k'/k)(\gamma_1 + \gamma_0) = 48.6$ MHz and to the width of the $2s_2 - 2p_4$ transition. For $2\Gamma_+$ we have 82.6 MHz. We note that the width of the two-photon backward scattering line is 104 MHz.

6.3.2 Experiment for Observation of Resonant SRS

The experiment on observation of the line shape of resonant SRS consisted in investigations of the line shape of the signal gain at $\lambda = 1.15\ \mu$m which arose under the action of a monochromatic signal at $\lambda = 1.52\ \mu$m. The scheme of an experimental arrangement is given in Fig. 6.5. Radiation of a high-power single-frequency laser 1 on the 1.52 μm line was directed into a discharge tube 3 (length 50 cm, diameter 2 mm) which was filled with pure isotope of ^{20}Ne. In order to measure the line shape of absorption or gain on the $2s_2 - 2p_4$ transition the radiation at $\lambda = 1.15\ \mu$m from a short scanned supermode He–Ne laser 2 was introduced into cell 3 either in the same direction as the strong wave (path of the

228   V. P. Chebotayev

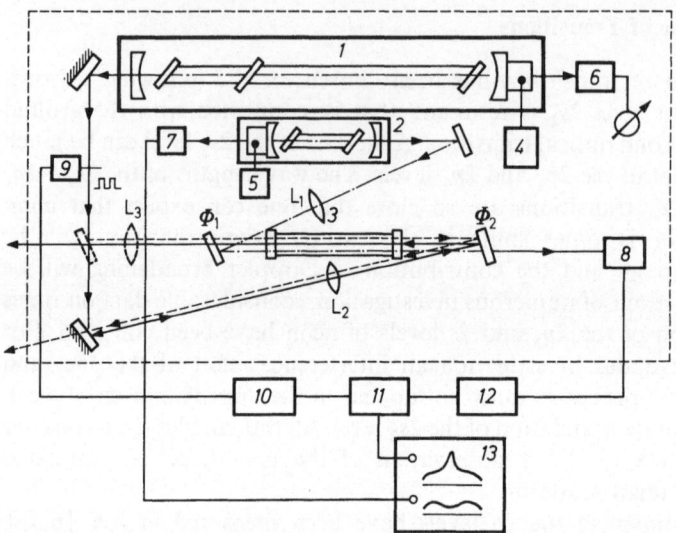

Fig. 6.5. Experimental arrangement for the observation of the line shape of resonant stimulated Raman scattering: forward $(k'k>0)$ (dotted lines) and backward $(k'k<0)$ in neon

1	He–Ne laser ($\lambda=1.52\,\mu$m, strong wave);	*6, 7, 8*	photodiodes;
2	He–Ne laser ($\lambda=1.15\,\mu$m, weak probe wave);	*9*	mechanical modulator;
		10	low frequency oscillator;
3	discharge cell;	*11*	phase-sensitive detector;
4, 5	units to control piezoceramics;	*12*	selective amplifier;
		13	automatic two-channel recorder

beam is marked by a dotted line in this case) or in the opposite direction. In the case of parallel waves, the strong wave at 1.52 μm must pass through a mirror φ_1 with the help of which the weak wave was introduced in the cell. Matching of wave fronts was made with the help of lenses which simultaneously provided focusing of the strong wave aimed for obtaining greater saturation in the cell. Non-coincidence of wave fronts influenced the signal size and the line shape. For the optical system used in the experiment, the influence of these factors might contribute to a width no more than 1 MHz.

A photodiode PD-3 was used as the photodetector 8 to record laser radiation at $\lambda=1.15$ μm after passing through the absorption cell 3. The strong wave at $\lambda=1.52$ μm was modulated with the help of an electromechanical modulator at a frequency of about 40 Hz. Percentage of modulation was about 100%. As a result of the nonlinear interaction, the modulation at the same frequency of 40 Hz arose in the laser radiation at $\lambda=1.15$ μm after passing through the cell. The a.c. signal of the photodetector was amplified with the help of a selective amplifier and supplied to the input of a synchronous detector.

The frequency of the weak wave was slowly scanned by supplying a sawtooth voltage to piezoceramics on which one of the mirrors of the laser cavity 2 was fixed. The laser 2 had a discharge tube of 1 mm in internal diameter and 7 cm in length of a discharge section. The tube was filled with a mixture of neon and helium isotopes in a ratio of $1:10$ at a total pressure of 3 Torr. The high He pressure in the tube permitted us to obtain an almost flat top on the dependence of output power on the frequency that provided minimum change of the weak wave intensity in the range of 300–400 MHz near the centre of the line. The distance between mirrors was 12.3 cm (the free spectral frequency range was 1220 MHz). The laser 2 radiation was supplied to the photo-detector 7 (PD-3) operating in the photovoltaic regime and the output of the photodetector was fed to the second channel of the automatic recorder 13. Thus, the second channel of the automatic recorder recorded the change of the weak wave amplitudes v.s. its frequency scanning. The a.c. signal at the output of the synchronous detector was proportional to the value of the weak wave. It is therefore clear that the ratio of the first spectrogram of the recorder 13 to the second one is proportional to the a.c. part of the absorption coefficient at the 1.15 μm neon line in the tube 3, this part being due to the strong external wave at $\lambda = 1.52$ μm.

The time of recording of one mode was about 40 seconds. Therefore a high stability of the probe frequency was required. The passive stabiliza-tion of the length of cavity 1 was used. For this purpose the experimental apparatus was placed on a massive steel plate which lay on an isolated concrete foundation. Massive steel heads with mirrors were attached to the plate. Spaces between windows sealing off discharge tubes and mirrors were carefully isolated from the surrounding medium. As a result of these arrangements an error in determining the width due to drift of the pump frequency for the time of recording of a narrow structure of the line shape amounted to a magnitude of less than 1 MHz. The relative short-time stability of the laser amounted to 10^{-9}.

In conclusion, let us discuss the method of eliminating the "back-ground" resulting from the dragging of the resonant radiation. This was necessary to improve the measurement accuracy. The presence of the "background" complicated the analysis of shapes of the narrow reso-nances (see curve 1 in Fig. 6.6). Since the shape of the "background" coincides with the absorption line, we could choose the amplitude and phase of the discharge-current modulation in such a way that the signal making up the "background" was completely cancelled out by the signal produced as a result of the modulation of the current, in the entire range of variation of the probe frequency. In practice, the selection was carried

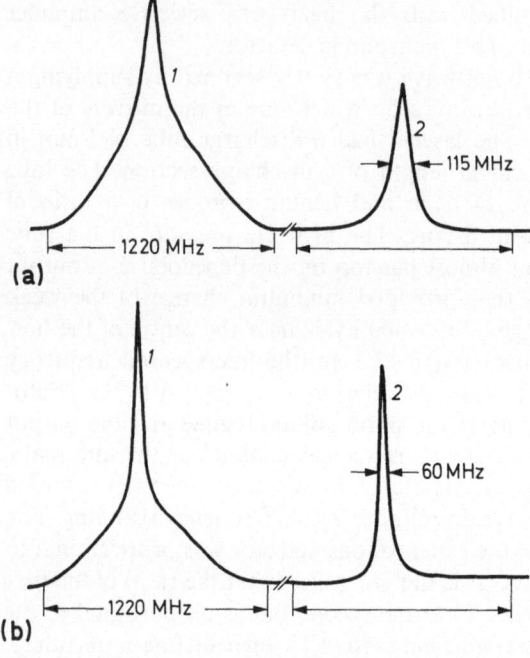

Fig. 6.6a and b. The line shape of resonant SRS in neon: a—backward scattering ($p = 0.5$ Torr, $I = 17$ mA), b—forward scattering ($p = 0.9$ Torr, $I = 15$ mA). Curve 1—SRS line without compensation of the Doppler "background", Curve 2—SRS line with compensation of the Doppler "background"

out in such a way that the amplitude of the a.c. signal was zero in the case of frequency deviation $\Omega' \gg \Gamma_-$ from the narrow peak. This indeed signified the cancellation of the Doppler "background" as shown by curve 2 in Fig. 6.6.

6.3.3 SRS Line Shape

Anisotropy of the Line Shape: Figure 6.6 shows the recorded stimulated emission (absorption) line shape for weak and strong waves traveling in parallel (Fig. 6b) and antiparallel (Fig. 6a) directions. The curve 1 shows that the gain line has a rather complicated shape. It constitutes a narrow peak on a much broader "background". The peak widths are much less than the Doppler width of the 1.15 μm line and are comparable with the radiative transition width. We note strong anisotropy of the line. For the waves moving in parallel ($k'k > 0$) the peak is much (by a factor of 2) narrower than for the oppositely moving ones ($k'k < 0$).

The "background", which has the Doppler width and is connected with the equilibrium velocity distribution, can be explained by the diffusion of the excitation in velocity space (see Sect. 6.6).

The character of the anisotropy agrees qualitatively with that predicted by the theory. For a probe wave propagating in the same direction as the pump wave ($k'k > 0$) the line has a width of 58 MHz at a pressure of 0.9 Torr which is approximately half the value that follows from consideration of hole-burning effects. For oppositely moving waves ($k'k < 0$) the width of the peak turns out to be 118 MHz and is almost equal to the sum of the width of the $2s_2 - 2p_4$ transition and the width in the velocity distribution of the atoms on the $2s_2$ level in units of k'.

The observed resonance width turned out to be close to that predicted by theory. Experimentally we observed the dependences of the resonance widths on the field intensities and gas pressures. The analysis of the results is simple when the fields are weak. In this case there is no need to take into account the transverse and longitudinal inhomogeneities of the optical fields. But it must be taken into account in the case of very weak fields that the signal magnitudes are small. Therefore the values of the width corresponding to very weak fields were obtained by extrapolating the dependence of the widths to zero field. Since in our experiments the saturation parameter did not exceed $\kappa = 0.5$, we could use linear extrapolation. Values of the widths $2\Gamma_+ = 86 \pm 3$ and $2\Gamma_- = 31 \pm 2$ MHz extrapolated to zero field and zero pressure proved to be in good agreement with those calculated by using known values of the level lifetimes. This permits us to draw a conclusion about the efficiency of this method for spectroscopic investigations.

Measurements of the Level Lifetimes. The difference between the backward and forward halfwidths, Γ_+ and Γ_-, is equal to the width of the common level: $\Gamma_+ - \Gamma_- = \gamma_0$.

In the presence of quenching or "strong" collisions which also reduce the atom-field interaction time, we have

$$\Gamma_+ - \Gamma_- = \gamma_0 + \nu_{col}$$

where ν_{col} is the frequency of quenching and strong collisions. The results of measurements of the widths of forward and backward scattering were used for measurements of the rate of decay of the $2s_2$ level. The dependence of the difference between the widths, $2\Gamma_+ - 2\Gamma_-$, on pressure is given in Fig. 6.7. The extrapolation to zero pressure yields the following values of the rate of decay of the level: $\gamma(2S_2) = 27$ MHz corresponding to the lifetimes $\tau = 6 \times 10^{-9}$ s, which is in good agreement with the results of the direct measurements by using the technique of delayed coincidences

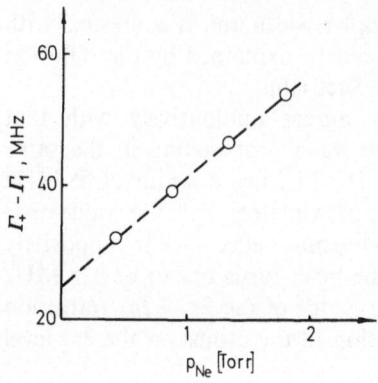

Fig. 6.7. Dependence of difference between the halfwidths of backward and forward scattering on neon pressure

and registration in a vacuum ultraviolet ($\tau(2S_2) = 7.8 \times 10^{-9}$ s [6.33]). We note that the larger the width of the common level (greater difference between the widths γ_0 and γ_1, γ_2), the more exact are measurements of the lifetime by the method of TLS. The lifetimes of the other levels can be obtained by using known values of γ_0 and the linewidths of the $1 \rightarrow 0$ and $2 \rightarrow 0$ transitions. The latter can be defined by observing a Lamb dip shape. For instance, the investigations of the Lamb dip in the 1.15 μm Ne laser [6.32] have given a value of natural halfwidth of the $2s_2 - 2p_4$ transition $\Gamma = 17 \pm 1$ MHz. Hence the corresponding lifetimes are $\tau(2p_4) = 2.2 \times 10^{-8}$ s, and $\tau(2p_1) = 1.25 \times 10^{-8}$ s. The agreement between the spectroscopic data and the results of the direct measurements of luminescence damping is rather convincing.

6.4 Polarization Characteristics of Resonant SRS and Investigations of Fine Structure of Levels

Up to now we have not taken into account of the polarization characteristics of resonant SRS which are directly related to degeneration of the levels 1, 0, and 2. Our investigations [6.34] have shown that the observed sharp structure is exceedingly sensitive to polarization of the optical fields E and E'.

Thus, Fig. 6.8 shows the recorded line shape of the scattering for a circularly polarized wave at $\lambda = 1.52$ μm. The weak wave at 1.15 μm also has a circular polarization, which in the case 1) is opposite to the direction of rotation of the polarization of the strong wave, while in the case 2) the directions coincide. The intensity ratio at the maximum of the scattering line is 6:1, which agrees (within the limits of measurement accuracy) with the intensity ratio of the Zeeman com-

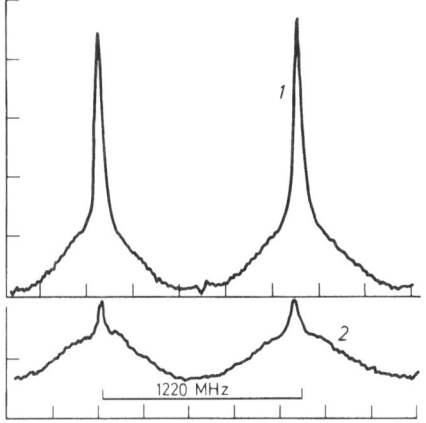

Fig. 6.8. Plot of the line shape
of resonant SRS in neon for
circularly polarized waves. Curves
1 and 2 correspond to the opposite
and the same circular polarizations with the pump wave,
respectively

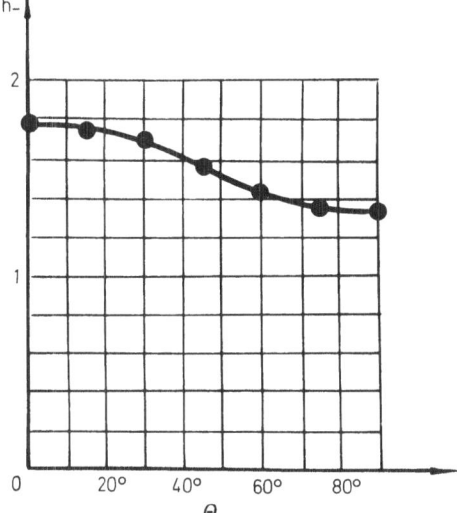

Fig. 6.9. Polarization anisotropy
of the intensity of resonant SRS
in neon

ponents calculated with the help of $6J$-symbols on the transition $J = 1 \rightarrow J' = 2$.

Investigations were also made of the degree of polarization of the
scattering line as a function of the azimuthal angle between the polarization planes of the probe and pump waves (Fig. 6.9). The observed
maximum degree of polarization of the scattering line is 25%. The
polarization anisotropy of the sharp structure of the line is described
with a sufficient accuracy by the ratio

$$\alpha/\alpha_0 = 1 - 0.25 \sin^2 \theta,$$

where α_0 is the gain for parallel polarizations.

In the investigated pressure range (0.3–1.5 Torr) the polarization properties of the scattering line remained practically unchanged. This indicates that the reorientation of the magnetic moments of an atom is accompanied by a simultaneous change in its velocity in its own gas under the conditions of resonant exchange of excitation. In the case of "weak" collisions the relaxation of orientation should lead to a depolarization of the sharp structure of the line. If the $2s_2 - 2p_2$ ($\lambda = 1.1767\,\mu\text{m}$) transition was used as a probe wave, the narrow resonances were not observed at linear polarizations of the external and probe waves. This can be easily understood. The transitions between the sublevels of the fine structure of the $2s_2$ ($J = 1$) and $2p_2$ ($J = 1$) levels are allowed for $M = \pm 1$. Simultaneously the wave at $\lambda = 1.52\,\mu\text{m}$ with the same linear polarization causes the transitions between the sublevels with $M = 0$. The interaction of the waves with circular polarizations leads to a sharp structure of the emission line at $\lambda = 1.1767\,\mu\text{m}$ in the presence of the wave at $\lambda = 1.52\,\mu\text{m}$.

The narrow lines of resonant SRS can be used for investigations of the fine structure of levels. Here we shall describe the investigations of the resonant SRS line in a magnetic field [6.34]. The longitudinal magnetic field removes degeneracy of the $2s$ and $2p$ levels. When we observe the $1.15\,\mu\text{m}$ line along the magnetic field H, we can detect three transitions with left circular polarization (LCP $\Delta M = \pm 1$) and three transitions with right circular polarization (RCP $\Delta M = \pm 1$). For the light with LCP the frequencies differ from the non-shifted frequency ω_{02} by values of $-\dfrac{g_0}{\hbar}\mu_0 H,\ -\dfrac{g_2}{\hbar}\mu_0 H,\ -\dfrac{2g_2 - g_0}{\hbar}\mu_0 H$, and for the light with RCP by values of $\dfrac{g_0}{\hbar}\mu_0 H,\ \dfrac{g_2}{\hbar}\mu_0 H,\ \dfrac{2g_2 - g_0}{\hbar}\mu_0 H$, where μ_0 is the Bohr magneton. Owing to the thermal motion, the Doppler widths of these transitions overlap in the region where splitting is linear to the field H, and the difference between g-factors is difficult to detect. Therefore for the sake of simplicity we consider $g_0 = g_2 = g$ for the $1.15\,\mu\text{m}$ line. The observed line along the field is the sum of two non-interacting Doppler-broadened lines which shift in the opposite directions as the magnetic field increases.

The Zeeman effect in a three-level system in the presence of resonant monochromatic light appears in a different way. In this case one can observe splitting of the resonance at weak magnetic fields, and this permits us to make exact measurements of g-factors of the levels and the intensities of the separate Zeeman components. The possibility of selective excitation of the magnetic sublevels and of the investigation of exchange between them is attractive from the point of view of the investigations of the Zeeman effect.

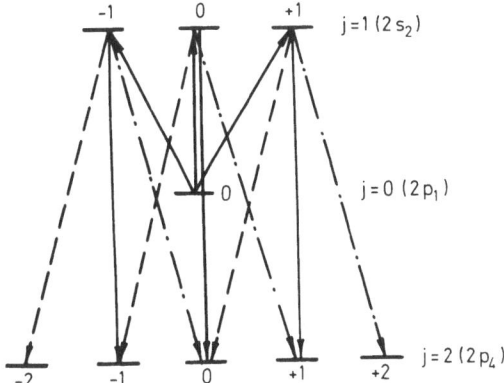

Fig. 6.10. The scheme of Zeeman sublevels for the $2s_2 - 2p_4$ and $3s_2 - 2p_1$ transitions in neon

Let us consider the Zeeman effect of the 1.15 μm line in the presence of the external monochromatic wave at $\lambda = 1.52$ μm in the scheme given in Fig. 6.10. The superposition of the longitudinal magnetic field leads to splitting of the absorption line of the strong wave into two σ-components. The linearly polarized wave may be resolved into two components with LCP and RCP with the same intensities. The absorption of the component with LCP leads to excitation of the atoms to the sublevel with $M = -1$, and with RCP to the sublevel with $M = +1$, whose velocities satisfy the resonance condition interacting with the wave: the Doppler-shifted frequency is equal to the detuning from the centre of the line. The atoms which are in exact resonance have the velocities: $kv_p = \Omega + \Delta_0$ on the sublevel with $M = -1$ and $kv^+ = \Omega + \Delta_0$ on the sublevel with $M = +1$. It has been taken into account that the lower level of the $2s_2 - 2p_1$ transition is not degenerated ($J = 0$) and $\Delta_0 = \dfrac{g_0}{\hbar} \mu_0 H$.

Two electric dipole transitions from each magnetic sublevel on the $2s_2 - 2p_4$ transition in a longitudinal magnetic field turn out to be resolved in accordance with the selection rule for dipole radiation ($\Delta M = \pm 1$) with the resonance frequencies $\omega_{02} + \Delta_0 - 2\Delta_2 (1 \rightarrow 2, \text{LCP})$, $\omega_{02} - \Delta_0 (1 \rightarrow 0, \text{RCP})$, $\omega_{02} - \Delta_0 (-1 \rightarrow 0, \text{LCP})$, $\omega_{02} - \Delta_0 + 2\Delta_2 (-1 \rightarrow -2,\ \text{RCP})$. These frequencies correspond to the emitting atoms which have different projections of velocities onto the direction of propagation of the external optical field. In the laboratory coordinate system the emission on the 1.15 μm line in the same direction as the external field will be concentrated near the frequencies:

$$\Omega'^{(1)} = \frac{k'}{k}\Omega + \left(1 - \frac{k'}{k}\right)\varDelta_0 - 2\varDelta_2 \qquad (1\rightarrow2, \text{LCP})$$

$$\Omega'^{(2)} = \frac{k'}{k}\Omega + \left(1 - \frac{k'}{k}\right)\varDelta_0 \qquad (1\rightarrow0, \text{RCP})$$

$$\Omega'^{(3)} = \frac{k'}{k}\Omega - \left(1 - \frac{k'}{k}\right)\varDelta_0 \qquad (-1\rightarrow0, \text{LCP})$$

$$\Omega'^{(4)} = \frac{k'}{k}\Omega - \left(1 - \frac{k'}{k}\right)\varDelta_0 + 2\varDelta_2 \qquad (-1\rightarrow-2, \text{RCP}) \qquad (6.58)$$

that is, the lines corresponding to the same circular polarizations are split.

The experimental investigations were given in the scheme described in Section 6.2. A difference was that an external absorption cell was placed in a longitudinal magnetic field. In order to eliminate polarization anisotropy, the external cell had windows perpendicular to the axis of the discharge tube and the direction of the magnetic field. It was possible to change arbitrarily the polarization of both the pump wave at $\lambda = 1.52\,\mu\text{m}$ and the probe at $\lambda = 1.15\,\mu\text{m}$. The circularly polarized light was obtained by introducing quarter-wave micaceous plates respectively selected for the wavelengths 1.15 μm and 1.52 μm into the beam.

Figure 6.11 shows (with compensation of the Doppler background which is due to the diffusion of excitation from the position of trapping of resonance radiation) the experimental recordings of the line shape of the stimulated emission on the 1.15 μm line in the presence of the plane-polarized moving wave at $\lambda = 1.52\,\mu\text{m}$ and the magnetic field $H = 80$ Oe for forward scattering.

The splitting of the magnetic sublevels is defined by the formula $\varDelta = 1.4\, gH$ (MHz), where H is the magnetic field in Oe. For $H = 100$ Oe and $g = 1.3$ we obtain a magnitude of $\varDelta \simeq 180$ MHz. The distance between the maxima of σ-components at the normal Zeeman effect on the $2s_2 - 2p_4$ transition must be equal to $2\varDelta = 360$ MHz which is appreciably less than the Doppler line width in a discharge, $\varDelta v_D = 800$ MHz. In the three-level system we have a well-resolved structure with a maximum distance between the components, 806 MHz. Besides two strongly shifted components with different circular polarizations, there are two components less shifted from the centre of the transition. These components are related to the transition to the $M = 0$ sublevel which does not shift in the magnetic field. It is obvious that at very close wavelengths, that is, at $(k'/k - 1 \ll 1)$ these components will not shift in a very strong field

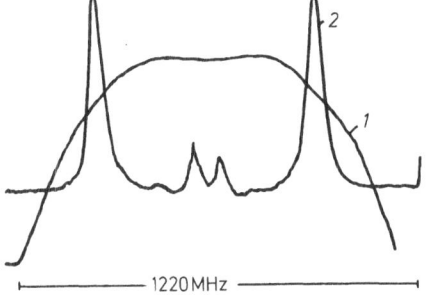

Fig. 6.11. Plot of the line shape of resonant forward SRS in a magnetic field ($H = 80$ Oe). Curve 1 corresponds the profile of probe signal, curve 2 is the SRS line shape

in spite of the fact that the common level in the magnetic field shifts greatly enough. This effect reflects the specificity of scattering in the three-level system. The frequency of scattered light with the same circular polarization as the external field does not change in the magnetic field whilst the change of polarization of scattered photons in the magnetic field is connected with the change of frequency of the scattered light.

The splitting of the stimulated emission of the $\lambda = 1.15\ \mu m$ line was used for exact determination of the ratio of g-factors of the $2s_2$ and $2p_4$ levels. The measurements have yielded a magnitude of the ratio of g-factors, $g(2p_4)/g(2s_2) = 1.035 \pm 0.02$. Using the previously obtained magnitude of $g(2p_4) = 1.301$ [6.35], we obtain $g(2s_2) = 1.26 \pm 0.03$. This magnitude agrees with that determined by the method of double resonance [6.36]. From these measurements one can obtain absolute magnitudes of g-factors provided that we measure exactly the magnitude of the magnetic field. Our measurements have yielded a magnitude of 1.30 ± 0.03 for $g(2p_4)$ which agrees with the above-mentioned value. We should point out that these measurements proved to be possible due to the narrowness of the peaks obtained in the stimulated forward scattering. The widths of the backward peaks are much wider. Therefore, the resolution is less when the ordinary population saturation effect is used.

6.5 Investigation of Resonance Radiation Trapping

The TLS methods can turn out to be rather effective at investigations of the processes of excitation transfer between magnetic sublevels, of the velocity change of excited atoms and so on. These processes can be of importance in the interaction of the strong light field with a gas, and also these can influence the emission characteristics of gas lasers. We shall show the use of the TLS methods by studying the effects of the resonant radiation trapping on the $2s_2$ level of neon. Early investigations of the Lamb dip on the 1.15 μm transition in the He–Ne laser have shown an appreciable homogeneity of saturation [6.37, 38]. It was assumed that this was caused by "strong" collisions at which the velocities of excited atoms changed within the range of thermal distribution. Further investigations of the influence of collisions [6.32] have shown that the collisions accompanying the phase change are of great importance, and the differences in the results on collisional broadening are to a large extent connected with the model of collisions which is assumed as a basis of the experimental data processing. We note that the resonant radiation trapping leads to the same effects as the strong collisions do. Therefore the investigations of the resonant radiation trapping are important for the investigations of collisions [6.32].

At the concentrations of atoms of neon ($p \sim 10^{-1}$ Torr) used in the operating conditions on He–Ne lasers a free path length of an ultraviolet photon is much shorter than the laser beam sizes. Therefore, complete trapping of the resonant radiation occurs in the volume, and the lifetime of the $2s_2$ level is defined by the decay to the underlying $2p$ levels only. The direct measurements of the lifetimes of the $2s_2$ level made on the spontaneous emission of the $2s_2 - 2p$ transitions have yielded values of about 10^{-7} s [6.31].

The influence of radiation trapping upon the character of field-atom interaction can be explained with the help of a simple illustration (Fig. 6.12). Let an atom 1 move in the direction perpendicular to the plane of the illustration. In this plane the atom emits with the frequency coincident with the centre of the line. The different directions of the photon emission are marked by dotted lines. The photons emitted by the atom can be absorbed by other atoms whose directions of motion are perpendicular to the direction of photon propagation. Let us take an atom 2, for instance. In the direction of an observer, it interacts with the field of the other frequency, and its velocity projection onto the z axis is not zero. Thus, the exchange of excitation occurs between the atoms with different velocities. After excitation of the atom, one can detect atoms with arbitrary velocities in the narrow velocity interval Δv due to emission and trapping of photons. On the other hand, the excitation transfer from the region of atomic velocities which do not interact with the photon to the region of velocities which are resonant can occur in a similar way. This process leads to the saturation homogeneity at the interaction with the strong light, and is similar to the strong collisions in its effect upon the character of the interaction with the light field.

The same scheme of neon levels is used below. The absorption at $\lambda = 1.52$ μm transfers atoms to the $2s_2$ level which is resonant. Since in the conditions of experiment and at gas pressure of about 1 Torr the homo-

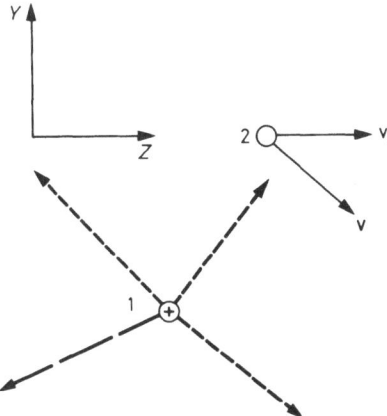

Fig. 6.12. Radiation trapping by moving atoms

geneous width is much smaller than the Doppler width, interaction with the atoms whose velocities meet the condition $v_p = \pm \Omega/k$ occurs. As a result of such interaction the peak arises in the velocity distribution of atoms on the $2s_2$ level due to the above-mentioned processes of the resonance trapping.

The non-equilibrium velocity distribution of atoms and the wide band Maxwell distribution may be recorded on the emission line shape from the $2s_2$ level. The observation of the line shape of the spontaneous emission on any adjacent transition may be used. It is easier to study the line shape of the stimulated emission by using a single-frequency laser which is tunable within the Doppler width. The $2s_2 - 2p_4$ transition on which lasing can be easily obtained in a wide range with the 1.15 µm He–Ne laser is the most convenient for these purposes. A non-equilibrium part in the velocity distribution of atoms will be responsible for the narrow resonance in the gain (or absorption) on the transition which arises under the pumping at $\lambda = 1.52$ µm. And the wide band signal has an ordinary Doppler contour of the gain (or absorption) line. The relation between the amplitudes of the sharp and wide components in the contour of the gain line indicates the role of resonance radiation trapping.

The first experiments were made on the observation of the line shape of the stimulated emission in a three-level laser. The scheme of this experiment is shown in Fig. 6.13. The output of a high-power single-frequency He–Ne laser at $\lambda = 1.52$ µm was focused in a cavity of the 1.15 µm laser. As a rule, the absorption at $\lambda = 1.52$ µm is observed in the discharge of pure neon. Both absorption and gain can take place at $\lambda = 1.15$ µm depending on the discharge current and Ne pressure. It is more convenient to make an investigation of the resonant radiation trapping in the conditions when the small absorption is observed at

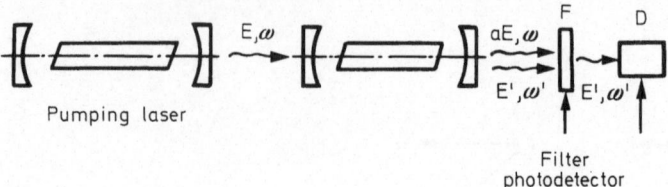

Fig. 6.13. A scheme of a three-level laser

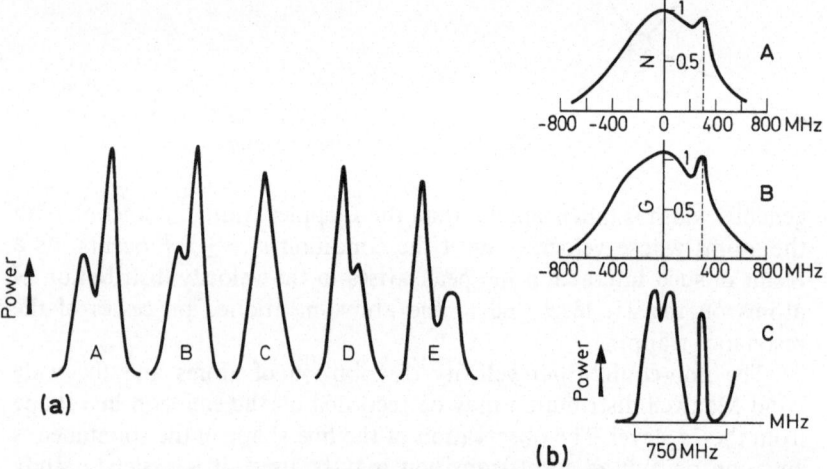

Fig. 6.14. (a) The power-frequency characteristics of the laser at $\lambda = 1.15\,\mu\text{m}$ under the pumping at $\lambda = 1.52\,\mu\text{m}$ with different values of detuning Ω. A: 150 MHz, B: 50 MHz, C: 0, D: -50 MHz, E: -160 MHz. (b) Velocity distribution of atoms on the $2s_2$ level (A), the gain (B) and the output power (C) at $\lambda = 1.15\,\mu\text{m}$ under the pumping at $\lambda = 1.52\,\mu\text{m}$

$\lambda = 1.15\,\mu\text{m}$. Then the produced gain and emission are fully due to the action of the pumping at $\lambda = 1.52\,\mu\text{m}$. The velocity distribution of atoms which arises on the level under the laser pumping at $\lambda = 1.52\,\mu\text{m}$ is shown in Fig. 6.14. Figure 6.14a shows the gain line shape in the 1.15 μm laser. It consists of two peaks of different intensities. Since the standing wave field in a laser can be represented by a superposition of two plane waves moving in opposite directions, the resonance conditions arise on two frequencies $\omega' = \omega_0 \pm \dfrac{k'}{k}\Omega$ at the interaction of the moving wave field at $\lambda = 1.52\,\mu\text{m}$ with two oppositely moving waves. Hence, two resonances arise in the standing wave field during the frequency scanning; each of them corresponds to the gain in each direction. The narrow resonance corresponds to the interaction of 1.15 μm and 1.52 μm waves moving in the same direction. A laser excitation threshold is selected so that the output radiation passes near resonances only, and the Doppler background does not appear. The emission has a complicated form (Fig. 6.14a) at the frequency detuning of $\lambda = 1.52\,\mu m$ and at the increase of its intensity. There is a sharp peak, and the dependence of its power on frequency shows the Lamb dip at the centre of the line which is characteristic of ordinary gas lasers. This dependence is due to the resonance radiation trapping. The Doppler "background" was observed in the neon pressure from 0.2 to 1.5 Torr. Changing the neon pressure by a factor of 8, the ratio of amplitudes of the peaks and the background changed slightly. An estimate has shown that at a neon pressure of 0.1 Torr the frequency of the strong collisions must be higher than $10^8\,s^{-1}$ in order to lead to an appreciable effect of the diffusion of excitation in the velocity space. On the other hand, at a pressure of higher than 1 Torr the sharp structure as well as the Lamb dip must vanish under the high collision frequency. These considerations and further investigations permit us to conclude that the observed diffusion of excitation is due to the resonance radiation trapping.

The quantitative investigations made with the help of the three-level laser are somewhat difficult because of complications of the description of its output characteristics. The detailed investigations of the radiation trapping and the 1.15 μm line broadening by using the three-level schemes were made on the line shape of the stimulated emission [6.39].

Figure 6.6 shows the recorded signal of the absorption line at $\lambda = 1.15\,\mu m$ under the pumping at $\lambda = 1.52\,\mu m$. Curves 2 and 1 correspond to the recordings of the signals with and without compensation of the Doppler background. The background-to-resonance amplitude ratio has a value of about 1. The relative amplitude of the background R increases with the pressure as shown in Fig. 6.15. However, the analysis shows that its relative increase is mainly connected with the narrow resonance broadening and not with the increase of the collision frequencies which leads to the formation of the background itself.

Convincing proof of the main part of the resonance radiation trapping in the formation of the background at the neon pressure of about 1 Torr consists in the fact that the extrapolation of the dependence of R on pressure to zero pressure gives a non-zero value. If the background were due to collisions only, the dependence of R on pressure would naturally start from zero. The obtained experimental dependence given in Fig. 6.15 depends slightly on pressure and it again confirms the small role of collisions. It is seen from the pressure dependence of R that at a pressure of about 1 Torr the contribution of collisions to the diffusion of excitation is about 20%. Thus, collisions will play a role comparable with the radiation trapping at pressures of over 5 Torr.

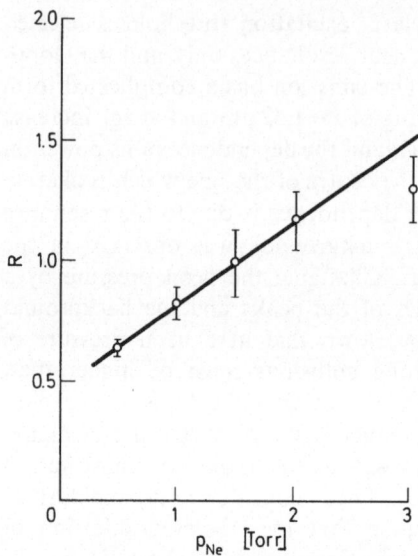

Fig. 6.15. The ratio R between the intensities of the background and the resonance components as a function of neon pressure

Thus, the method of observation of resonance Raman scattering permits us to study directly the processes of excitation relaxation as a function of the velocity of excited atoms.

6.6 Influence of Collisions upon Resonant SRS Line

Since the line shape on the adjacent transition is sensitive to the velocity distribution of atoms on the common level, one can study the change of the atomic velocities at the collisions by the change of the line shape. The SRS line shape which is due to the two-quantum transition turns out to be sensitive to the mechanism of the line broadening by collisions. The difference in the collisional broadening of the line between forward and backward scattering permits us to clear up contributions of different levels in the line broadening. Such information could not be received earlier with the conventional spectroscopic methods.

The above-described experiments on the observation of the radiation trapping are related to the observation of the diffusion of excitation over velocities within the average velocity $\bar{v}$. It is obvious that the method can be successfully used in the cases when the velocity diffusion occurs in a small fraction of the average velocity. For example, it may occur when the Doppler shift at scattering by an angle θ, equal to $k\bar{v}\theta$, is larger than the width of the emitted line γ. Then the corresponding background has the line width $k\bar{v}\bar{\theta}$. From the experimental data and the theory of collisions the angle θ is $1°$ in the order of magnitude, that corresponds to the background line width of about 10 MHz in the optical range. Since this magnitude is commensurate with the resonance widths, its direct observation is difficult in such conditions. The analysis of the results requires a very careful performance of the experiments and the processing of their results. Up to now such experiments have not yet been made. However, even a simple analysis of the collisional broadening of SRS resonance gives much information on the mechanism of collisions. The character of the resonance broadening turns out to be rather critical to the type of the interaction of colliding particles, to the contribution of different levels into the line broadening. Therefore, the investigations of collisions on resonant SRS lines turn out to be much more informative as compared with those on the Lamb dip broadening.

The simplest way is to use a phenomenological model of collisions for the description of their influence upon the resonant SRS line. The system of Eqs. (6.25) for a density matrix permits us to describe the influence of the phase-perturbing and quenching collisions by the corresponding replacement of the relaxation constants Γ_{ik} and γ_i.

As should be expected, the most interesting features arise in forward scattering [6.39]. When the phase-perturbing collisions are available, there occur not only the change of the resonance parameters, but also the qualitative changes of its form: the phase-perturbing collisions lead to the formation of the background line with the width of the step-by-step transition. Its intensity is proportional to the collision frequency. The similar phenomenon arises in resonance fluorescence under collisions [6.40].

The resonant SRS line shape for $k'/k > 1$ and $N_0 \simeq 0$ is of the form

$$
\alpha_- = \alpha'_0 \left[\frac{\Gamma_0 + \Gamma_{02} - \gamma_0 - \Gamma_{12}}{\Gamma_{01} + \Gamma_{02} - \Gamma_{12}} \frac{\Gamma_-}{\left(\Omega' - \frac{k'}{k}\Omega\right)^2 + \Gamma_-^2} \right.
$$

$$
\left. + \frac{\gamma_0}{\Gamma_{01} + \Gamma_{02} - \Gamma_{12}} \frac{\Gamma_+}{\left(\Omega' - \frac{k'}{k}\Omega\right)^2 + \Gamma_+^2} \right]. \qquad (6.59)
$$

Here

$$
\Gamma_{ik} = \frac{\gamma_i + \gamma_k}{2} + \frac{\nu_i + \nu_k}{2}
$$

$$
\Gamma_- = \Gamma_{12} + \left(\frac{k'}{k} - 1\right)\Gamma_{01}, \qquad \Gamma_+ = \Gamma_{02} + \frac{k'}{k}\Gamma_{01},
$$

γ_i is the rate of decay of the level i allowing for collisions, $v_i = v_0 n_0 \bar{v} \sigma_i$, σ_i is the total elastic scattering cross section of the atom in the level i at the collisions with surrounding particles, n_0 is the particle density,

$$\alpha_0' = \frac{8\pi^{3/2} \, k' |p_{02}|^2 \, |G|^2 \, N_1}{k \bar{v} \gamma_0}.$$

One can see from (6.59) that the collisions on the upper level only lead to the formation of the background with the width Γ_+. Therefore we transform (6.59) to the form

$$\alpha_- = \alpha_0' \left[\frac{v_0}{\gamma_0 + v_0} \frac{\Gamma_-}{\left(\Omega' - \dfrac{k'}{k}\Omega\right)^2 + \Gamma_-^2} + \frac{\gamma_0}{\gamma_0 + v_0} \frac{\Gamma_+}{\left(\Omega' - \dfrac{k'}{k}\Omega\right)^2 + \Gamma_+^2} \right]$$

$$\alpha_+ = \alpha_0' \left[\frac{\Gamma_+}{\left(\Omega' + \dfrac{k'}{k}\Omega\right)^2 + \Gamma_+^2} \right]. \tag{6.60}$$

The backward scattering line has a Lorentzian shape. The resonance width depends on the line broadenings on the $0 \rightarrow 1$ and $0 \rightarrow 2$ transitions.

a) Line Broadening

Figure 6.16 shows the dependence of $2\Gamma_+$ and $2\Gamma_-$ on pressure. The backward line width consists of the broadening of the $2s_2 - 2p_1$ transition (it defines the Bennett hole broadening in the velocity distribution of atoms) by taking account of the wavenumbers k' and k and the broadening of the $2s_2 - 2p_4$ transition. Both transitions belong to the same electron configuration and have the common level which must give the main contribution to the line broadening. Therefore one can consider the line broadenings to be equal. Hence, the collisional broadening of the neon $2s - 2p$ transition is $\simeq 10$ MHz/Torr. The obtained values of the broadening agree with those measured in Ref. [6.32]. Using the early measurements of the broadening of the $2s_2 - 2p_4$ transition on the Lamb dip (8.8 MHz/Torr), we obtain 10 MHz/Torr for the broadening of the $2s_2 - 2p_1$ transition, which is in good agreement with the results of the measurement of the power resonance in a laser with nonlinear absorption [6.41].

b) Forbidden Transition Broadening

The forward line broadening consists of the broadening of the $2p_1 - 2p_4$ forbidden transition and the broadening of the Doppler part of the SRS

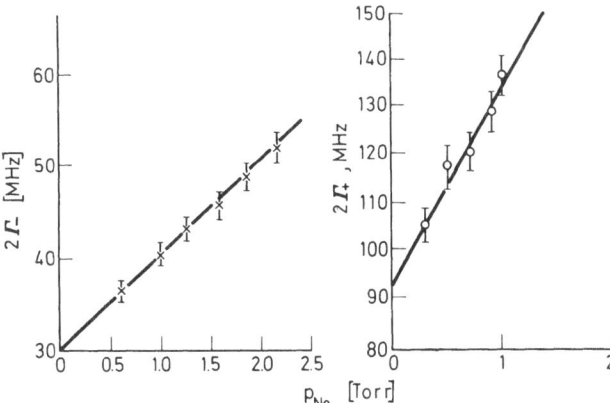

Fig. 6.16. Dependence of the linewidths of resonant forward (left) and backward (right) SRS on neon pressure in the cell

line. The collisions with the phase perturbation lead to the formation of the background line.

As we have pointed out, the direct observation of the broad background component is difficult because of its small amplitude. Let us estimate its influence upon the effective linewidth of the scattering, by assuming $v_0 \ll \gamma_0$. The change of the halfwidth which is due to the background is equal to

$$\Delta\Gamma \sim \Gamma_+ \frac{\Gamma_-^2}{\Gamma_+^2 + \Gamma_-^2} \cdot \frac{v_0}{\gamma_0} \frac{\Gamma_+}{\Gamma_-}. \tag{6.61}$$

This change may be taken as an additional broadening of the resonance. The tentative estimation of the magnitude may be made by using the data of the measurement of the collisional broadening of the 1.15 μm line. Assuming that the contribution of the phase-perturbing collisions amounts to the maximum possible magnitude of 8 MHz/Torr, we obtain from (6.61) an upper estimation for the magnitude of the additional resonance broadening of 2.5 MHz/Torr. This magnitude is within the limits of an experimental accuracy; therefore, the influence of collisions with the phase perturbation will be neglected. From the broadening of the 1.52 μm and the 1.15 μm lines we obtain a magnitude of 5 ± 2.5 MHz/Torr for the broadening of the forbidden transition.

c) Relaxation of Individual Levels

As we have already pointed out, one can determine the frequency of quenching and strong collisions on the level 0 by the difference of $\Gamma_+ - \Gamma_-$. The broadening of the common level is 6.5 ± 3 MHz/Torr. The difference between the broadenings of the 1.15 μm and 1.52 μm lines and those of the $2s_2$ level give the broadenings of the $2p_1$ and $2p_4$ levels to be 1.5 MHz/Torr and 3.6 MHz/Torr, respectively. Their sum gives the broadening of the forbidden transition obtained above. We note that the obtained values of the broadening of the $2p_1$ and $2p_4$ levels agree with the data obtained from the measurements of relaxation of the magnetic moment (the measurements were made for the $2p_9$ level) by collisions [6.42].

6.7 Resonant Raman Gas Lasers

The narrow gain lines which arise on the coupled transitions can be used for production of lasers. The gain width in such lasers may be much narrower than the Doppler linewidth. The saturation behaviour of such lasers may differ from that of ordinary lasers. No wonder therefore that Raman lasers have emission features which differ from those of ordinary lasers.

The difference between the widths and the values of gain as a function of the direction of light propagation may be used for production of unidirectional amplifiers and optical valves [6.15, 26]. The production of tunable single-frequency dye lasers opens up new possibilities of transformation of radiation in a gas at near resonant SRS (see Ref. [6.43]).

The first gas laser with the narrow gain line which was obtained in a three-level scheme was realized on the $2s_2 - 2p_4$ and $3s_2 - 2p_4$ transitions of neon [6.44]. The optical scheme of such a laser is similar to that described above (see Fig. 6.13). The pumping at a wavelength of 0.63 μm reduces the population of atoms in the narrow velocity interval on the $2p_4$ level. As a result a peak arises in the population difference between the $2s_2$ and $2p_4$ levels in the velocity distribution of atoms; and the maximum of gain arises due to this peak. Since the lifetime of the common level $2p_4$ is longer than the lifetimes of the $2s_2$ and $3s_2$ levels, the gain line shape is defined by the population effect, and the anisotropic features of the gain line are produced only slightly. As is known, the standing wave in a Fabry-Perot resonator can be represented by a superposition of two oppositely moving waves. Therefore, at the detuning of the pumping frequency Ω from the centre of the line, the effective gain line

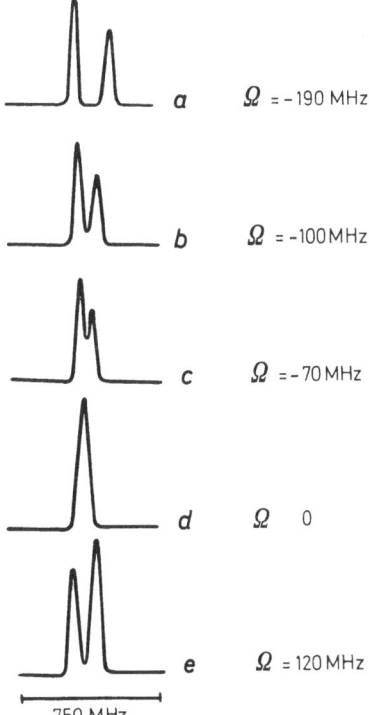

a $\Omega = -190\,\text{MHz}$

b $\Omega = -100\,\text{MHz}$

c $\Omega = -70\,\text{MHz}$

d Ω 0

e $\Omega = 120\,\text{MHz}$

|← 750 MHz →|

Fig. 6.17a–e. Dependence of the output power on frequency at $\lambda = 1.15\,\mu\text{m}$ at different frequency detunings Ω of pumping at $\lambda = 0.63\,\mu\text{m}$

has two sharp maxima with different intensities and widths at frequencies $\omega' = \omega_{02} \pm \dfrac{\Omega k'}{k}$. The gain line in the standing wave field will be the sum of contours, α_+ and α_-, corresponding to the interaction of two waves.

Figure 6.17 shows the dependence of the output power of the $\lambda = 1.15\,\mu\text{m}$ laser under pumping by radiation at $\lambda = 0.63\,\mu\text{m}$ at different frequency detunings from the centre of the 0.63 μm line. The dependence of output power on detuning is asymmetrical. It is connected with anisotropy of the gain line.

The lasers in which the two-quantum process plays a main role are of great interest by their features. Such resonant Raman lasers have features interesting for spectroscopy. The three- and four-level schemes can be of interest for the production of short-wave radiation up to the γ-ray region [6.45].

The resonant Raman laser worked on the $2s_2 - 2p_4$ and $2s_2 - 2p_1$ transitions of neon. It was briefly described in Section 6.4. Using special conditions, one can get absorption to be observed on the $\lambda = 1.15\,\mu\text{m}$ line in the absence of pumping at $\lambda = 1.52\,\mu\text{m}$. Under pumping at

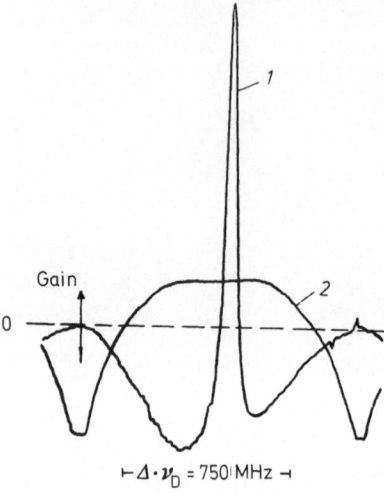

Gain

0

⊢ $\Delta \cdot \nu_D$ = 750 MHz ⊣

Fig. 6.18. Curve *1* is line shape of signal gain and absorption at $\lambda = 1.15\,\mu m$ under the pumping at $\lambda = 1.52\,\mu m$. Curve *2* shows absorption at $\lambda = 1.15\,\mu m$ in the absence of the pumping at $\lambda = 1.52\,\mu m$

$\lambda = 1.52\,\mu m$, the gain at $\lambda = 1.15\,\mu m$ arises. The gain line shape for this case is shown in Fig. 6.18 [6.15, 46]. The gain line broadening and anisotropy show the features which differ from those of ordinary lasers.

Figure 6.14a shows the characteristic line shape of the three-level gas laser at $\lambda = 1.15\,\mu m$ obtained at different fixed detunings of the $\lambda = 1.52\,\mu m$ laser with respect to the centre of the line. It is easy to notice that the line shape considerably differs from those which are observed in gas lasers and are caused by the interaction of the electromagnetic standing wave field with the inhomogeneously broadened Doppler line of gain. The gain line width depends on the excess of the gain over the threshold.

The output amplitude reduction and the broadening are first observed at the frequency detuning of the pump from the centre of the absorption line. Then the peak bifurcates, and a curve of the output power at $\lambda = 1.15\,\mu m$ as a function of frequency is asymmetrical. The sign of asymmetry is defined by the sign of detuning Ω. When $\Omega > 0$, the peak with a large amplitude is observed on the high-frequency side of the line, and for $\Omega < 0$, on the low-frequency side.

As a new phenomenon, a sharp dip of a small amplitude with a width of about 10 MHz (Fig. 6.19) arises by increasing the pumping intensity near the centre of the line. The theoretical and experimental investigations of this phenomenon [6.19] are of interest for spectroscopy. This resonance arises when the difference between the frequencies of incident and scattered radiation is equal to that of the forbidden transition. In contrast to the two-quantum resonance it arises in higher orders of the perturbation theory. It may be considered as a saturation effect

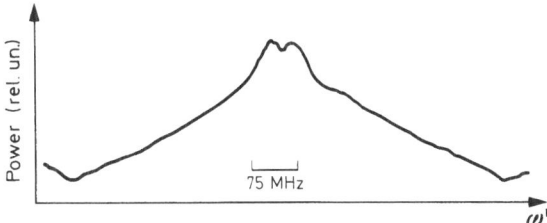

Fig. 6.19. Dip of the output power in a Raman laser at $\lambda = 1.15\,\mu m$ pumped by a laser at $\lambda = 1.52\,\mu m$

in multiphoton transitions. The resonance dip can arise when the resonant SRS line undergoes inhomogeneous broadening due to the motion of atoms. Following Ref. [6.19], we shall give an explanation of the obtained result.

The transition probability of an atom from the level 1 to the level 2 is given by the expression of (6.23). In the case of the standing wave field on the $0 \rightarrow 2$ transition for the transition probability of the atom having the velocity projection v, we have

$$W_{1\to 2}(v) = W_{1\to 2}^{+} + W_{1\to 2}^{-} . \tag{6.62}$$

Later on we shall need the case $\gamma_0 \gg \gamma_1, \gamma_2$, for which (6.62) is of the form

$$W_{1\to 2}(v) = \frac{2|G|^2 |G'|^2}{(\gamma_0/r)^2 + (\Omega - kv)^2} \cdot \frac{1}{\gamma_1}$$
$$\cdot \left[\frac{\gamma_{12}}{\gamma_{12}^2 + (\Omega' - k'v - \Omega + kv)^2} + \frac{\gamma_{12}}{\gamma_{12}^2 + (\Omega' - \Omega + k'v - kv)^2} \right] . \tag{6.63}$$

When the Doppler width is much broader than γ_0, the ensemble of atoms with Maxwell velocity distribution emits with a Lorentzian line shape, depending on the direction of observation: forward or backward.

When $\gamma_0 \gg \gamma_1$ and $k' - k \sim k$, the inhomogeneous width of the SRS line is much broader than the homogeneous width of the two-quantum process. According to (6.63), the contribution to absorption of the standing wave on the $0 \rightarrow 2$ transition is provided by the atoms moving with the velocities near

$$v_{\pm} = \frac{\Omega' - \Omega}{\pm k' - k},$$

with the width

$$\Delta v_\pm = \frac{\gamma_{12}}{|k' \mp k|} .$$

The inhomogeneous broadening must lead to an inhomogeneous character of saturation with increasing field intensities.

When $\Omega' - \Omega = 0$, both waves interact with the same atom. As a result, saturation in multi-quantum transitions is larger, and the resonance dip with the width of the forbidden transition arises in the saturated SRS line in the standing wave field when $\omega' - \omega = \omega_{12}$.

In addition to the method of two-quantum resonance it can be used for the precise frequency measurement of the forbidden transitions and so on.

Acknowledgements. "I would like to thank Prof. K. Shimoda for valuable remarks, Dr. Ye. Baklanov and Dr. Ye. Titov for discussions and help".

References

6.1 V. S. Letokhov, V. P. Chebotayev: *Principles of Non-linear Laser Spectroscopy* (Izd. "Nauka" 1975)

6.2 K. Shimoda: In *Laser Spectroscopy*, ed. by R. G. Brewer and A. Mooradian (Plenum Press, New York 1974)

6.3 B. Decomps, M. Dumont, M. Ducloy: In *Topics in Applied Physics*, Vol. 2, ed. by Walther (Springer Berlin, Heidelberg, New York 1976) pp. 283—347

6.4 W. R. Bennett, Jr.: Phys. Rev. **126**, 580 (1962)

6.5 W. E. Lamb, Jr.: Phys. Rev. **134**A, 1429 (1964)

6.6 W. R. Bennett, Jr., V. P. Chebotayev, J. W. Knunson: Proc. of 5th Internat. Conf. Phys. Electronic and Atomic Collisions, Leningrad, 1967

6.7 R. H. Cordover, R. A. Bonczyk, A. Javan: Phys. Rev. Lett. **18**, 730 (1967)

6.8 W. E. Schweitzer, Jr., M. M. Birky, J. A. White: J. Opt. Soc. Am. **57**, 1226 (1967)

6.9 H. R. Schlossberg, A. Javan: Phys. Rev. **150**, 267 (1966)

6.10 G. E. Notkin, S. G. Rautian, A. A. Feoktistov: JETP **52**, 1673 (1967)

6.11 H. K. Holt: Phys. Rev. Lett. **19**, 1275 (1967)

6.12 V. B. Berestetsky, Ye. M. Lifshitz, L. P. Pitayevsky: Relativistic Quantum Theory, Ch. I, Nauka, 1968 (in Russian)

6.13 I. M. Beterov, Yu. A. Matyugin, V. P. Chebotayev: JETP **64**, 1495 (1973)

6.14 T. W. Hänsch, P. Toschek: Zeitschr. f. Phys. **236**, 213 (1970)

6.15 I. M. Beterov, V. P. Chebotayev: Three Level Systems and their Interaction with Radiation. In *Progress in Quantum Electronics*, Vol. 3, part 1 (Pergamon Press 1974)

6.16 M. S. Feld, A. Javan: Phys. Rev. **177**, 540 (1969)

6.17 N. Skribanowitz, M. S. Feld, R. E. Francke, M. J. Kelley, A. Javan: Appl. Phys. Lett. **19**, 161 (1971)

6.18 T. Ya. Popova, A. K. Popov, S. G. Rautian, R. I. Sokolovsky: JETP **57**, 850 (1969)

6.19 YE. V. BAKLANOV, I. M. BETEROV, B. YA. DUBETSKY, V. P. CHEBOTAYEV: Report on IV Vavilov Conference on Non-linear Optics, Novosibirsk, June 12–14, 1975

6.20 L. S. VASILENKO, V. P. CHEBOTAYEV, A. V. SHISHAYEV: JETP Lett. **12**, 161 (1970)

6.21 Physics Today **17** (July 1974)

6.22 K. SHIMODA: Appl. Phys. **9**, 239 (1976)

6.23 T. HÄNSCH, R. KEIL, A. SHABERT, C. H. SCHMELZER, P. TOSCHEK: Zeit. f. Phys. **226**, 293 (1969)

6.24 A. K. POPOV: JETP **58**, 1623 (1970)

6.25 B. J. FELDMAN, M. S. FELD: Phys. Rev. A**5**, 899 (1972)

6.26 N. SKRIBANOWITZ, M. J. KELLEY, M. S. FELD: Phys. Rev. A**6**, 2302 (1972)

6.27 S. H. AUTLER, C. H. TOWNES: Phys. Rev. **100**, 703 (1955)

6.28 YE. V. BAKLANOV, V. P. CHEBOTAYEV: JETP **61**, 922 (1971)

6.29 H. K. HOLT: Phys. Rev. Lett. **20**, 410 (1968)

6.30 I. M. BETEROV, V. P. CHEBOTAYEV: JETP Lett. **9**, 216 (1969)

6.31 W. R. BENNETT, JR., P. J. KINDLMANN: Phys. Rev. **149**, 38 (1966)

6.32 YU. A. MATYUGIN, A. S. PROVOROV, V. P. CHEBOTAYEV: JETP **63**, 2043 (1971)

6.33 G. V. LAWRENCE, H. S. LISZT: Phys. Rev. **269**, 340 (1968)

6.34 I. M. BETEROV, YU. A. MATYUGIN, V. P. CHEBOTAYEV: Preprint No. 21, Inst. of Semicond. Physics, Siberian Branch, USSR Acad. of Sci., 1971

6.35 W. CULSHAW, J. KANNELAUD: Phys. Rev. **133** A, 691 (1964)

6.36 T. O. CARROL, C. J. WOLGA: IEEE J. QE-2, 456 (1966)

6.37 A. SZÖKE, A. JAVAN: Phys. Rev. Letters **10**, 521 (1963)

6.38 A. SZÖKE, A. JAVAN: Phys. Rev. **145**, 137 (1966)

6.39 I. M. BETEROV, YU. A. MATYUGIN, S. G. RAUTIAN, V. P. CHEBOTAYEV: JETP **58**, 1243 (1970)

6.40 YE. V. BAKLANOV: JETP **65**, 2203 (1973)

6.41 V. N. LISITSYN, V. P. CHEBOTAYEV: Opt. i Spectr. **26**, 856 (1969)

6.42 C. G. CARRINGTON, A. CORNEY: Opt. Commun. **1**, 115 (1970)

6.43 P. P. SOROKIN, in: *Laser Spectroscopy* Conf., Megeve, France, June 1975, ed. S. HAROCHE (Springer Berlin, Heidelberg, New York)

6.44 I. M. BETEROV, V. P. CHEBOTAYEV: JETP Lett. **9**, 216 (1969)

6.45 YE. V. BAKLANOV, V. P. CHEBOTAYEV: JETP Lett. **21**, 286 (1975)

6.46 I. M. BETEROV, YU. A. MATYUGIN, V. P. CHEBOTAYEV: Opt. i Spectr. **28**, 357 (1970); JETP Lett. **10**, 296 (1969)

7. Quantum Beats and Time-Resolved Fluorescence Spectroscopy

S. HAROCHE

With 25 Figures

There are many ways lasers may be used to perform high-resolution spectroscopy. Various techniques, which are described in other chapters of this book, rely on different specific properties of laser fields, such as monochromaticity, temporal coherence and intensity. The very simple method we deal with in this chapter consists in making use of the pulsed character of the light excitation provided by some lasers in order to induce a transient fluorescence signal whose time-resolved analysis provides spectroscopic information about the atomic system of interest. The most evident information one gets from such types of experiment is of course the lifetime of the excited levels. But, certainly, the most interesting and dramatic feature of the fluoresence transient following a pulse excitation is the possibility of observing a time modulation in the light signal. This modulation arises from the fact that the light pulse may prepare the atomic system in a coherent superposition of excited substates of slightly different energies. Provided the fluorescence light is detected with a proper polarization, the evolution of this atomic excited state coherence may reveal itself by oscillations in the fluorescence light at the Bohr frequencies corresponding to the excited state splittings. These oscillations are called quantum beats since they may be interpreted as arising from a quantum interference effect between the different fluorescence channels originating from the various excited states coherently prepared by the light pulse. The analysis of the beats provides, as we will see, a convenient way of determining with a very high resolution small energy structures in excited atomic or molecular states. In particular, the observed beat frequencies are, as we will show, practically not subjected to a Doppler spread, which is a big asset for spectroscopic applications in the gas phase.

The quantum beat effect is essentially of the same nature as other various atomic coherence effects in fluorescence experiments which have been developed as very useful tools in nuclear, atomic and molecular spectroscopy many years ago. The Hanle effect [7.1], level crossing [7.2], modulated pumping fluorescence experiments [7.3, 4], and photon correlation experiments in γ-ray cascades [7.5] are all techniques which consist of exciting in some way the atomic system in a coherent superposition of substates and of detecting the evolution of this coherence as a change in the polarization properties of the fluorescence light.

Among all these effects, which are reviewed in several articles [7.5–7], the quantum beat phenomenon is no doubt conceptually the simplest and the most basic one, since all the above-mentioned effects may be derived from it in a straightforward manner. It may thus seem at first sight strange that quantum beats have been explicitly discussed and experimentally demonstrated after the other more sophisticated steady state atomic coherence fluorescence effects mentioned above. Although the grounds for the theory of quantum beats may be found in papers dating from the early days of quantum mechanics such as that of BREIT [7.8], it was not until the late fifties and early sixties that some theoretical papers have explicitly considered the possibility of getting light modulations in the spontaneous emission following a light pulse and that a detailed analysis of the phenomenon was made [7.3, 9]. This paradoxical situation has obviously technical reasons, for until recently it had been very difficult to detect fast modulation signals and also to produce the short and intense pulses needed for an efficient excitation of the atomic system. It was only when technology had made sufficient progress to overcome these difficulties that the quantum beat effect emerged from what could be considered a long lethargy. About ten years ago, various quantum beat experiments using as light sources shuttered spectral lamps were performed by ALEXANDROV [7.10] and by DODD et al. [7.11, 12]. Figure 7.1 shows a typical beat signal observed in the resonance fluorescence of mercury [7.12]. The beat arises from the Zeeman splitting of the emitting level in a magnetic field of a few Gauss. Such experiments were rather difficult to perform, since the pulses had a very low peak power, so that long averaging times were necessary to extract the signal from the noise. They were performed more as demonstration experiments than as true spectroscopic investigations. A big improvement in this kind of experiments was made when the weak spectral lamp optical excitation was replaced by the more effective excitation provided by short collision processes. After HADEISHI and NIERENBERG [7.13] showed that quantum beats could be induced by the impulsive excitation produced by electron bombardment, beam foil techniques were used to produce quantum beats after passage of a fast ion beam through a thin carbon foil [7.14]. This technique has led to many interesting spectroscopic measurements, but is obviously restricted to some species and suffers from a lack of excitation selectivity.

The advent of tunable pulsed lasers a few years ago renewed the interest in optically induced quantum beat experiments. These lasers allow an efficient and selective impulsive excitation of optical transitions in the visible, near UV and infrared ranges, so that quantum beat ex-

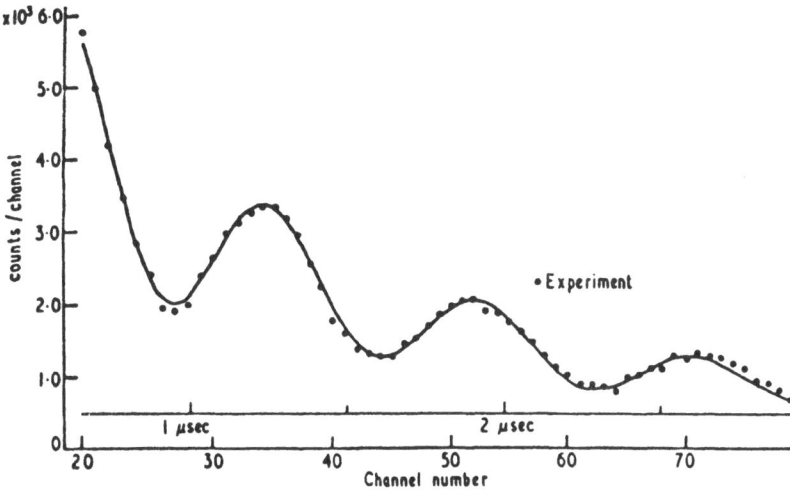

Fig. 7.1. Zeeman quantum beats observed in the fluorescence of the 6^3P_1 level of mercury after excitation by a shuttered spectral lamp (from [7.12])

periments have begun to develop into a very useful tool in high-resolution Doppler-free laser spectroscopy. A very attractive feature of this technique is the fact that it does not require very monochromatic laser excitation, in contrast with other high-resolution techniques such as saturation or two-photon spectroscopy. This advantage is offset by the fact that the technique is obviously limited to the detection of small frequency energy splittings and cannot in particular be used for the absolute determination of optical frequency intervals.

We intend in this chapter to review both the theoretical and experimental aspects of quantum beat experiments. As this book is devoted to laser spectroscopy techniques, we will restrict our interest to light-pulse-induced, and more specifically laser-pulse-induced quantum beat studies, excluding beam foil type of experiments. Of course, a very large part of the theoretical considerations about quantum beats is common to both laser and beam foil-induced beats. However, beam foil experiments pose specific problems concerning the description of the collision excitation processes [7.15], which are quite different from and much less well known than the light excitation ones. On the other hand, the technology of these two kinds of experiments is quite different. Specific information about beam foil-induced quantum beats may be found in articles [7.14] and review papers [7.16]. Our purpose in this chapter is of a pedagogical nature. On the one hand, we hope that theoretical considerations about quantum beat interpretation and calculation will be of some use to experimental physicists. On the other hand, we wish that the brief description of recently performed experiments will be of interest to those more theoretically oriented who would nevertheless like to know how basic effects in atomic fluorescence may be actually applied to laser spectroscopy.

In Section 7.1, we will describe quantum beats as a quantum interference effect in atomic fluorescence by making use of the quantum electrodynamics formalism. This basic approach will allow us to understand the main features of the quantum beat signal. We will then (Section 7.2) show how this signal may be quite generally calculated with the help of the density matrix formalism. In Section 7.3, we will compare the quantum beat technique with other types of atomic fluorescence experiments such as level crossing and double resonance techniques which have also been renewed by the advent of tunable laser sources. The last three sections will be devoted to the technology of laser-induced quantum beat spectroscopy and to the description of some spectroscopic measurements made by using this technique.

Some comments concerning our bibliography are needed at the end of this introduction. We have not tried to be complete in our references to previous works. As far as theory is concerned, it is obviously difficult to quote all the papers dealing with the subject, especially since there are so many connections between quantum beats and other atomic coherence phenomena which have been the object of intensive studies in so many fluorescence and optical pumping works. We have quoted only what we think to be basic articles or textbooks relevant to the subject. As for experiments, we have tried to be exhaustive in the quotation of all the laser quantum beat experiments performed up to now. We have not attempted to be complete in the quotation of the lifetime measurements using the pulsed laser technique, since these experiments are much more numerous and have been performed in various fields of atomic and molecular spectroscopy.

7.1 The Quantum Beat Phenomenon: A Quantum Interference Effect in Atomic Fluorescence

Quantum beats may be observed in the intensity of the light emitted by impulsively excited atoms when the fluorescence may occur through several undistinguishable channels. It is in fact a very good example of a quantum interference effect which may be thoroughly understood with simple physical arguments based on the "quantum theory of measurement". As this effect involves spontaneous emission, it is preferable to discuss it within the frame of Quantum Electrodynamics Theory (QED).

In this section, we intend to present a simple QED treatment of this phenomenon which will allow us to understand its main characteristics and to analyze it in several experimental situations. We will deal successively with quantum beats arising from a single-atom system (Subsection 7.1.1) and with those which are collectively emitted by a many-atom system (Subsection 7.1.2). At the end (Subsection 7.1.3), we will point out the differences between these two kinds of beats and emphasize the advantages of single-atom quantum beats for sub-Doppler spectroscopy of atoms or molecules in low-pressure gaseous samples.

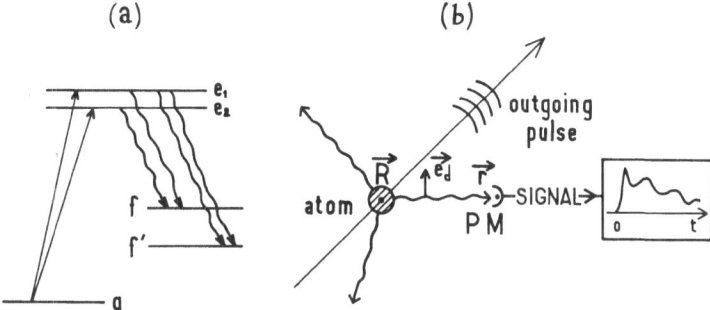

Fig. 7.2a and b. Single-atom quantum beats. (a) Energy diagram showing the transitions involved in the preparation of the excited state coherences (straight line arrows) and in the subsequent fluorescence (wavy lines). (b) Symbolic sketch of the experiment showing the emitting atom located at point R and the detector at point r

7.1.1 Quantum Beats in a Single-Atom System

Let us consider an atom with the energy diagram sketched in Fig. 7.2a. A typical Quantum Beat experiment involving this atom is symbolized in Fig. 7.2b. The atom, located at point R and initially in its ground state $|g\rangle$, is suddenly excited at time $t=0$ by a short pulse of broadband light irradiation into a coherent superposition of excited states $|e_i\rangle$ having slightly different energies E_i. Immediately after the excitation, the atom starts emitting light by spontaneous emission (wavy lines on Fig. 7.2b) and decays back to various final states $|f\rangle$, $|f'\rangle$ whose energies are $E_f, E_{f'} \ldots$ (among these states one finds of course the initial state $|g\rangle$). With a broadband filter and a polarizer one selects the light components emitted with a given polarization e_d to a set of final states $|f\rangle$. The light thus filtered is detected by a photomultiplier P.M. located at point r. The signal $S(e_d, r, t)$ detected in such an experiment may be readily derived from QED.

a) QED Derivation of the Quantum Beat Signal for a Single-Atom System

We assume that at time $t = 0$, the system is prepared by the light pulse in the state

$$|\psi(0)\rangle = \sum_i \alpha_i |e_i, 0\rangle \quad (i = 1, 2). \tag{7.1}$$

In this expression $|e_i, 0\rangle$ represents the atom in substate $|e_i\rangle$ with no photon present. The α_i are the probability amplitudes that the light pulse has prepared the atom in state $|e_i\rangle$. These amplitudes depend of course on the characteristics of the pulse used to excite the atoms. We will suppose that they are known quantities throughout this section. At time t, $|\psi(0)\rangle$ has evolved into

$$|\psi(t)\rangle = \sum_i \alpha_i e^{-iE_it/\hbar} e^{-\Gamma t/2} |e_i, 0\rangle + \sum_{f, k\varepsilon} C_{f, k\varepsilon}(t) |f, k\varepsilon\rangle. \tag{7.2}$$

The initial states $|e_i, 0\rangle$ have been damped at the rate Γ of spontaneous emission ($\tau = 1/\Gamma$ is the common radiative lifetime of all the e_i substates). $C_{f k\varepsilon}(t)$ is the probability amplitude to find at time t the atom in the final state f with a photon of wave vector k and polarization ε. This amplitude may be readily obtained from the Wigner-Weisskopf theory of spontaneous emission. One finds

$$C_{f k\varepsilon}(t) = \sum_i C_{f k\varepsilon}^{(i)}(t) \tag{7.3}$$

with

$$C_{f k\varepsilon}^{(i)}(t) = \alpha_i \mathscr{E}_k \langle f | \varepsilon \cdot \boldsymbol{D} | e_i \rangle e^{-ik \cdot \boldsymbol{R}} \frac{e^{-i(E_f + \hbar ck)t/\hbar} - e^{-iE_it/\hbar} e^{-\Gamma t/2}}{\hbar ck - (E_i - E_f) + i\hbar\Gamma/2}. \tag{7.4}$$

In this expression $\mathscr{E}_k$ is the electric field of a photon at frequency $\hbar ck$ and $\boldsymbol{D}$ is the electric dipole operator of the atom. The probability amplitude to find a photon $|k\varepsilon\rangle$ appears as the sum of two terms, each corresponding to the emission from a given excited state $|e_i\rangle (i = 1, 2)$. Each of these terms exhibits a resonance centered around $\hbar ck = E_i - E_f$ with a widht $\hbar\Gamma$, in agreement with both energy conservation and Heisenberg uncertainties relations. At resonance, each amplitude $C_{f k\varepsilon}^{(i)}(t)$ is modulated at the Bohr frequency $E_i/\hbar$ of the corresponding excited state.

The average photon counting rate of the detector located at point $\boldsymbol{r}$ is equal to the expectation value at that point of the operator $\mathscr{E}_d^-(\boldsymbol{r}) \mathscr{E}_d^+(\boldsymbol{r})$ representing the product of the positive and negative frequency parts of the electric field component along the direction e_d

$$S(e_d, \boldsymbol{r}, t) = \langle \psi(t) | \mathscr{E}_d^-(\boldsymbol{r}) \mathscr{E}_d^+(\boldsymbol{r}) | \psi(t) \rangle. \tag{7.5}$$

Expanding $\mathscr{E}^+(r)$ and $\mathscr{E}^-(r)$ in terms of the normal modes of the electromagnetic field

$$\mathscr{E}_d^+(r) = \sum_{k\varepsilon} \mathscr{E}_k \varepsilon_d a_{k\varepsilon} e^{ik \cdot r}$$
$$\mathscr{E}_d^-(r) = \sum_{k'\varepsilon'} \mathscr{E}_{k'} \varepsilon_d a_{k'\varepsilon'}^+ e^{-ik' \cdot r} \tag{7.6}$$

($a_{k\varepsilon}$ and $a_{k\varepsilon}^+$ are the annihilation and creation operators in mode $k\varepsilon$), one obtains the following expression for the signal:

$$S(e_d, r, t) = \sum_{\substack{k, k' \\ \varepsilon, \varepsilon'}} \sum_f \sum_{i,j} \mathscr{E}_k \mathscr{E}_{k'} C_{fk\varepsilon}^{(i)}(t) C_{fk'\varepsilon'}^{(j)*}(t) \varepsilon_d \varepsilon_d' e^{i(k-k') \cdot r} . \tag{7.7}$$

Replacing in (7.7) the $C_{fk\varepsilon}^{(i)}(t)$ by their expression given by (7.4) and carrying out all the angular and energy summations, one finally gets the signal

$$S(e_d, r, t) = \frac{1}{(4\pi\varepsilon_0)^2} \frac{k_0^4}{r_0^2} \sum_f \sum_{ij} \langle f|e_d D|e_i\rangle \alpha_i \alpha_j^* \langle e_j|e_d D|f\rangle$$

$$\cdot \theta\left(t - \frac{r_0}{c}\right) e^{-i\omega_{ij}\left(t - \frac{r_0}{c}\right)} e^{-\left(t - \frac{r_0}{c}\right)} . \tag{7.8}$$

In this expression, $r_0 = |R - r|$ is the distance between the atom and the detector, $k_0 = (E_e - E_f)/\hbar c$ is the average wavenumber of the detected optical transition, and $\omega_{ij} = (E_i - E_j)/\hbar$ is the Bohr frequency corresponding to the splitting between states e_i and e_j. $\theta(t - r_0/c)$ is the ordinary Heaviside function, equal to 1 if $t > r_0/c$ and to 0 if $t < r_0/c$, which allows for the propagation between the emitter and the detector. The summation over f in (7.8) is restricted to the set of final states selected by the broadband filter placed in front of the detector. Depending on the experiment, this summation may or may not include the initial ground state $|g\rangle$.

The expression of the signal given by (7.8) is a quite general one and applies to a situation where there is an arbitrary number of emitting excited states $|e_i\rangle$. It clearly describes an exponential curve modulated by sinusoidal functions with periodicity $2\pi/\omega_{ij}$. We shall try now to provide some physical insight concerning these modulations.

b) Physical Interpretation of the Quantum Beat Signal

In the experiment sketched in Fig. 7.2, the impinging light pulse is scattered by the atom through two different channels, corresponding each to a given excited state $|e_i\rangle$. These channels are symbolized by the diagrams of Fig. 7.3. When the atom has re-emitted a photon, there is no way to distinguish between the two possible channels and to tell whether the photon has been scattered through level $|e_1\rangle$ (with an amplitude $C^1_{fk\varepsilon}$) or through level $|e_2\rangle$ (with an amplitude $C^2_{fk'\varepsilon'}$). As a general postulate of quantum mechanics, the amplitude corresponding to these two indistinguishable processes must be added and their sum must be squared to yield the expression of the signal. This leads to interfering term such as $C^{(i)}_{fk\varepsilon} C^{(j)}_{fk'\varepsilon'}$ in the expression (7.7) of $S(e_d, r, t)$. After integration over the photon energy, these interfering terms produce the modulations in the observed signal. As $C^{(i)}_{fk\varepsilon}$ and $C^{(j)}_{fk'\varepsilon'}$ are at resonance modulated, respectively, at frequencies $E_i/\hbar$ and $E_j/\hbar$, their cross product does in-

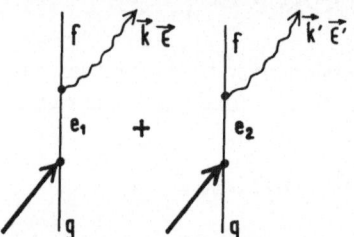

Fig. 7.3. Diagrams symbolizing two interfering amplitudes in a single-atom quantum beat experiment. The detected photon (wavy line) may have been emitted either by excited state e_1 or by excited state e_2 (the light excitation process is symbolized by a solid line arrow)

deed exhibit a modulation at frequency $\omega_{ij} = (E_i - E_j)/\hbar$. Quantum beats are thus a typical quantum interference effect quite similar to the Young's double-slit experiment. In complete analogy with this latter problem, it is quite clear that any attempt to perform an experiment in order to determine through which channel the photon has been scattered will result in a disappearance of the beat pattern. For example, we may try to make use of the polarization selection rules and detect the light emitted with a polarization e_d which can be emitted by only one level ($|e_1\rangle$, for example). In that case however, the matrix element $\langle e_2|e_d D|f\rangle$ is equal to zero and there are obviously no beats in (7.8). Another way of determining the emission channel would be to put in front of the detector a narrow band filter centered, for example, around the $(E_1 - E_f)/\hbar$ frequency. As a result of this filtering, the summation over k and k' in (7.7) is now restricted to a small frequency interval excluding the $(E_2 - E_f)/\hbar$ frequency. In that interval the $C^{(2)}_{fk\varepsilon}(t)$ amplitude remains very small, and the beats are again lost.

The above qualitative analysis also explains another very important feature of single-atom quantum beats. These beats can arise only from an excited state splitting and will never reveal the structure of the lower atomic states. This result can be read directly from the expression for the signal and arises from the fact that $S(e_d, r, t)$ appears as the sum of the probabilities of finding the atom in the various possible final states (and not as the square of the sum of the corresponding amplitudes). This is again a straightforward consequence of the quantum theory of measurement. The scattering channels ending in two different states $|f\rangle$ and $|f'\rangle$ are not indistinguishable in the sense that nothing forbids one, at least theoretically, to observe the atom long after the emission of the photon and to determine whether it is left in state $|f\rangle$ or $|f'\rangle$. Thus the corresponding amplitudes cannot interfere, and no beats can appear in that case at the lower states Bohr frequencies.

7.1.2 Quantum Beats in a Many-Atom System

We have so far analyzed a situation in which only one atom is involved in the scattering of the impinging light pulse. In an actual experimental situation, there is of course a large number of atoms in the sample, and one should thus consider physical situations in which the photon may have been "collectively" scattered by several atoms. We will see in this subsection that new types of quantum beats are actually observed when the scattered photon may have been emitted by many indistinguishable atoms, and we will analyze the main characteristics of these collective beats whose properties are completely different from those arising from a single-atom system.

For the sake of simplicity, we will restrict our attention to a two-atom system and discuss only qualitatively the various possible situations. For a more quantitative analysis, one could read the theoretical paper by CHOW et al. [7.17] in which the many-atom quantum beat signal is studied in detail.

a) Cooperative Quantum Beats Arising from a Splitting in the Atomic Upper State

Let us consider two identical atoms A and B, each of them having the energy diagram of Fig. 7.2a (small energy splitting in the upper state). These atoms are located at points R_A and R_B, respectively. The atomic system, which is initially in state $|gg\rangle$ representing both atoms in their ground state $|g\rangle$, is suddenly excited by a short broadband light pulse which prepares the system in a coherent superposition of excited states belonging to both atoms. Immediately after the light excitation, the system is assumed to be left in the state

$$|\psi(0)\rangle = \sum_i \alpha_{A_i}|e_i, g; 0\rangle + \sum_j \alpha_{B_j}|g, e_j; 0\rangle . \tag{7.9}$$

In this expression α_{A_i} and α_{B_j} are, respectively, the probability amplitudes that the pulse has excited atom A in state $|e_i\rangle$ or atom B in state $|e_j\rangle$.

The calculation of the signal emitted by the system at time $t > 0$ in a given direction with a given polarization follows the same lines as in Subsection 7.1.1. At time t, the system has evolved into

$$|\psi(t)\rangle = \sum_i \alpha_{A_i} e^{-iE_it/\hbar} e^{-\Gamma t/2} e_i, g; 0\rangle + \sum_j \alpha_{B_j} e^{-iE_jt/\hbar} e^{-\Gamma t/2} |g, e_j; 0\rangle$$
$$+ \sum_f \sum_{ij} (\sum_{k\varepsilon} C^{A_i}_{fk\varepsilon}(t)) |f, g; k\varepsilon\rangle$$
$$+ \sum_{k'\varepsilon'} C^{B}_{fk'\varepsilon'}(t) |g, f; k'\varepsilon'\rangle). \qquad (7.10)$$

In this formula, the $C^{A_i}_{fk\varepsilon}$ and $C^{B_j}_{fk\varepsilon}$ are given by (7.4), in which we have replaced R by R_A or R_B and α_i by α_{A_i} or α_{B_j}. They represent the partial amplitudes that the photon has been scattered from atom A in state $|e_i\rangle$ or from atom B in state $|e_j\rangle$. We should at this stage consider separately the case of inelastic and elastic scattering.

Suppose at first that one detects the photon emitted in a transition back to a final state $|f\rangle$ different from $|g\rangle$ (inelastic scattering). It is clear in that case that the amplitudes $C^{A_i}_{fk\varepsilon}$ and $C^{B_j}_{fk'\varepsilon'}$ correspond to distinguishable channels, since they end up in different atomic states, namely $|fg\rangle$ and $|gf\rangle$. As a result, these amplitudes cannot interfere, and the only interfering terms in the signal will be of the form $C^{A_i}_{fk\varepsilon} C^{A_i}_{fk'\varepsilon'}$ or $C^{B_i}_{fk\varepsilon} C^{B_i}_{fk'\varepsilon'}$. Only single-atom beats will be observable in this inelastic scattering process, and no collective effects are to be expected.

Suppose, on the other hand, that one observes the light emitted in the $|e\rangle \leftrightarrow |g\rangle$ transition back to the initial state $|g\rangle$ (elastic scattering). In that case, both amplitudes $C^{A_i}_{gk\varepsilon}$ and $C^{B_j}_{gk'\varepsilon'}$ correspond to the same $|g, g\rangle$ atomic final state and the two corresponding channels are now indistinguishable. These channels are represented by the two diagrams of Fig. 7.4. As a result, interference terms such as $C^{A_i}_{gk\varepsilon} C^{B_j*}_{gk'\varepsilon'}$ now appear in the signal. These interatomic interference terms give rise to a beat pattern at frequency ω_{ij} which adds to the single-atom beat signal. The two-atom situation discussed here can obviously be extended to an n-atom system [7.17]. The fundamental difference between the elastic and inelastic scattering cases, which did not exist for the single-atom quantum beat phenomenon, is quite similar to the one encountered in other kinds of collective scattering experiments (e.g., neutron scattering). In the elastic process, the atom which scatters the photon comes back to the initial state and cannot be distinguished from the other ones. The emission amplitudes from the different atoms do interfere with each other, and interatomic beats may occur. In the inelastic case, on the contrary, the scattering atom ends up in a different state and could, at least theoretically, be traced in the atomic sample. One must, in that case, add up the field intensities emitted by different atoms (and not the field amplitudes), so that collective interference effects are no longer observable.

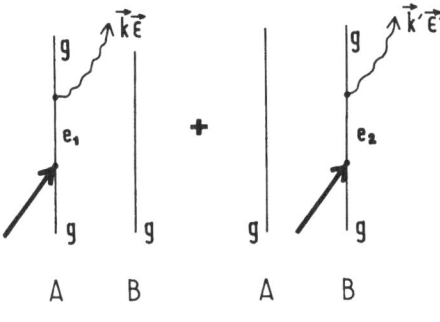

Fig. 7.4. Diagrams symbolizing two interfering amplitudes in a many-atom quantum beat experiment corresponding to an upper state energy splitting. The detected photon may have been emitted either by atom A excited in state e_1 or by atom B excited in state e_2

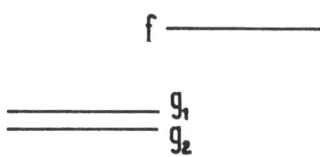

Fig. 7.5. Energy diagram of an atomic system which may exhibit cooperative quantum beats corresponding to a lower state energy splitting

b) Cooperative Quantum Beats Arising from a Splitting in the Atomic Lower States

Let us now consider two identical atoms A and B, each of them having the energy diagram of Fig. 7.5. We assume that the excited state is not split anymore, but that the initial ground state $|g\rangle$ is split in two closely spaced substates $|g_1\rangle$ and $|g_2\rangle$. As noted in Subsection 7.1.1, such atoms will not exhibit single-atom quantum beats. Cooperative quantum beats may, however, be emitted by these atoms as will be shown in this subsection.

Let us suppose that the atomic system is initially in the state $|g_1, g_2\rangle$ representing atom A in level $|g_1\rangle$ and atom B in level $|g_2\rangle$. Assume that this system is irradiated by a short broadband light pulse which can excite either atom A or atom B to level $|e\rangle$. As a result, the system immediately after the pulse will be prepared in state

$$|\psi(0)\rangle = \beta_A |e, g_2; 0\rangle + \beta_B |g_1, e; 0\rangle \tag{7.11}$$

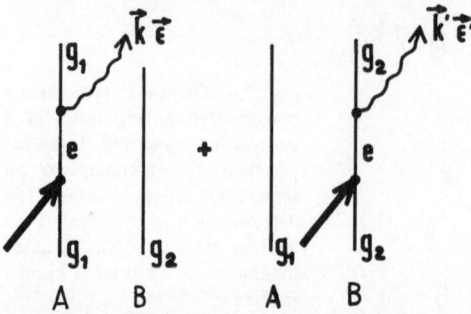

Fig. 7.6. Diagrams symbolizing two interfering amplitudes in a many-atom quantum beat experiment corresponding to a lower state energy splitting. The detected photon may have been emitted either by atom A back to state g_1 or by atom B back to state g_2

where β_A and β_B are complex amplitudes. At time t, the system will have evolved into

$$|\psi(t)\rangle = \beta_A e^{-iE_e t/\hbar} e^{-\Gamma t/2}|e, g_2; 0\rangle + \beta_B e^{-iE_e t/\hbar} e^{-\Gamma t/2}|g_1, e; 0\rangle$$
$$+ (\sum_f \sum_{k\varepsilon} C^A_{fk\varepsilon}(t)|f, g_2; k\varepsilon\rangle + \sum_{f'} \sum_{k'\varepsilon'} C^B_{f'k'\varepsilon'}(t)|g_1, f'; k'\varepsilon'\rangle) . \quad (7.12)$$

Here again, one should distinguish between elastic and inelastic processes. If one detects the light emitted on a transition back to a set of states $|f\rangle$ different from $|g_1\rangle$ and $|g_2\rangle$ (inelastic scattering) the two channels corresponding to $C^A_{fk\varepsilon}$ and $C^B_{f'k'\varepsilon'}$ will be distinguishable since they end in the two different atomic final states $|fg_2\rangle$ and $|g_1 f'\rangle$. There will be no possible interference between these channels and no quantum beats at all. On the other hand, if one detects the light emitted back to the $|g_1\rangle$ and $|g_2\rangle$ states, it is clear that the probability amplitudes $C^A_{g_1 k\varepsilon}$ and $C^B_{g_2 k'\varepsilon'}$ correspond to identical final atomic states. Terms such as $C^A_{g_1 k\varepsilon} C^{B*}_{g_2 k'\varepsilon'}$ will appear in the signal, corresponding to the interference between the two emission processes originating from different atoms which are represented by the diagrams of Fig. 7.6. Collective quantum beats at frequency $\omega_{g_1 g_2} = (E_{g_1} - E_{g_2})/\hbar$, due to the splitting of the atomic lower state $|g\rangle$, will then be observable. As pointed out in [7.17], this result is qualitatively different from the single-atom situation, for which only excited-state-splitting beats do exist.

7.1.3 Comparison Between Single- and Many-Atom Beat Signals

We compare, in Table 7.1, the main features of single- and many-atom beat signals. The discussion of the previous section has clearly shown some very important differences between the two kinds of beat effects, con-

Table 7.1. Compared features of single and many atom beat signals

	Single-atom quantum beats	Many-atom quantum beats
Type of scattering process	Elastic scattering *and* Inelastic scattering	Elastic scattering *only*
Observed beat frequencies	*Upper state* Bohr frequencies only	Upper *and* lower state Bohr frequencies
Minimum atomic density	Zero	~ 1 atom per λ^3
Sensitivity to Doppler effect	No	Yes
Observation	In numerous direct time resolved fluorescence experiments	No direct observation as yet

cerning, in particular, the type of scattering experiments in which they occur and the beat note frequencies which can be observed in each case. Other important distinctions between the two kinds of beats should be emphasized.

Let us first consider the case of fixed atoms (e.g., atoms in a crystal). The many-atom beat signal is in that case very sensitive to the relative distances between the emitting atoms and the detector. The amplitudes of emission from atoms A and B are proportional to $e^{ik \cdot R_A}$ and $e^{ik' \cdot R_B}$, respectively. Each term such as $C^A_{fk\varepsilon} C^B_{fk'\varepsilon'} e^{i(k-k')r}$ which appears in the expression of the signal contains a factor $e^{ik(R_A - r)} e^{-ik'(R_B - r)}$ allowing for the propagation from atoms A and B to the detector located at point r. If the two atoms are at a distance $|R_A - R_B|$ much larger than $\lambda = 2\pi/k$ from each other, the path difference $|R_A - r| - |R_B - r|$ will vary very rapidly with the position of the detector, and two closely spaced points r_1 and r_2 will see interatomic beat patterns completely out of phase with each other. In an actual experiment, the photomultiplier receives the photon on a finite surface area, so that the interatomic beat pattern produced by two atoms whose distance is much larger than λ will be almost completely washed out in the detector. Consequently, the atoms have to be located at a mutual distance of the order of, or smaller than λ, in order to yield an appreciable cooperative beat signal. This is of course not the case for single-atom quantum beats, for which the relative distances of emitting atoms is irrelevant and which can be observed at densities much lower than one atom per λ^3.

The difference between single- and many-atom beat signals is also very important when considering moving atoms.

For single-atom beat signals, the interfering amplitudes $C^{(i)}_{fk\varepsilon}$ and $C^{(j)}_{fk'\varepsilon'}$ which corresponds to the emission by the same atom are both Doppler shifted in frequency by almost the same amount. This shift is consequently cancelled out in the beat note so that the beat patterns emitted by atoms having different velocities will add up in the signal.

In the many-atom beat case, however, the interfering amplitudes $C^A_{gk\varepsilon}$ and $C^B_{gk'\varepsilon'}$ corresponding to different atoms will be frequency Doppler shifted by different amounts so that the interatomic beat note frequency will depend upon the relative velocities of atoms A and B. Thus, in general, it will be impossible to observe directly a collective beat signal in a macroscopic sample, in which the atoms have a thermal velocity distribution. In contrast, single-atom beats will be observable in this condition. This is a very important advantage of the single-atom quantum beat effect, which makes it, as we will see, very attractive for sub-Doppler spectroscopy.

The general conclusion we may derive from the above discussion is that single-atom quantum beats are by far much easier to produce and to detect than many-atom beat signals. Being obliterated by the Doppler effect, many-atom beat signals could only be observed in the fluorescence from closely spaced fixed atoms in a crystal or from atoms in a monocinetic atomic beam. In fact, to our knowledge, such collective atomic beats so far have not been observed directly in atomic fluorescence experiments. Collective emission effects somewhat related to the phenomena described above have, however, been recently detected in photon-echo type of experiments. Sinusoidal modulations in the intensities of echoes involving atoms with nearly degenerate excited or ground states have been observed by various groups [7.18–20]. These experiments are very different from fluorescence experiments, since they imply, in contrast with the latter ones, the preparation of a macroscopic optical dipole in the atomic medium. Nevertheless, the appearance of excited or ground-state splitting beats in such experiments may certainly be interpreted in terms of interference processes analogous to the one responsible for the collective beats predicted in pure fluorescence experiments. The description of these photon-echo experiments is, however, out of the scope of this review, which will be restricted to the study of single-atom fluorescence quantum beats and to the description of their applications to high-resolution excited-state spectroscopy.

Remark: If, instead of using QED, which we believe to be the right description for spontaneous emission, we had tried to describe the beat effects discussed in this section with the help of the neoclassical theory of radiation (NCT) we would have reached completely different conclusions as shown in [7.17] and [7.21]. NCT, in particular, predicts single-atom quantum beats produced by lower state splittings in contradiction with QED. Quantum beat experiments have been proposed to test NCT versus QED, although the results of such tests do not seem to be dubious since all experiments of this kind have so far unambiguously supported QED.

7.2 Quantum Beat Theory in the Density Matrix Formalism

Two problems remain to be solved in order to calculate the single-atom quantum beat signal from (7.8). At first, one must determine the α_i, α_j excitation amplitudes which depend on the light pulse characteristics (polarization, intensity, spectral profile). Once the α_i, α_j are known, one must carry out explicitly all the state summations included in the expression, which may be a rather lengthy and tedious task if not performed with some precautions. We show in this section that these two calculation steps may be greatly simplified if one describes the quantum beat signal in terms of the density matrix of the atomic excited state. The density matrix formalism, which has been widely used in most papers and textbooks dealing with optical pumping problems [7.22, 23], may be briefly summarized as follows. The optical signals in a fluorescence experiment turn out to be proportional to the mean value of some atomic observable in the excited state e, which can be very easily expressed as a combination of components of the density matrix $\varrho_e(t)$ of this state. The evolution of $\varrho_e(t)$ due to the light excitation process, to the precession of the coherences in the atomic excited state and to spontaneous emission is adequately described by a set of linear differential equations (the so-called optical pumping equations). The solution of these equations yields $\varrho_e(t)$ and allows the explicit calculation of the atomic fluorescence signal as a function of time. Furthermore, the atomic density matrix $\varrho_e(t)$ may be represented as an expansion over a set of spherical tensor operators among which only the scalar, dipolar and quadrupolar terms affect the fluorescence light. This results in a considerable simplification in the calculation of the detection signal, especially for high angular momentum excited states of atoms or molecules.

   Of course, we will not derive again here all the details of the theory which has been popularized by many articles and optical pumping works. We will only recall its main features and insist upon the specific problems which are relevant for the description of a laser-induced quantum beat experiment.

   Starting from (7.8), we will at first derive very briefly the expression of the quantum beat signal in terms of the excited state density matrix

(Subsection 7.2.1). We will thus make the connection between the quantum interference approach developed in the previous section and the density matrix point of view. We will see that the quantum beat signal depends on the density matrix $\varrho_e(0)$ induced at time $t=0$ in the atomic excited state by the light pulse, and we will in Subsection 7.2.2 show how $\varrho_e(0)$ may be derived from the equations of the optical pumping theory. Finally we will recall briefly (Subsection 7.2.3) how the use of a multipole expansion of the atomic density matrix may greatly simplify the analysis of the fluorescence signal.

7.2.1 The Quantum Beat Signal Expressed in Terms of the Excited State Density Matrix

The product of amplitudes $\alpha_i \alpha_j^*$ in (7.8) is nothing but the matrix element between states $|e_i\rangle$ and $|e_j\rangle$ of the excited state density matrix ϱ_e, evaluated at time $t=0$

$$\alpha_i \alpha_j^* = \langle e_i | \varrho_e(0) | e_j \rangle . \tag{7.13}$$

With this new notation, (7.8) may be rewritten as

$$S(e_d, t) = C \sum_{i,j} (e^{-i\omega_{ij}t} \langle e_i | \varrho_e(0) | e_j \rangle e^{-\Gamma t}) (\sum_f \langle e_j | e_d D | f \rangle \langle f | e_d^* D | i \rangle) \tag{7.14}$$

with

$$C = \frac{1}{(4\pi\varepsilon_0)^2} \frac{k_0^4}{r_0^2} .$$

As we will not explictly consider anymore the r dependence of the signal, we have dropped the r variable in $S(e_d, r, t)$. We have also, for the sake of simplicity, assumed that the detector is close enough to the emitting atom so that the retardation term r_0/c may be neglected and $t - r_0/c$ replaced by t.

The first bracket in (7.14) is in fact the density matrix element of ϱ_e between states $|e_i\rangle$ and $|e_j\rangle$ at time t

$$\langle e_i | \varrho_e(t) | e_j \rangle = e^{-i\omega_{ij}t} \langle e_i | \varrho_e(0) | e_j \rangle e^{-\Gamma t}, \tag{7.15}$$

so that (7.14) may be finally expressed in the much more compact form

$$S(e_d, t) = \text{Tr}[\varrho_e(t) \mathscr{L}(e_d)] \tag{7.16}$$

with

$$\mathscr{L}(e_d) = C \sum_f e_d D | f \rangle \langle f | e_d^* D . \tag{7.17}$$

The fluorescence signal thus appears as the expectation value in the atomic excited state of a "detection" operator $\mathscr{L}(e_{\mathrm{d}})$. This detection operator is essentially proportional to the component corresponding to the $e - f$ transition of the square of the atomic dipole projected along the detection polarization e_{d}.

Equation (7.16) allows one to see the beat phenomenon with a point of view somewhat different from the one developed in Section 7.1. The modulations in the atomic fluorescence now result from the time evolution of the atomic density matrix "coherences" $\langle e_i|\varrho_e(t)|e_j\rangle$. These coherences, which are prepared by the light pulse in the excited state, start precessing at time $t=0$ at the Bohr frequencies ω_{ij} corresponding to the energy splittings between the various sublevels. This precession entails a modulation of the square of the atomic dipole along the e_{d} direction and the atom thus behaves as a lighthouse which emits bursts of light in a given direction with a periodicity corresponding to its own eigenfrequencies.

In order to calculate explictly the quantum beat signal given by (7.16), one has to know $\varrho_e(t)$, that is in fact $\varrho_e(0)$, the density matrix just after the pulse excitation, from which $\varrho_e(t)$ may be deduced by (7.15).

7.2.2 Preparation of the Excited Density Matrix by the Light Pulse

In order to describe the evolution of the atomic system under the light excitation process, one must know precisely the characteristics of the pulse itself. Two very important limiting cases may be considered.

(i) The electric field pulse has well-defined frequency and phase, with a slow varying amplitude. In this case, the pulse is said to be Fourier limited since its spectral width Δ is the reciprocal of its duration T

$$\Delta = \frac{1}{T}. \tag{7.18}$$

Such "coherent" pulses are required in experiments such as free induction decay [7.24] or photon echoes [7.18] in which a macroscopic optical dipole has to be induced in the atomic system by the light excitation. The evolution of the atomic system is in this case described by generalized "Bloch equations" coupling together the excited and ground-state density matrices and the optical coherences as well.

(ii) The electric field pulse has a large spectral width corresponding to a correlation time $\tau = 1/\Delta$ much shorter than the pulse duration T. The pulse is then not Fourier limited, and the broadline condition

$$\Delta \gg 1/T \tag{7.19}$$

amounts to considering that the electric field pulse is a random function of time in a time interval of the order of T. In this case, the effect of the light irradiation is equivalent to a time-dependent stochastic process acting on the atoms. Provided the pulse is not too intense, the evolution of the system is then described by a set of linear differential rate equations, the so-called optical pumping equations [7.25–29], which couple together the ground and excited states density matrices. In contrast with case (i), no optical coherences appear in these equations, since no optical macroscopic dipole may be generated by an incoherent pulse excitation. In a quantum beat experiment, the signal is not sensitive to optical coherences, but only to atomic coherences induced in the excited state. Incoherent broadband pulses fulfilling the condition of (7.19) are thus well suited for these experiments which do not require at all Fourier-limited pulses of the (i) type. We will thus restrict our study to the broadband pulse excitation and not consider any more the case of a Fourier-limited pulse. We will not in this subsection derive in detail the optical pumping equations which may be found elsewhere in the literature [7.27–29]. We will only recall briefly the general conditions the light pulse must fulfill for these equations to be valid and give their solutions in the limiting cases of weak and strong pumping irradiation.

a) Description of the Optical Pulse in the Broadline Excitation Case

Three time parameters are very important for the description of the pulse: its duration T (the pulse is assumed to interact with the atoms between time $t = -T$ and $t = 0$), its correlation time $\tau = 1/\Delta$ and its pumping time $T_p(t)$. The pumping time is inversely proportional to the instantaneous spectral density $u(\omega_0, t)$ of the pulse at the frequency ω_0 of the optical transition, and to the oscillator strength of the transition. It may be defined by the relation

$$\frac{1}{T_p(t)} = \frac{\pi}{\varepsilon_0 \hbar^2} u(\omega_0, t) |\langle e||D||g\rangle|^2 \tag{7.20}$$

(where $\langle e| |D| |g\rangle$ is the radial part of the electric dipole matrix element between states $|e\rangle$ and $|g\rangle$). $T_p(t)$ represents the instantaneous average time between two successive photon absorptions from the pulse.

In order to be able to derive rate equations for the evolution of the atomic system, the pulse should fulfill the broadline condition (7.19) together with the "weak coupling" condition

$$\Delta \gg \frac{1}{T_p(t)} \tag{7.21}$$

which sets up an upper limit to the allowed pulse intensity and expresses the fact that the atomic system does not evolve appreciably under the pulse excitation in a time interval of the order of the field phase memory time τ. We will suppose that conditions (7.19) and (7.21) are fulfilled throughout the remainder of this section. They are in fact not too restrictive and likely to be satisfied in a standard quantum beat experiment [if one uses a N_2-laser pumped pulsed dye laser of a few hundred watts peak power, one has in general $\Delta \approx 10^{11}$ Hz, $\frac{1}{T} \approx 10^9$ Hz, $\frac{1}{T_p(t)} \approx 10^{10}$ Hz, and both conditions (7.19) and (7.21) are fulfilled].

In addition to these essential conditions and unless otherwise specified, we will also consider that the following inequalities are satisfied:

$$1/T \gg \Gamma , \tag{7.22}$$

$$1/T \gg \omega_{ij}, \tag{7.23}$$

$$\Delta \gg \omega_{ij}. \tag{7.24}$$

First (7.22) means that one can neglect the effects of spontaneous emission during the pulse itself and well separate in the quantum beat experiment the preparation phase (between $t = -T$ and $t = 0$) from the detection phase ($t > 0$). Inequality (7.23) means that the pulse is short enough so that the atomic coherences do not have the time to precess during the pulse excitation. Finally (7.24) expresses the fact that the pulse bandwidth is large enough to entirely cover the structure of the studied excited state. We must emphasize that conditions (7.22), (7.23) and (7.24), in contrast to (7.19) and (7.21), are not essential ones and could be disregarded at the expense of a slight modification in the expression of the quantum beat signal. An example of these modifications, when condition (7.23) is not fulfilled, will be given below [see (7.31, 32)].

b) The Weak Pumping Limit

In most quantum beat theories, it is assumed that the pulse used to excite the atoms is weak enough so that it interacts linearly with the atomic system. In terms of photon processes, this means that at most one photon from the pulse is absorbed during time T, which amounts to assuming the weak pumping condition

$$\frac{T}{T_p(t)} \ll 1 . \tag{7.25}$$

Such an assumption is natural for the description of the excitation by the light pulses produced by weak spectral lamps or by lasers on a weakly allowed optical transition. In such cases, the evolution of the excited state density matrix is described by a very simple rate equation [7.22]:

$$\frac{d}{dt} \varrho_e = \frac{1}{T_p(t)} \frac{1}{G_g} P_e e_0 D P_g e_0^* D P_e \tag{7.26}$$

where e_0 is the polarization of the pulse, P_e and P_g are the projectors into the excited and ground states, respectively, and G_g is the degeneracy of the ground state. This equation may be readily integrated to yield the excited state density matrix just after the pulse.

$$\varrho_e(0) = \frac{K_0}{G_g} P_e e_0 D P_g e_0^* D P_e \tag{7.27}$$

where

$$K_0 = \int_{-T}^0 \frac{dt}{T_p(t)} \tag{7.28}$$

is the time-integrated pumping rate.

Equation (7.27) is valid when the atomic ground state is not oriented prior to the pulse excitation. It merely expresses the fact that the atomic density matrix components in the excited state are obtained as products of two amplitudes proportional to the atomic dipole matrix elements between the ground state and the relevant excited substates. Replacing $\varrho_e(0)$ by its expression (7.27) in (7.16), one gets for the quantum beat signal the following very simple and well-known expression, which is quite symmetric with respect to the polarizations of the ingoing pulse and outgoing fluorescence:

$$S(e_d, t) = \frac{C K_0}{G_g} \sum_f \sum_{i,j} \sum_g \langle e_i | e_0 D | g \rangle \langle g | e_0^* D | e_j \rangle \langle e_j | e_d D | f \rangle$$

$$\cdot \langle f | e_d^* D | e_i \rangle e^{-\Gamma t} e^{-i\omega_{ij}t} \tag{7.29}$$

where $\sum_{f,g}$ bears on all the relevant initial and final substates.

In some experiments, it may happen that the ground state exhibits some anisotropy before the pulse excitation. This is the case for example in stepwise processes, when the excited state is prepared via an intermediate state g which is itself oriented or aligned by a first polarized light pulse exciting the atom from a lower state. In that case, (7.27) has to be modified to account for the anisotropy of the atom in state g and becomes

$$\varrho_e(0) = K_0 P_e e_0 D P_g \varrho_g(-T) P_g e_0^* D P_e \tag{7.30}$$

where $\varrho_g(-T)$ is the density matrix in state g at time $-T$ prior to the pulse excitation. [To derive (7.30), it is assumed that the ground state has no time to evolve between the times $-T$ and 0, so that $\varrho_g(-T)$ could in fact be replaced by $\varrho_g(0)$.]

At the expense of a small complication, (7.27), (7.29), and (7.30) may at last be generalized to allow for a time evolution in the excited state when condition (7.23) is no more fulfilled (long pulse case). Equation (7.27), for example, becomes

$$\langle e_i | \varrho_e(0) | e_j \rangle = \frac{K_{0_{ij}}}{G_g} P_e e_0 D P_g e_0^* D P_e \tag{7.31}$$

where

$$K_{0_{ij}} = \int_{-T}^{0} \frac{e^{-i\omega_{ij}t}}{T_p(t)} dt \tag{7.32}$$

is the Fourier transform at frequency ω_{ij} of the pulse profile. The width of $K_{0_{ij}}$ is of the order of $1/T$, which entails that the coherence $\langle e_i | \varrho_e(0) | e_j \rangle$ is washed out if the pulse duration becomes larger than the Bohr period $1/\omega_{ij}$. This is a quite natural result which merely expresses the fact that one can observe a beat note only if the pulse used to excite the atom is shorter than the corresponding period.

c) Saturation Effects Induced by a Strong Pumping Pulse

In the case of a laser pulse excitation, it may very often happen that the pulse is strong enough to saturate the optical transition during its duration so that condition (7.25) is no longer fulfilled (a peak power of the order of a few watts is sufficient to saturate in a nanosecond a well-allowed atomic transition). In that case, the pulse interacts in a highly nonlinear way with the atoms and the simple rate equation (7.26) is no longer valid. One must then take into account not only the initial absorption process which brings the atom from the ground state into the upper state, as was done in the previous subsection, but also the subsequent stimulated emission processes which may, during the pulse, carry the atoms back to the ground state. Virtual transitions induced by the off-resonant components of the light pulse may in addition shift the atomic levels and affect the atomic evolution during the pulse duration.

For a complete theory of laser-induced quantum beats, one should analyze all these phenomena which come together into play for the preparation of the atomic excited state density matrix. Such an analysis may be found in [7.27] in which the evolution of the system is quite generally described by a set of coupled linear differential rate equations taking into account all the possible interaction processes between the pulse and the atom. We will just indicate here how the saturation effects modify the results of the simple weak pumping limit. Resolution of the complete set of equations yields the following results:

i) The effects of light shifts induced by the pulse are in general completely negligible on $\varrho_e(0)$.

ii) The stimulated emission processes which bring the atoms back to the lower state $|g\rangle$ may change the ratios between the various excited state density matrix components. These changes are however in general rather small and entail a modification of at most a few percent in the relative weights of the various frequency components in the quantum beat signal.

iii) For long pulses ($T \gtrsim 1/\omega_{ij}$), the beat washing-out effect observed in the weak pumping limit [see (7.32)] may be reduced and the modulation at frequency ω_{ij} may be enhanced. This effect is related to a slowing down of the atomic excited state precession due to saturation by the light pulse.

All the saturation effects are however likely to be very small and difficult to observe in an actual experimental situation since they require in general a very large saturation parameter T/T_p. Although the excitation by a strong broadband light pulse is a very complex process, we thus reach the conclusion that the excited state density matrix just after the pulse is in general adequately described by the weak pumping solutions (7.28) or (7.30). We will assume these solutions to be valid throughout the remainder of this chapter.

In conclusion, we may say that the preparation stage of an optically induced quantum beat experiment may be quite generally described by rate equations which allow the accurate expression of the atomic excited state density matrix just after the pulse as a function of the various pulse parameters. This is a big advantage over other kinds of quantum beat experiments, such as beam foil experiments, in which the excitation is achieved by a complicated collision process which can be only phenomenologically described and for which no accurate expression of the atomic density matrix can be derived.

Remark: We have restricted here our description to pulses fulfilling conditions of (7.19) and (7.21). If the pulse is too narrowband or too intense for these conditions to be satisfied, but not monochromatic enough to be Fourier limited, neither the Bloch equations nor the optical pumping rate equation are valid anymore. The description of the excitation process is then very complicated, and one has to take into account in the calculations the higher-order correlation functions of the optical field. Such calculations have been recently performed by AVAN and COHEN-TANNOUDJI [7.30].

7.2.3 The Multipole Expansion of the Quantum Beat Signal

Let us consider in this subsection the specific case of a fine structure quantum beat experiment. The relevant energy levels of the atomic system are sketched in Fig. 7.7. L_g, L_e, and L_f are the respective angular momenta of states g, e and f which all have the same spin momentum S. Each of these states is split by the fine structure interaction into fine structure levels, each having a given total angular momentum (labelled J_g, J_e, or J_f). Each of these fine structure levels has, furthermore, a $(2J+1)$ degeneracy corresponding to the possible values of the magnetic quantum number m_J. The quantum beat signal emitted by such an atom on the $e - f$ transition appears as a sum of contributions corresponding to all these states. The explicit and direct calculation of such a fine structure quantum beat signal would thus be a rather formidable task. It is possible to bypass a great deal of summations and calculations if one represents the atomic density matrix ϱ_e in terms of a multipole basis operators. This representation allows one to consider only those combinations of the atomic density matrix components which appear explicitly in the signal and to expand it in a form which shows clearly the signal symmetries and angular properties.

a) The Spherical Tensor Basis

Consider the atomic excited state multiplicities $|J_e, M_e\rangle$ and $|J'_e, M'_e\rangle$. We can define ([7.31, 32]) a spherical basis of operators for the atom in state $|e\rangle$ by the relation

$$
{}^{J_e J'_e} T^k_q = \sum_{M_e, M'_e} \sqrt{2k+1} (-1)^{J_e - M_e} |J_e M_e\rangle \langle J'_e M'_e|
$$
$$
\cdot \begin{pmatrix} J_e & k & J'_e \\ -M_e & q & M'_e \end{pmatrix}
\tag{7.33}
$$

in which the brackets ( ) denote an ordinary $3j$ symbol.

These operators form an orthonormal basis system in state $|e\rangle$ in the sense that they satisfy the condition

$$
\mathrm{Tr}\{{}^{J_e J'_e} T^k_q {}^{J''_e J'''_e} T^{k'}_{q'}\} = \delta_{J_e J''_e} \delta_{J'_e J'''_e} \delta_{kk'} \delta_{qq'} .
\tag{7.34}
$$

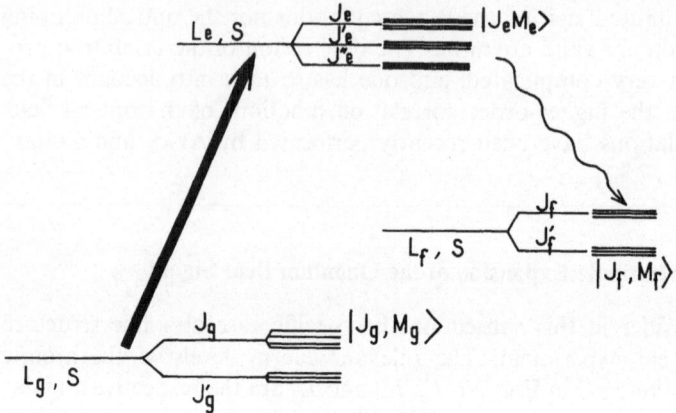

Fig. 7.7. General energy level scheme of an atomic system which may exhibit fine structure quantum beats

They are transformed under a rotation by the following relation:

$$\mathcal{R}(\alpha, \beta, \gamma)^{J_e J'_e} T^k_q \mathcal{R}^{-1}(\alpha, \beta, \gamma) = \sum_{q'} R^{(k)}_{q'q}(\alpha, \beta, \gamma)^{J_e J'_e} T^k_{q'}. \tag{7.35}$$

Each operator $^{J_e J'_e} T^k_q$ has matrix elements only between states J_e and J'_e. When no external field is applied to the atom, the time evolution of $^{J_e J'_e} T^k_q$ is thus given by

$$e^{-i\mathcal{H}_0 t/\hbar} \, {}^{J_e J'_e} T^k_q e^{i\mathcal{H}_0 t/\hbar} = e^{-i\omega_{J_e J'_e} t} \, {}^{J_e J'_e} T^k_q \tag{7.36}$$

(where $\mathcal{H}_0$ is the atomic Hamiltonian).

b) Multipole Expansion of the Density Matrix and Detection Operator in Terms of the $^{J_e J'_e} T^k_q$

The excited state density matrix $\varrho_e(0)$ and the detection operator $\mathcal{L}(e_d)$ may be readily expanded in terms of the $^{J_e J'_e} T^k_q$ according to the following relations:

$$\varrho_e(0) = \sum_{J_e J'_e} \sum_{k, q} {}^{J_e J'_e} \varrho^k_q(0) \, {}^{J_e J'_e} T^k_q, \tag{7.37}$$

$$\mathcal{L}^k_q(e_d) = \sum_{J_e J'_e} \sum_{k, q} {}^{J_e J'_e} \mathcal{L}^k_q(e_d) {}^{J_e J'_e} T^k_q. \tag{7.38}$$

The $^{J_e J'_e} \varrho^k_q$ and $^{J_e J'_e} \mathcal{L}^k_q$ coefficients can be calculated from the coupling of the angular momenta involved in the various levels. For example, when the detection does not distinguish between the final J_f states, the $^{J_e J'_e} \mathcal{L}^k_q$

coefficients are given by the following expression:

$$
{}^{J_e J'_e}\mathscr{L}^k_q(e_Q) = (-1)^Q \delta_{q0} \begin{pmatrix} k & 1 & 1 \\ 0 & Q & -Q \end{pmatrix} \begin{Bmatrix} k & 1 & 1 \\ L_f & L_e & L_e \end{Bmatrix}
$$

$$
\cdot \begin{Bmatrix} L_e & k & L_e \\ J'_e & S & J_e \end{Bmatrix} |\langle e| |D| |f \rangle|^2 \tag{7.39}
$$

($|\langle e| |D| |f \rangle|$ is the reduced matrix element of D between states e and f).

In (7.39), Q is a number equal to ± 1, 0 which defines the type of detection polarization ($Q = \pm 1$ corresponds to $e_{\pm 1} = e_x \pm i e_y$ circular polarization and $Q = 0$ to $e_0 = e_z$ linear polarization). When $\varrho_e(0)$ is given by the simple (7.27), which is quite similar to (7.17) for $\mathscr{L}(e_d)$, the ${}^{J_e J'_e}\varrho^k_q$ coefficients are given by an expression quite analogous to (7.39) in which L_f is replaced by L_g. When the atomic lower state g is oriented prior to the laser pulse, the ϱ^k_q coefficients will be given by more complicated expressions, since the excited state multipole components result from the coupling of the lower state spherical tensors with the vector operators $P_e e_0 D P_g$ and $P_g e_0 D P_e$ describing the light excitation [see (7.30)]. We will not give here the explicit form for the ϱ^k_q coefficients in this general case, which involves $9j$ coefficients.

Some very simple and general properties may be deduced from symmetry considerations or read directly from expression (7.39):

i) The only nonvanishing $\mathscr{L}^k_q$ coefficients are those with $k=0$, $k=1$ and $k=2$ [the $3j$ coefficient in (7.39) obviously vanishes for $k>2$].

ii) For a linear polarization ($Q=0$), only the $k=0$ and $k=2$ components are different from zero [the $3j$ coefficient in (7.39) vanishes for $k=1$ and $Q=0$].

iii) Finally, the sum of the ${}^{J_e J'_e}\mathscr{L}^k_q(e_Q)$ coefficients over three orthogonal detection polarizations ($Q = +1$, -1 and 0) yields a null result when $k \neq 0$.

From these properties, one can deduce immediately the main features of the quantum beat signal.

c) The Scalar, Vector and Tensor Parts of the Quantum Beat Signal

The multiple expansion of the quantum beat signal may be readily derived from (7.16), (7.37) and (7.38). One finds

$$
S(e_d, t) = \sum_{J_e J'_e} \sum_{k=0}^{2} \sum_q {}^{J_e J'_e}\varrho^k_q(0) \, {}^{J_e J'_e}\mathscr{L}^{k*}_q(e_d) e^{-i\omega_{J_e J'_e} t} e^{-\Gamma t}. \tag{7.40}
$$

The signal thus appears as the sum of a scalar part ($k=0$), a vector part (or "orientation") corresponding to $k=1$ and a tensor part (or "alignment") corresponding to $k=2$.

The scalar part has only $J_e = J'_e$ terms and is therefore not modulated. The beats appear only on the vector and tensor parts of the signal. When linear polarizations are used, the orientation ($k=1$) part of the signal vanishes. Finally, when summing the signal over three orthogonal detection polarizations, the vector and the tensor parts each vanish identically, and the beat disappears. In other words, quantum beats appear in phase opposition for different states of detection polarization. This feature is very often used, as we will see later, to improve the signals in a quantum beat spectroscopy experiment.

In this subsection, we have chosen for the sake of definiteness to describe the case of fine structure quantum beats. The same tensor formalism may be used to discuss hyperfine structure beats, or Zeeman beats. In all cases, the expansion of the signal in terms of multipole moments results in a considerable simplification and reduces the quantum beat calculation to the determination of a small number of ϱ_q^k and $\mathscr{L}_q^k$ coefficients. Most summations over initial, excited or final states are taken care of in the tabulated $6j$ or $9j$ symbols which appear in these coefficients. This simplification is most obvious in the case of molecular fluorescence, since the angular momenta involved in the molecular excited states are generally very large.

The tensor formalism has other very interesting advantages. It makes indeed very clear all the angular properties of the signal and allows simple calculations of how it changes when one rotates the polarization of the incoming pulse or outgoing photon. The spherical tensor transformation under such a rotation is described by (7.35) and depends on rotation matrix elements of angular momentum $k=1$ and $k=2$. These matrix elements may be expressed in terms of spherical harmonics or Legendre polynomials of cosine functions. The angular dependence of the signal is thus simply expressed as a combination of such angular functions. When the excited state is prepared by a stepwise process via an intermediate state implying two successive exciting pulses, the quantum beat signal depends on the respective polarization of these two pulses and on the polarization of the emitted photon. Here again, the tensor formalism allows one to express in a very elegant and concise way the signal as a function of these respective polarizations [7.27].

The tensor formalism is also very useful when collisions in the atomic excited state have to be taken into account. We have so far supposed that the only damping mechanism is spontaneous emission. In the presence of a buffer gas, the emitting excited level may be affected by collisions which contribute to the damping of its population and coherences. It may quite generally be shown that these collision processes result in a unique damping constant Γ_k for each tensor rank k, Γ_k being independent of q, J_e, J'_e. In the presence of collisions, the beat signal expressed in the form of (7.40) thus remains valid if one replaces Γ in the scalar, vector and tensor parts, respectively, by Γ_0, Γ_1, and Γ_2. It is worth noticing that the analysis of the quantum beat signal allows one to measure at the same time these three damping constants, while level crossing or double resonance experiments are not sensitive to Γ_0.

7.3 Connection Between Quantum Beats and Other Types of Atomic Fluorescence Experiments

The detection of the single-atom quantum beat signal described in the previous section appears to be especially suitable for high-resolution spectroscopy of small energy intervals in atomic or molecular excited states. As such, the quantum beat detection is very closely related to classical techniques in atomic fluorescence which also make use of broadband optical excitation and, nevertheless, allow one to overcome the Doppler effect (level-crossing, double-resonance, modulated fluorescence experiments). In this section, we briefly show how these various techniques are related to quantum beat detection.

7.3.1 Relation to Level-Crossing Experiments

In a level-crossing experiment [7.1, 2], the excited state atomic coherences are continuously prepared by a broadband light beam with spectral width Δ. Such a light beam may be described as a succession of pulses randomly distributed in time. The duration of each pulse is inversely proportional to Δ. This interpretation accounts for the fact that the atoms in the source incoherently emit bursts of radiation at random times. To calculate the amount of fluorescence light emitted by the atoms under this optical excitation, we have to sum over time the elementary signal produced by each irradiating pulse. It is convenient to write the signal in the form of (7.14). The integration is then straightforward to perform. We find that the average level $\bar{S}(e_d)$ of fluorescence is constant and given by

$$\bar{S}(e_d) = \int_0^\infty S(e_d, t)dt$$
$$= C \sum_f \sum_{i,j} \frac{\langle e_i|\varrho_e(0)|e_j\rangle \langle e_j|e_d D|f\rangle \langle f|e_d^* D|e_i\rangle}{\Gamma + i\omega_{ij}} . \tag{7.41}$$

This formula yields the well-known level-crossing signal which has been derived by FRANKEN [7.33] in a way quite similar to the one developed in this section. This signal contains non-resonant "diagonal" terms $(i=j)$ which come from light "incoherently" emitted by the various excited sublevels $e_i, e_j \ldots$ and which correspond to the contributions of the non-modulated part of each elementary signal. We find also in the signal various "off-diagonal" terms $(i \neq j)$ with energy dependent denominators. These terms are important only if there is a Bohr splitting ω_{ij} of the order of or smaller than Γ, that is in the vicinity of a level crossing in the atomic excited state. This level-crossing behaviour is easy to understand in the quantum beat picture. Each elementary light pulse in the beam will produce its own beat pattern in the fluorescence. However, after successive pulses, contributions to the interference term generally average to zero because the pulses are randomly distributed in time. It is only when two levels are degenerate or quasi-degenerate that the zero or near-zero frequency beat signal will add up and give a non-zero contribution to the fluorescence. The signal will then exhibit a resonant behavior when the level energies are swept through the degeneracy value (by the scanning of a magnetic or an electric field for example). In summary, the effect of the quantum beats vanishes completely under continuous excitation unless they arise in the emission from degenerate excited states and they occur at zero frequency. In this respect, a very descriptive name for level crossing would be zero frequency quantum beats.

7.3.2 Relation to Modulated Pumping Experiments

Instead of exciting the atoms with pulses randomly distributed in time, let us consider now an excitation by pulses succeeding each other at a given fixed time interval T_0. It is clear that the elementary beat patterns produced by these pulses will reinforce each other and add up constructively whenever the beat frequency is equal to a multiple of the pulse rate $1/T_0$. More generally, if the intensity I of the exciting light beam is a periodic function of time with a fundamental frequency $\omega = 2\pi/T_0$, one may expand it in a Fourier series according to the relation

$$I = \sum_k I_k e^{-ik\omega t} . \tag{7.42}$$

One then obtains the fluorescence signal $\tilde{S}(e_d, t)$ emitted by the atom under this optical excitation by summing over time elementary signals. One gets, after a straightforward integration,

$$\tilde{S}(e_d, t) \propto \sum_k I_k e^{-ik\omega t} \sum_f \sum_{i,j}$$
$$\times \frac{\langle e_i | \varrho_e(0) | e_j \rangle \langle e_j | e_d \mathbf{D} | f \rangle \langle f | e_d^* \mathbf{D} | e_i \rangle}{\Gamma + i(\omega_{ij} - k\omega)} . \tag{7.43}$$

The atomic fluorescence light will thus exhibit a resonant component modulated at the Bohr frequency ω_{ij}, whenever the condition $\omega_{ij} = k\omega$ is fulfilled. This is the basis for a modulated pumping fluorescence experiment [7.3, 4]. One excites the atom or the molecule with a light beam whose intensity (or polarization) is time modulated and one sweeps the modulation frequency ω until a resonance is observed in the fluorescence light at a multiple of ω. The measurement of ω at resonance yields the energy splitting ω_{ij} between the excited levels. The modulated pumping fluorescence experiment appears thus as the steady state counterpart of a transient quantum beat experiment. Instead of exciting the free precession of the atomic coherence with a single impulsion, one tries to force it into resonance by a continuous succession of impulsions regularly distributed in time. The signal versus frequency one obtains is, for $k = 1$, nothing but the Fourier transform of the time resolved quantum beat signal.

7.3.3 Relation to Double Resonance Experiments

In a double resonance experiment [7.34] the atoms are excited by a broadband continuous light irradiation which prepares only diagonal matrix elements in the excited states (population pumping). A radiofrequency field (frequency ω) is simultaneously applied to the atoms, which transforms the population differences into atomic coherences. These coherences, which evolve at frequency ω, undergo a resonant variation when the condition $\omega = \omega_{ij}$ is fulfilled. As a result, one may observe resonant modulations at frequency ω in the fluorescent light emitted by the atomic sample. Observation of these modulations provides a very convenient way of detecting the double resonance signal [7.35]. Described in that way, the double resonance experiment appears as a special kind of atomic coherence experiment in which the coherences are driven in a steady state by the simultaneous action of the light and the rf field. In this respect, double resonance may also be considered as the steady state version of a quantum beat experiment. The frequency response obtained in double resonance is again essentially the Fourier transform of the time dependent signal provided by the atoms in a quantum beat experiment.

7.3.4 Advantages and Limitations of Quantum Beat Spectroscopy as Compared to Steady State Fluorescence Techniques

For the study of a given atomic or molecular excited state, one has in principle to choose between the steady state methods mentioned above (level crossing, modulated pumping and double resonance) and their transient counterpart (quantum beats). It is therefore interesting to compare the advantages and limitations of these various methods as spectroscopic tools.

Level-crossing and double resonance types of experiments using spectral lamps have been successfully applied, over the last 20 years, to investigate a very large number of atomic excited states. Interest in this type of experiments has also been renewed recently with the advent of tunable continuous lasers which allow one to reach an even wider range of excited state levels [7.36]. The most evident advantage of the steady state techniques is their much higher duty cycle, the fluorescent light being recorded continuously instead of in small time intervals only a few times per second as is the case in a transient fluorescence experiment. This generally results in a much simpler detection procedure, for the steady state fluoresence experiment than for quantum beat experiments, which necessitate more or less sophisticated fast response detection systems (see Subsection 7.4.2). This advantage of the steady state techniques becomes the more obvious the shorter the lifetimes and the larger the frequency intervals to be measured. The quantum beat experiments are eventually limited by a cutoff frequency which depends on the laser excitation duration and detection rise time. As shown in Section 7.4, this cutoff is in most experiments of the order of a few hundred MHz, with a possible extension in the GHz range in some special cases (laser-beam experiments). The level-crossing and double resonance techniques do not suffer from these limitations and allow the determination of much higher frequency intervals up into the microwave range. However, for small enough structures, the quantum beat techniques have several interesting advantages over the steady state methods.

The first asset of quantum beat detection is related to the present development of lasers. As shown in Section 7.4, the tunability range of pulsed dye lasers (2400 Å to about 9000 Å) is very large and for the time being much wider than the range of cw dye lasers which are restricted to the visible part of the spectrum, and in most cases, to the gain curve of rhodamine $6G$ (5500–6100 Å). As a result, a lot of levels can be reached with pulsed sources, and not with a steady state laser excitation. Thus in some cases (especially for transitions in the near UV or in the blue part of the spectrum) it is more difficult to develop level-crossing and double resonance techniques than quantum beat spectroscopy. Using a N_2 laser pumped pulsed dye laser (see Subsection 7.4.1), it is possible to excite a very large number of UV, visible and near infrared transitions. These transitions do not have to be strongly allowed, since the high power available during the pulse permits the saturation of even very weak transitions. If one combines two successive laser pulses in a stepwise excitation process, one can also reach levels of the same parity as the ground state and extend the technique to a large number of highly excited states. In molecular spectroscopy too, the pulsed dye lasers are very useful for selective pulsed excitation of rotation-vibration levels which would be very difficult to study otherwise.

The second advantage of the quantum beat technique is the simplicity and versatility of the method. In such an experiment, one has just to ring the atom with the optical pulse and observe it while it displays its own spectrum. There is no scanning or tuning of a static field or of an rf field and no complicated interpretation of line structure, as may be the case in level crossing or double resonance. Nor is it necessary to know before the measurement the approximate value of the structure to be studied. The quantum beat technique is thus a very universal and versatile one which is particularly well suited for a first and systematic determination of atomic lifetimes or energy splittings in a series of levels.

The last advantage is of a more conceptual nature. In a quantum beat experiment, one observes the free decay of atoms or molecules after the pulse excitation. However intense or saturating this excitation may be, it does not affect the atomic evolution during the measurement time, and the measured frequencies are thus neither broadened nor shifted by the electromagnetic field, as may be the case in a level-crossing, a modulated fluorescence or a double resonance experiment. (In Subsections 7.3.1 and 7.3.2, we have assumed for simplicity that the optical fields interacted linearly with the atomic medium and neglected the radiative effects. For strong cw laser light fields, the signals are in fact not given any more by the simple Eqs. (7.41) or (7.43), but by more complicated expressions taking into account the saturation effects induced by the light irradiation itself). While a level-crossing or a double resonance experiment using lasers often necessitates an extrapolation procedure to vanishing light intensity, one can in a quantum beat experiment perform directly the measurement with a very strong light irradiation. This may be a very strong conceptual advantage, which makes, in particular, the technique very valuable for the study of highly excited states which may be very sensitive to all kinds of electromagnetic perturbations.

In this review of advantages of time-resolved spectroscopy, one should also quote the possibility of obtaining in certain cases a resolution slightly improved over the natural linewidth. This is however a rather controversial point which will be discussed in more detail in Section 7.6.

7.4 Technology of Time-Resolved Fluorescence Spectroscopy Using Lasers

Before describing in the next section various transient fluorescence experiments, we will briefly discuss the technology of these experiments which have been made possible by the recent development of pulsed laser sources and fast signal detecting systems. Most transient fluorescence

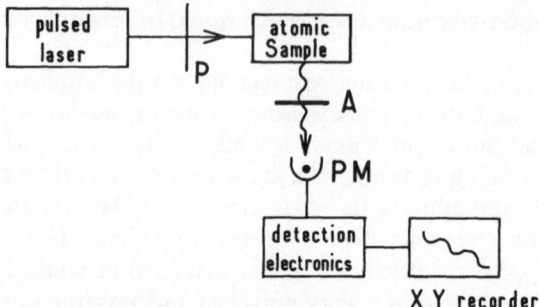

Fig. 7.8. Sketch of a "quantum beat laser spectrometer"

lifetime and quantum beat measurements are performed with a "quantum beat laser spectrometer" which may be sketched as shown in Fig. 7.8. A pulsed laser source is used to excite selectively atomic or molecular species in the sample (which may be a resonance cell or an atomic beam). The fluorescent light emitted by the sample is received by a photomultiplier tube (PM) connected to a fast detection electronics device. The output of the detection system is sent to an XY recorder which displays the fluorescence pattern versus time. For quantum beat modulations observation, a polarizer P and an analyzer A are generally used to conveniently polarize the incoming light pulse and emitted fluorescence. These polarizers are in general omitted for simple lifetime measurements. In Subsections 7.4.1 and 7.4.2, we provide some information concerning, respectively, the most widely used pulsed laser sources and fast response detection systems. To be complete, we briefly describe in Subsection 7.4.3 a different kind of laser transient fluorescence spectrometer that has also been used in some lifetime and quantum beat experiments; this consists of a continuous laser impulsively exciting atomic species moving in a fast atomic beam (beam-laser technique).

7.4.1 Pulsed Laser Sources

In a transient fluorescence experiment, the laser excitation has to be shorter than the atomic or molecular excited state lifetime (if one wants to measure only the radiative damping of the level), and also to be shorter than the excited state splittings Bohr frequencies (if one wants to detect quantum beat modulations). The laser excitation must also be efficient and carry enough energy to bring as large a number of atoms or molecules as possible into the excited state during the short pulse duration. In practice, for well-allowed transitions and energy splittings of the order of a few hundreds of megahertz, these requirements imply that the laser pulse

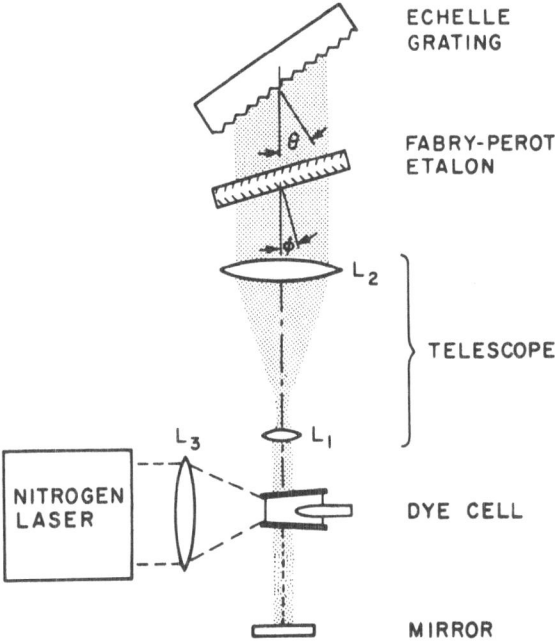

Fig. 7.9. Sketch of a Hänsch design NLPD laser. The telescope and Fabry-Perot in the cavity are optional and may be removed if a narrow output is not needed

should be at most a few nanoseconds long and have a peak power of several watts. (Of course, these figures have to be modified if one wants to detect faster beats or to excite very weak transitions).

The N_2-laser-pumped-pulsed-dye-laser [7.37] (NLPD laser) fulfills these general requirements and is the most widely used type of lasers for lifetime and quantum beat experiments. The NLPD laser may deliver several kilowatts peak power pulses with a pulse duration of 2 to 10 nanoseconds and a spectral bandwidth in the range 0.001–10 Å depending on the cavity design. Several types of cavity may be used, one of the most popular being the HÄNSCH design [7.38] sketched in Fig. 7.9. The dye cell is transversally pumped by a UV pulsed light beam delivered by a powerful N_2 laser. The dye laser cavity is closed at one end by a poorly reflective output mirror and at the other end by an echelle grating which is tilted for wavelength selection. A telescope may be placed between the dye cell and the grating to expand the beam waist on the grating and improve the resolution. A tilted Fabry-Perot etalon may be added between the grating and the telescope to further narrow the laser bandwidth. With etalon and telescope in the cavity, a typical bandwidth of about 500 MHz– 1 GHz is achieved. This bandwidth range is generally well adapted for most experiments. It is indeed narrow enough to permit selective excitation of well-resolved transitions in complicated atomic or molecular structures, and nevertheless wide enough to allow efficient excitation of the whole Doppler profile in most species. When high resolution is not needed, removal of the etalon increases the bandwith to about 10–20 GHz. Removing the telescope further broadens the laser output to about 2 to 10 Å. The wavelength tunability of such a laser is very large. By a convenient choice of dye solutions one may cover the whole spectral range between about 3600 Å and 9000 Å. Frequency doubling with a proper nonlinear crystal extends the wavelength range in the UV down to about 2400 Å. The laser repetition rate depends on the type of N_2 laser used for the pump and varies from a few pps up to 10^3 pps. This laser is thus a very efficient and versatile light source which may be easily adapted to each particular study.

Other types of pulsed dye lasers may also be used for time-resolved spectroscopy. Flashlamp lasers are in general less convenient than NLPD lasers, owing to their rather long pulse duration and their low repetition rate. A most attractive possibility would be to use the train of pulses emitted by a mode-locked continuous dye laser [7.39]. The pulse length is then only of the order of picoseconds, which would allow the excitation of quantum beats in the GHz range.

7.4.2 Fast Response Detection Systems

The fluorescent light following the light pulse is detected by a short rise time photomultiplier tube, wired for fast response. Nanosecond rise time tubes are commonly available and are well adapted for the detection of the beats—up to a few hundred MHz—induced by an NLPD laser. For faster beat patterns such as those which could be produced by mode-locked lasers, subnanosecond rise time photomultiplier tubes should be used. The technology of these tubes is now making rapid progress.

The way the photocurrent is time analyzed depends on the actual experimental situation. If the signal-to-noise ratio is large enough (high atomic density and efficient pulse pumping of the excited state), the photocurrent may be recorded directly on a fast oscilloscope and the beats detected on a single sweep.

For lower light intensity, the beats are obliterated by the quantum noise in the photocurrent or by electrical noise, so that some kind of signal averaging procedure has to be used. Before the recent advent of fast transient digitizers, the beats could be averaged using a sampling technique. The atoms or molecules were excited by a train of pulses. Using a sampling oscilloscope, each transient signal was detected at a time slightly delayed with respect to the previous one, so that the whole signal was reconstructed after a few hundred laser shots. The signal was then stored in a multichannel analyzer and averaged for many runs. The technique was not very efficient. Due to the pulse to pulse fluctuations of the laser output, the sampling added its own specific noise which had to be averaged out. Furthermore, most of the information was lost by using this method, since each transient was detected only in a small time interval. Closely related to the sampling method is the boxcar technique. The signal is integrated and averaged piece by piece in small preset time gates. This method basically suffers from the same drawbacks as the previous one.

Much better is the technique which makes use of a fast transient digitizer. This recently marketed electronics device allows one to record, to read out and to store in digital form the whole transient signal in real time. It may, for the fastest models, respond to frequencies up to 500 MHz. The number of channels in the transient trace, which limits the resolution of the technique, varies according to the type of apparatus, from about 20 to several hundreds. Using this technique, the signal is no longer sensitive to the fluctuations of the laser output. All the information is used, which also results in a considerable gain in time averaging.

When the light intensity is very low, the photomultiplier receives photons one at a time. The detection then resorts to the photocounting technique. In a standard procedure, the signal is conveniently amplified and sent through a discriminator which eliminates spurious pulses and standardizes those that are acceptable. The output of the discriminator is applied to the stop input of a time-to-amplitude converter whose start signal is initiated by the laser pulse. This converter changes the signal in a pulse whose height is proportional to the time delay between the laser excitation and the photon emission. The pulses are stored according to their amplitude in a channel analyzer. After a large number of pulse detections, the spectrum of pulse heights averaged by the analyzer represents the probability of photon emission versus time. This technique is obviously restricted to very low signal with at most one photon per light pulse, since the converter can receive only a single photon before being initiated again by the next light pulse.

7.4.3 A Variant Technique: The Beam-Laser Experiment

When the species to be studied may be produced in fast moving atomic beams, a variant laser technique may be used which allows one to bypass most of the difficulties connected with the problems of fast signal detection. The method consists of exciting the atoms with a laser crossing the beam at a given angle. Figure 7.10 shows a typical beam-laser spectrom-

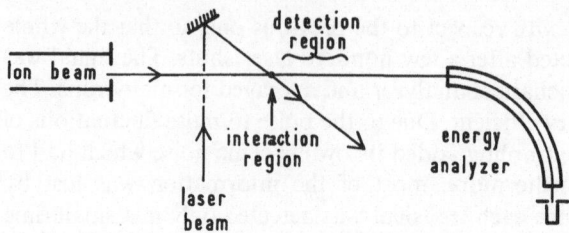

Fig. 7.10. Sketch of a "laser-beam spectrometer"

eter setup. The technique is very similar to the beam-foil method [7.14], the laser beam replacing the thin carbon foil which excites the atoms or ions in this latter type of experiment. For fast-moving atoms ($v \approx 10^3$ km/s) and well-focused laser beams (0.1 mm waist size), the interaction time may be as short as 100 ps, which allows the excitation of beats in the GHz range. Tuning to the resonance line may be quite easily performed by tilting the laser beam direction with respect to the atomic beam and taking advantage of the Doppler effect. Ba^+ ions have been excited in this way with an argon ion laser [7.40] (see Subsection 7.5.3). As in beam foil experiments, the excited species emit along their path and the time modulation of the fluorescence reveals itself as a spatial modulation down the atomic beam. For a good time calibration, the velocity of the beam should of course be accurately determined (with an ion energy analyzer). With a spatial resolution of a tenth of a mm and a beam velocity of 1000 km/s, one can reconstruct time dependent beat patterns at frequency up to 2 GHz, one order of magnitude larger than is possible for a standard quantum beat experiment performed with an NLPD laser. While conserving the high time resolution of beam foil spectroscopy, the beam-laser technique has some important advantages over this latter technique. It provides indeed a much more selective excitation which is very important in order to avoid, for example, undesirable cascading effects. The beam-laser methods thus appear as very promising for the spectroscopy of various ion species.

7.5 Review of Time-Resolved Fluorescence Spectroscopy Experiments

During the last three years, many experiments have been performed which have demonstrated the feasibility of quantum beat detection as a spectroscopic technique and have also provided some new interesting information about excited atomic and molecular states. From the damping constant of the transient fluorescent signal, radiative and collisional lifetimes have been measured. From quantum beat modulations, fine and hyperfine structures, g factors and atomic polarizabilities have been deduced. Interesting information has been obtained about levels which were very difficult to study otherwise, such as molecular excited states or atomic Rydberg states.

In this section, we present a review of these experiments. Besides quantum beat experiments, we include also the closely related lifetime measurements which have recently been performed using pulsed laser excitation. The experiments have been ordered according to the spectroscopic parameters measured.

7.5.1 Lifetimes

The most obvious transient fluorescence experiment is the direct determination of lifetimes by observing the exponential decay of selectively excited states. The experimental technique is somewhat simpler than for quantum beat detection, since one does not have to excite coherently the system and to observe modulations in the light signal following the pulse excitation. However, the technology of such experiments is sufficiently close to quantum beat spectroscopy to justify that they should be described here.

The pulsed laser technique for lifetime measurements competes with other techniques such as Hanle effect, double resonance, phase shift and beam foil spectroscopy. Its main advantage over some of its competitors results from its high efficiency and selectivity which allows one to measure, even in complicated spectra, the lifetimes of well-resolved and -defined levels. Its selectivity is also useful to eliminate cascading effects from upper states which might in some of the other techniques perturb the lifetime measurements.

Of course some precautions have to be taken in order to avoid systematic errors. Among the most important sources of errors, is non-linearities in the detectors which may affect the shape of the exponential decay curve. Extreme care must be taken to avoid ringing effects or over loading of the photomultiplier which should always be adequately wired for fast response. Radiation trapping in the atomic sample may also modify the decay curves and cause an increase of the effective measured lifetime. These effects can always be suppressed by lowering the atomic density.

a) Atoms and Ions

As for many other types of experiments with dye lasers, the Na atom has been a test species for various lifetime measurements. ERDMANN et al. [7.41] have measured the lifetime of the Na $3P$ state, which was well

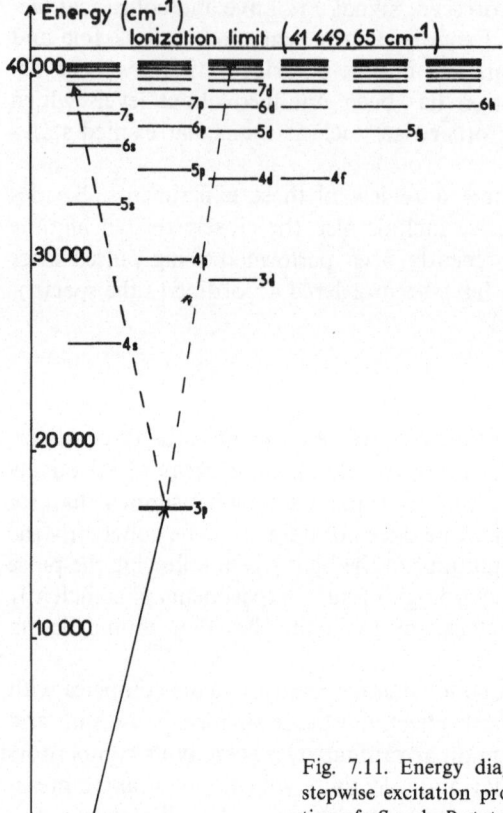

Fig. 7.11. Energy diagram of Na atom showing the stepwise excitation processes which allow the preparation of nS and nD states

known from double resonance and level-crossing studies. They made use of a flashlamp pumped dye laser which was switched off in a time much shorter than the 16 ns lifetime, by acting on the intracavity Lyot filter which determines the wavelength.

More recently, two different experimental groups (in Germany and United States) have independently measured the lifetime of a whole series of nS and nD states of Na (with n up to 13) [7.42–44]. These states were excited by a stepwise excitation process involving two pulsed dye lasers. The first laser excited the $3S - 3P$ transition and the second the $3P - nS$ or $3p - nD$ one (see Fig. 7.11). In one experiment (GALLAGHER et al. [7.44]), both lasers were NLPD lasers pumped by the same N_2 laser which ensured a good synchronization of both excitations. In the other one (GORNIK et al. and KAISER [7.42, 43]), the first step was performed by a flashlamp pumped dye laser and the second by an NLPD laser triggered by the light from the first laser pulse. Figure 7.12 gives an example of

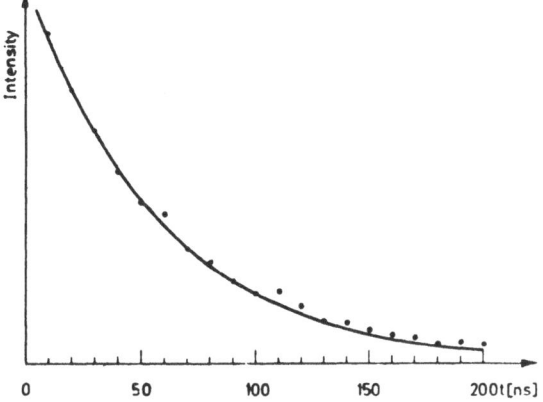

Fig. 7.12. Exponential decay of the fluorescence from the $4D$ level of Na (from [7.42])

fluorescence signal corresponding to the decay of the $4D$ level as observed by GORNIK et al. [7.42]. Using a different kind of stepwise process, the same German group has measured the lifetime of the $4s5s\,^3S_1$ and $4S4d\,^3D_1$ levels of Ca [7.45]. In a first step, the atoms were excited by a low Penning discharge into the metastable $4s4p\,^3P_2$ level. An NLPD laser in a second step brought the atoms in the levels of interest.

One must also quote the very interesting experiments performed by FIGGER et al. [7.46], SIOMOS et al. [7.47] and HELDT et al. [7.48] in the excited states of Fe and Ni. These atomic species were produced in an atomic beam and were excited by a single pulse or by a stepwise process using synchronized NLPD lasers. Radiative lifetimes of single resolved fine structure levels of several multiplets in highly excited atomic configurations were for the first time measured with an accuracy of about 1 %. Figure 7.13 shows a typical decay curve corresponding to the $3d^8 4s\,(a^2F)$ $4py\,^3D_2^0$ level of Ni. The data have been recorded using photo-counting techniques. In these experiments, the use of pulsed dye lasers represents a major improvement over previous techniques. The measured lifetimes yield, in particular, accurate values for the oscillator strengths of several transitions which are of astrophysical interest (they are useful for the determination of abundances of these elements in solar and stellar atmospheres).

The beam laser technique [7.40] has also been recently used to measure with great accuracy the lifetimes of various excited states in atomic or ionic species such as Ba^+ [7.49–50], Ne [7.50], Sr [7.51] and Rb^+ [7.52]. In some of these measurements, the excited state of interest has been directly prepared by laser excitation from the ground state [7.49]. In others [7.50–52], it has been excited by a two-stage process

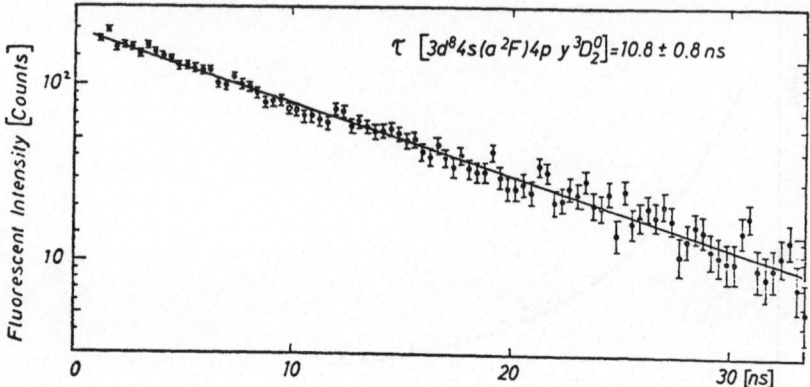

Fig. 7.13. Decay of the fluorescence from the $3d^8 4s(a^2F)4py\,3D_2^0$ level of Ni, in a logarithmic scale (from [7.48])

involving in a first step a non-selective beam-gas interaction populating excited levels and in a second step a selective laser irradiation. This combination of beam-gas and beam-laser techniques permits the excitation of states optically inaccessible from the ground state, as in a regular beam foil experiment, but in contrast with this latter technique, it allows one to avoid the effects of undesirable cascading effects induced by the non-selective first step collision. To suppress these effects, one may-perform a difference measurement, measuring the intensity of the decay curve from the level of interest with and without the laser selective excitation [7.50]. The beam-laser technique, with or without the beam-gas one, seems thus to be a very attractive and promising method for lifetime measurements.

b) Molecules and Radicals

If it is true that the development of lasers has been an important break-through for atomic physics, it is even more so for molecular spectroscopy. Due to the weakness of molecular transitions and to the complexity of molecular spectra, very little was known about molecular excited states until a few years ago. The advent of lasers, especially the tunable ones, has suddenly changed the situation and made available a lot of information about molecular excited states. Lifetimes data, in particular, have been obtained by numerous transient experiments.

Measurement of lifetimes of rovibronic levels in the $B^3\Pi_{0_u}^+$ state of molecular iodine has in particular received much attention. Three relaxation processes mainly contribute to the decay of the excited state

molecules: spontaneous emission, non-radiative spontaneous predissociation and self-quenching collisions with ground states molecules. Collisional contributions may be isolated by studying the variation of lifetimes as a function of gas density. The experiment yields the self-quenching cross section for the level of interest. Extrapolation to zero pressure
allows one to isolate the total contribution of radiative and predissociative processes. Predissociation arises from the coupling of the $B^3\Pi_{0_u}^+$
state with an unbound ungerade state through an off-diagonal rotational
term in the molecular Hamiltonian. This term should strongly depend
on the vibration and rotation number v' and J' of the level. Radiative
lifetimes must on the other hand depend on v', but not on J'. A systematic
study of lifetimes of individual rovibronic levels is thus necessary in
order to test in detail the mechanisms of the relaxation processes and to
be able to separate the radiative from predissociative contributions.

Using an NLPD laser tuned from 4995 to 6400 Å, CAPELLE and
BROIDA [7.53] have measured the lifetimes and self-quenching cross
sections of a large number of vibration levels (with v' varying from 6 to
70). However, the laser they used had a bandwidth of about 3 to 5 Å, too
large to allow the resolution of the various rotation levels, so that the
measured lifetimes were, in fact, averages over many different J' values.
Using a very narrowband NLPD laser, PAISNER and WALLENSTEIN [7.54]
have been able to resolve the rotational structures and to measure the
lifetimes of single rovibronic levels corresponding to v' values ranging
from 21 to 62. More recently, BROYER et al. [7.55] have systematically
extended the measurements to lower vibration quantum numbers and,
for each v' value, studied the variation of the lifetime as a function of J'.
They have in that way been able to separate completely the contributions
of the radiative from that of the predissociative processes. For v' values
between 14 and 18, the predissociative contribution has been found to be
negligibly small. For $v' < 14$ or $v' > 18$, the predissociation is appreciable,
and its contribution to the relaxation rate scales as $J'(J'+1)$, which dem-
onstrates the rotational nature of the term responsible for predissociation in the molecular Hamiltonian. To be complete with this review of I_2
transient fluorescence studies, one should quote miscellaneous measurements performed on some particular levels whose excitation is made
possible by a coincidence between a molecular and a gas laser line.
SHOTTON and CHAPMAN [7.56] have used a cavity dumped He–Ne laser
and KELLER et al. [7.57] a mechanically chopped Ar ion laser to study
by pulse excitation some well-identified rovibronic states. Results of
these lifetime measurements are in good agreement with those obtained
in the more general studies quoted above.

Many other molecular or radical state lifetimes have been measured
using NLPD lasers. Without giving details and without trying to be ex-

haustive, let us mention the experiments performed on ICl by HOLLEMAN and STEINFELD [7.58], on the alkaline earth monohalides by DAGDIGIAN et al. [7.59], on CN by JACKSON [7.60], on OH and OD by GERMAN [7.61] and on BaO by PRUETT and ZARE [7.62].

7.5.2 Zeeman Beats and g-Factors

We describe in the next subsections true quantum beat experiments involving the detection of coherence modulations in excited states. The simplest beat patterns which can be observed correspond to Zeeman splittings in a small magnetic field. They have been observed in both atomic and molecular cases.

a) Atoms

The first experiment demonstrating the feasibility of quantum beat experiments using pulsed dye lasers was performed by GORNIK et al. [7.63] on a thermal beam of ytterbium atoms. The $6s6p^3P_1$ resonant state was excited by a pulse light at 5556 Å from an NLPD laser. The excited state was split by a magnetic field of a few Gauss in three Zeeman components $m = \pm 1, 0$. The polarization of the exciting light was chosen to be perpendicular to the magnetic field so that the $m = \pm 1$ sublevels of the $^3P_1$ state were excited coherently. The fluorescent light in the direction of the magnetic field was detected through a second polarizer, and Zeeman beats at the Bohr frequency corresponding to the energy separation between the $m = +1$ and $m = -1$ substates were detected by a photomultiplier connected to a fast oscilloscope. Figure 7.14 shows a quantum beat modulation at a few MHz produced by a single shot of excitation. The advantage of the laser technique over conventional light source excitation is evident, since one obtains this data in only 5 µs. When spectral lamps were used, this measurement required several hours of averaging. From this experiment, lifetime and g-factors of the $6s6p^3P_1$ state of Yb were deduced. Similar Zeeman beats have been observed in levels 6^1P_1 of Ba and Ca by SCHENCK et al. [7.64] and on Na $4D$ state by SCHENCK and PILLOFF [7.65].

b) Molecules

A Zeeman beat experiment analogous to the one described above has been performed on molecular I_2 by WALLENSTEIN et al. [7.66]. The molecules were excited by a narrowband NLPD laser which selectively prepared individual rovibronic levels. The light pulse was linearly polarized perpendicular to the static magnetic field and detection was made with a polarization either perpendicular or parallel to the field. The signal,

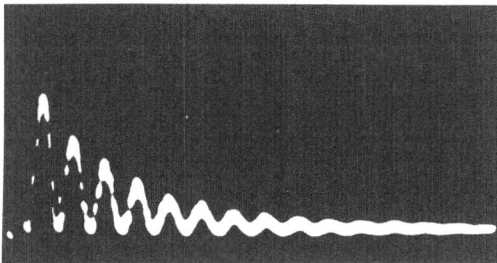

Fig. 7.14. Zeeman quantum beats produced in ytterbium by a single shot of NLPD laser excitation (from [7.63])

about 10^6 times weaker than in a standard atomic experiment, had to be amplified, sampled and averaged for many thousands of runs in order to get a good signal-to-noise ratio. Figure 7.15 shows a typical beat pattern observed in the fluorescence from the $J' = 79$, $v' = 40$ level. The curve (σ) shows a modulation detected with a polarization perpendicular to the field which clearly shows the presence of Zeeman beats. The curve (π) which has been recorded with a polarization parallel to the field is not sensitive to excited state coherence and appears as purely exponential. The difference between the two signals more clearly displays the beat structure. The modulation is washed out after a few oscillations, which indicates a complicated excited state structure. Interpretation of the Zeeman beat pattern is indeed much more complicated than in the previously discussed atomic case, due to the hyperfine interaction between the electronic and the nuclear spins, which splits each rotation vibration level into several unresolved components. These components have different Landé factors, which results in a superposition of several frequencies in the beat pattern. The signal may be theoretically calculated assuming a given value for the electronic Landé factors g_J and for the effective nuclear factors g_e ($g_e = g_I + g_1$ where g_I is the true nuclear g-factor and g_1 the chemical shift correction). A best computer fit to the experimental curves yields both g_J and g_1 values. These parameters have in that way been determined for several rotation-vibration levels.

7.5.3 Hyperfine Structure Beats

Hyperfine structure beats following laser excitation were observed for the first time by HAROCHE et al. [7.67] in the fluorescence light emitted by the $7P_{3/2}$ level of Cs. The relevant energy levels of Cs are shown in Fig. 7.16. The light from an NLPD laser excites the $6^2S_{1/2} - 7^2P_{3/2}$ transition at 4555 Å. The ground state $6^2S_{1/2}$ is split into two hyperfine

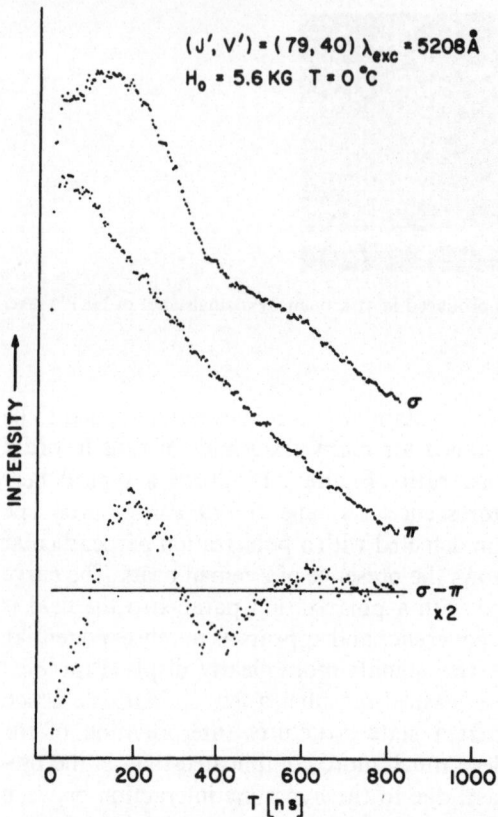

$(J', V') = (79, 40)$ $\lambda_{exc} = 5208\text{Å}$

$H_0 = 5.6\text{ KG}$ $T = 0\,^\circ\text{C}$

INTENSITY

σ

π

$\sigma - \pi$
$\times 2$

| 0 | 200 | 400 | 600 | 800 | 1000 |

T [ns]

Fig. 7.15. Zeeman quantum beats observed in molecular iodine (fluorescence from the $J' = 79$, $v' = 40$ level of the $B^3\Pi_{0_u}+$ state). The upper trace corresponds to a polarization perpendicular to the applied magnetic field; the middle trace corresponds to a polarization parallel to the field and the lower trace represents the difference signal (from [7.66])

levels ($F = 3$ and $F = 4$) separated by the hyperfine interval $\Delta g = 9193$ MHz which is larger than the laser bandwidth (about 1000 MHz). On the other hand, the excited state is split into four hyperfine levels whose separations ν_{54}, ν_{43} and ν_{32} (whose values are given on the figure) are much smaller than the laser width. As a consequence, the laser may be tuned to two distinct sets of transitions a and b starting, respectively, from ground states $F = 3$ and $F = 4$ and exciting two different coherent superpositions of excited states hyperfine structure levels. In each case (a) or (b) the atoms are excited to a superposition of the three hyperfine states which, according to the selection rule $\Delta F = 0, \pm 1$, are coupled to the relevant hyperfine ground state. Thus, when the laser is tuned to transitions (a), one must observe beats at frequencies ν_{43}, ν_{32}, and $\nu_{42} = \nu_{43} + \nu_{32}$. On

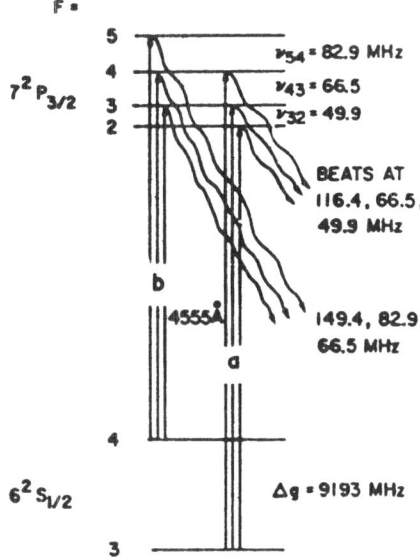

Fig. 7.16. Hyperfine structure of the $6^2S_{1/2}$ and $7^2P_{3/2}$ levels of Cs. Transitions starting, respectively, from levels $F=3$ and $F=4$ in the ground states are labelled by letters (a) and (b). The sets of quantum beat frequencies expected in each case (a) and (b) are indicated (from [7.67])

the other hand, when the laser is in resonance with transitions (b), beats at frequencies v_{54}, v_{43}, and $v_{53} = v_{54} + v_{43}$ must be detected. The laser radiation is linearly polarized and the fluorescent light emitted by the atoms at a right-angle is observed with a polarization either parallel (I_π) or perpendicular (I_σ) to that of the exciting pulse. Detection is made by a sampling technique with an average over a few hundred runs. Figure 7.17 shows the I_π and I_σ beat signals corresponding to resonances (a) and (b). One can compare them to the corresponding theoretical patterns shown under each picture. Those theoretical curves have been plotted from a calculation of the beats using the tensor algebra developed in Subsection 7.2.3 [7.68]. One sees that the modulation patterns in the different signals correspond very well to the theoretical predictions. Note in particular that the modulated parts of I_π and I_σ are opposite in phase, the modulation in I_π being twice as big as the one in I_σ. This result follows from the fact that the quantum beat effect must vanish if one detects the sum of the light intensities emitted along three orthogonal polarizations, this sum being obviously equal to $I_\pi + 2I_\sigma$ (see Subsection 7.2.3). This phase opposition relationship between I_π and I_σ may be used in order to enhance the signal-to-noise ratio of the experiment. By plotting the difference $I_\pi - I_\sigma$, one gets indeed a beat pattern with enhanced modulation depth and reduced noise level. In order to get from this pattern the energy splittings, one performs a Fourier analysis of the $I_\pi - I_\sigma$ signal. The result corresponding to cases (a) and (b) is plotted in Fig. 7.18. In each spectrum, one observes, besides the zero frequency peak

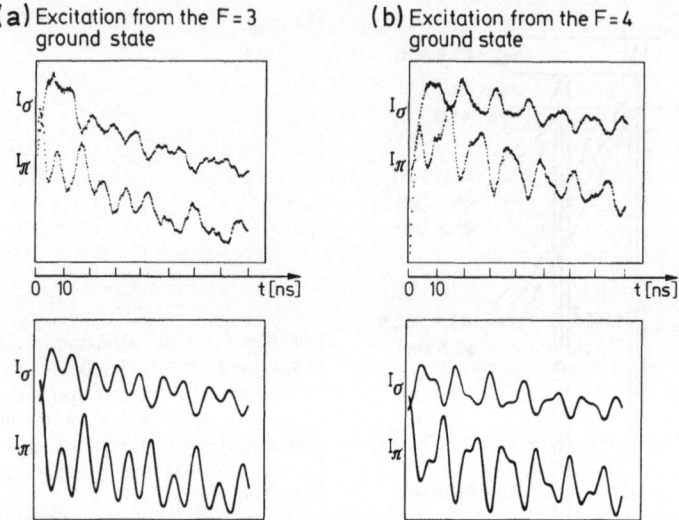

Fig. 7.17a and b. Hyperfine quantum beat signals observed in the fluorescence from the $7P_{3/2}$ level of Cs. (a) The atom is excited from the $F=3$ hyperfine ground state. (b) The atom is excited from the $F=4$ hyperfine ground state. The lower and upper trace in each picture correspond, respectively, to a detection polarization parallel or perpendicular to that of the exciting pulse. The corresponding theoretical plots of the beats are shown under each picture (from [7.67])

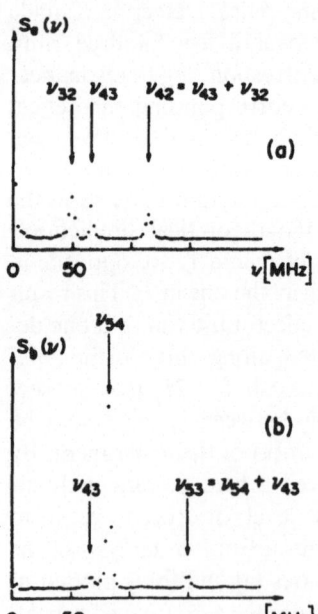

Fig. 7.18a and b. Fourier spectra of the hyperfine quantum beat signals observed in the $7^2P_{3/2}$ level of Cs. (a) Excitation starting from the $F=3$ hyperfine ground state. (b) Excitation starting from the $F=4$ hyperfine ground state (from [7.67])

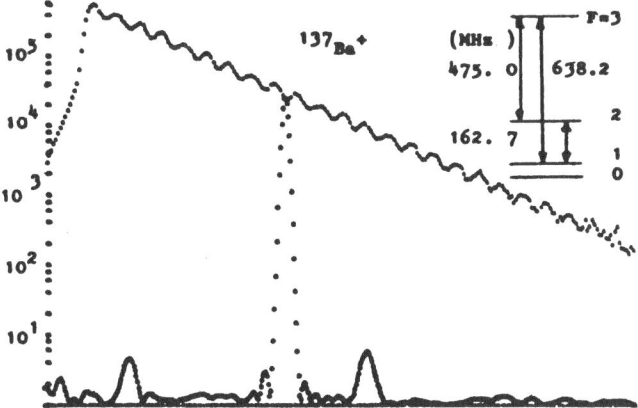

Fig. 7.19. Quantum beat signal and Fourier spectrum of the beats corresponding to the hyperfine structure of the $6^2P_{3/2}$ level of Ba$^+$, as observed in a laser-beam experiment. The energy diagram indicates the three expected beat notes (from [7.40])

corresponding to the unmodulated part of the fluorescent light, three peaks corresponding to the three eigenfrequencies expected in each beat signal. The beat frequencies are measured with an accuracy of about 1 % and are in good agreement with the hyperfine interval measurements made in previous works.

Quantum beat spectroscopy has also been used more recently by DEECH et al. [7.69] to determine the hyperfine structures of the 8, 9, and $10^2D_{3/2}$ levels of Cs. The levels are prepared by stepwise excitation using as sources a conventional Cs lamp for the first step ($6^2S_{1/2} - 6^2P_{1/2}$ transition) and an NLPD laser for the second step ($6^2P_{1/2} - n^2D_{3/2}$ transition). Detection of the fluorescent light is made by photo-counting. Each beat pattern is, as in the experiment described above, the superposition of several note frequencies corresponding to the coherent excitation of hyperfine states belonging to the $n^2D_{3/2}$ level. Analysis of the beats yields the values of the hyperfine structure constants of the 8, 9, and $10^2D_{3/2}$ levels. The values obtained are in good agreement with those obtained recently by level-crossing spectroscopy. In addition, the experiment has permitted the first determination of the radiative lifetimes of these states.

Using the beam-laser technique, ANDRÄ et al. [7.40] have also observed hyperfine structure quantum beats in the $6^2P_{3/2}$ level of Ba$^+$. Excitation of the ions was performed as described in Subsection 7.4.3. The laser beam and the fluorescent light were linearly polarized. The beat pattern was observed as a spatial modulation downstream from the excitation region. Figure 7.19 shows the beat signal and its Fourier spectrum which is very similar to those observed in Cs. The modulations, however, occur at much higher frequencies corresponding to the Ba^+ $6^2P_{3/2}$ hyperfine intervals (peaks in the Fourier spectrum are centered around 162.7, 475 and 638.2 MHz). The same group has recently observed in a similar experiment the hyperfine structure beats emitted by the resonance $6^2P_{3/2}$ level of Na [7.70].

7.5.4 Fine Structure Beats

The study of fine structure intervals in nD states of alkali atoms reveals interesting features for spectroscopists. In some of these states, the structure is inverted while in others it is positive, as it should be according to the ordinary fine structure theory of one electron atom. In Na for example, the fine structures of the $3D$ to $6D$ levels were found to be inverted by classical spectroscopy. In order to study how these structures vary when the principal quantum number n is increased, a series of fine structure quantum beat measurements in the nD states of Na was performed by HAROCHE et al. [7.71] and FABRE et al. [7.72]. The excitation scheme is the same as the one used by GORNIK et al. [7.42] and GALLAGHER et al. [7.44] in their experiments. Two synchronized NLPD lasers pumped by the same N_2 laser excite the nD state via the $3^2P_{3/2}$ intermediate level, and the fluorescence light is detected on its transition back to the $3P$ states (see Fig. 7.11). The difference with the lifetime measurements follows of course from the fact that one must induce fine structure coherences in the excited state and detect fast beats in the photocurrent. The experimental setup is sketched in Fig. 7.20a, and the diagram of relevant energy levels is recalled in Fig. 7.20b. The two laser pulses propagate in opposite directions for maximum overlapping in the Na resonant cell. Two linear polarizers P_1 and P_2 are used to polarize these pulses along directions e_1 and e_2. By rotating these polarizers, one can vary the amount of alignment induced in the intermediate $3^2P_{3/2}$ level and the alignment in the excited $n^2D_{3/2,\,5/2}$ states as well. Calculations using the spherical tensor formalism introduced in Subsection 7.2.3 [7.27] show that the two polarizers have to be crossed at right angles for maximum atomic fine structure coherence in the excited level. The fluorescent light emitted perpendicular to the laser beams is passed through a linear analyzer A which selects a detection polarization e_d and through a filter F which isolates the blue line corresponding to the transition back to the $3P$ level. The light signal is analyzed by a real time transient digitizer and a signal averager. Figure 7.21 shows a typical beat pattern observed in the $9D$ level. One observes a single beat note frequency, since there are only two fine structure levels $D_{3/2}$ and $D_{5/2}$. (No complication arises from the hyperfine structure of the levels, which is smaller than the natural linewidth and completely negligible in this

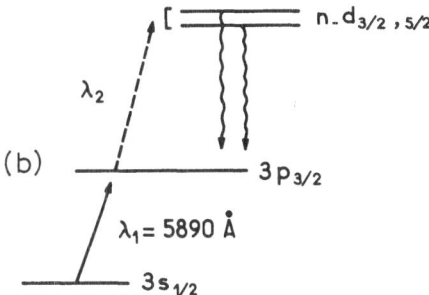

(b)

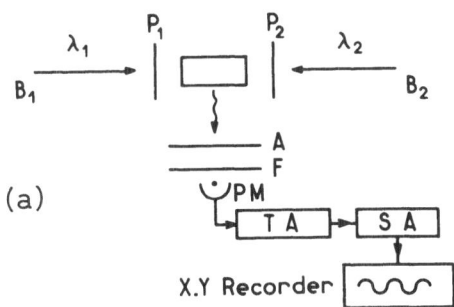

(a)

Fig. 7.20a and b. Fine structure quantum beat experiment in Na nD states. (a) Schematics of the experimental setup. P_1, P_2: polarizers of beams B_1, B_2; A (detection analyzer) and F (filter) in front of PM (photomultiplier). TA and SA transient digitizer and signal averager, respectively. (b) Partial level diagram of Na showing the relevant states for the experiment (from [7.71])

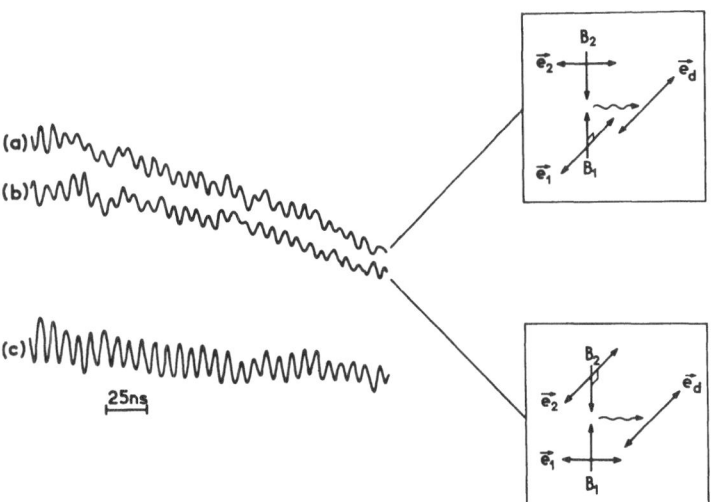

Fig. 7.21. Recording of fine structure beats in Na $9D$ level. Trace a: signal obtained with configuration of polarizers $e_1 \perp e_2, e_d /\!/ e_1$ as show in inset. Trace b: signal obtained with $e_1 \perp e_2, e_d /\!/ e_2$ as shown in inset. Trace c: result of subtracting trace b from trace a (from [7.71])

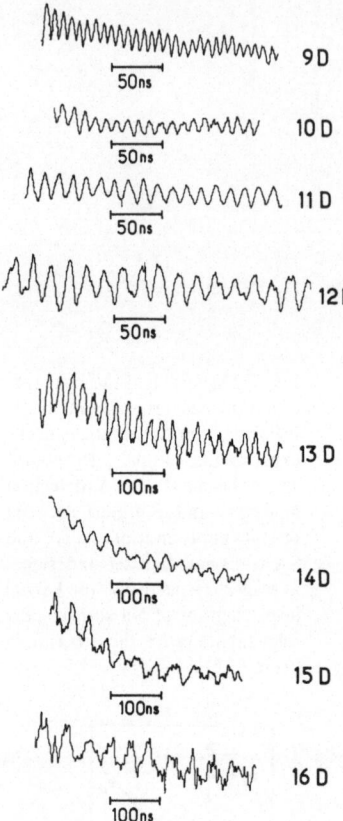

Fig. 7.22. Fine structure quantum beats observed in levels 9D to 16D of Na (from [7.72])

experiment). Two detection polarization schemes have been used, with e_d parallel either to e_1 (trace a) or to e_2 (trace b). Beats are found in phase opposition as expected from theory. Subtraction of the beat signals (a) and (b) once more yields a beat pattern with improved signal-to-noise ratio (trace c). Figure 7.22 shows similarly subtracted beat patterns observed up to level 16D. The continuous slowing down of the modulation frequency reveals decrease of the absolute value of the fine structure intervals as one climbs in the energy diagram. Measurements of the frequencies of these beat patterns have yielded the first experimental determination of fine structure intervals of these levels (with range from 124 MHz to 23 MHz). In Fig. 7.23, we have plotted the variation of these intervals as a function of the effective quantum number of the level. Values for $n=3$ to $n=6$ are taken from other types of experiments (interferometry [7.73] and Doppler-free two-photon spectroscopy [7.74]). As for the sign of the

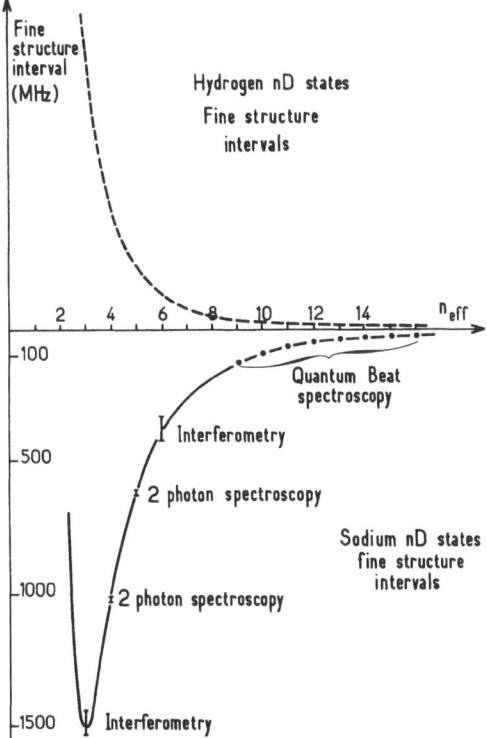

Fig. 7.23. Fine structure intervals in level nD of Na as a function of n_{eff}. The full line curve corresponds to the empirical function $-a/n^3 + b/n^5$ with $a = 96\,500$ MHz and $b = 498\,500$ MHz. The dashed line curve gives, for comparison, the variation with n of the normal fine structure of hydrogen nD level (from [7.72])

fine structure intervals, the quantum beat experiment described above does not determine it. However this sign is known to be negative [7.73] for $n = 3$ to $n = 6$, and it has furthermore been found to be also negative in the $n = 10, 11, 12$ levels by quantum beat Stark effect measurements described in the next subsection [7.75]. One can therefore conclude that all the nD levels of Na have negative fine structure intervals. Variation of the experimental fine structure intervals as a function of n are very well fitted by an empirical function which is the sum of a negative n^{-3} and a positive n^{-5} contribution (solid line curve in Fig. 7.23). For comparison, the positive n^{-3} function giving the normal fine structure of a hydrogen d electron has been plotted in dashed line in the same figure. It is quite remarkable to note that even for very high nD levels the fine structure is so much different in Na and in H. This puzzling result is now being investigated by theoretical spectroscopists.

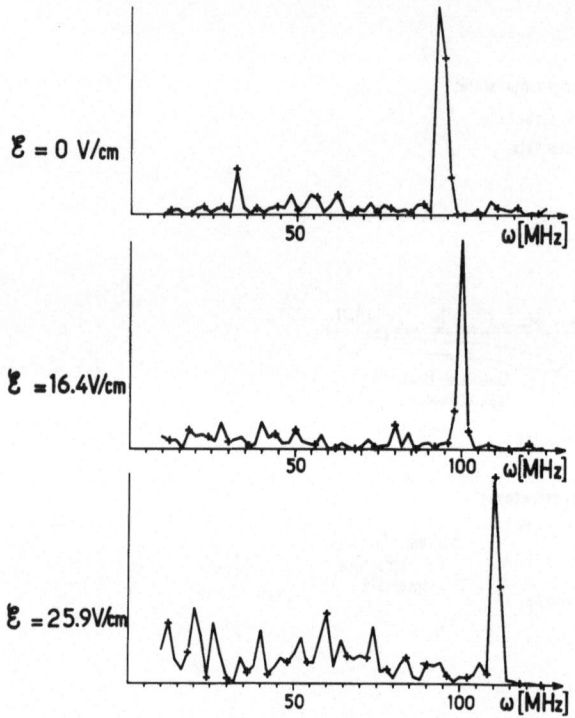

Fig. 7.24. Fourier spectra of the fine structure beat signal in level 10D of Na for three values of the electric field $\mathscr{E}$ (from [7.75])

7.5.5 Polarizability Measurements

Quantum beat spectroscopy may also be used to investigate Stark effects and to perform atomic polarizability measurements. When an electric field is applied to the excited states, the various sublevels are shifted and split apart by the Stark effect so that the atomic quantum beat pattern should be deeply altered. By studying the modification of the quantum beat signal emitted by Na nD states, FABRE and HAROCHE [7.75] have measured the very large tensor polarizabilities of these states. The experiment is performed with the experimental setup described in the previous subsection. A small electric field of a few volts per cm is applied to the atoms which Stark shifts the various $|JM_J\rangle$ substates in the $nD_{3/2}$ and $nD_{5/2}$ fine structure levels. The detection polarization is parallel to the electric field, so that only $\Delta M_J = 0$ interval frequencies can be detected. The fine structure beat spectrum should thus be the superposition of two beat notes corresponding to $M = 3/2 \leftrightarrow M = 3/2$ and $M = 1/2 \leftrightarrow M = 1/2$ intervals. However, calculation shows that the amplitude of beat note $M=1/2 \leftrightarrow M=1/2$ is ten times smaller than the other one, so that there should be essentially one single note in the beat pattern. This beat note is frequency shifted with respect to the zero electric field fine structure quantum beat frequency by an amount proportional to the square of the electric field and to the atomic tensor polarizability. This shift should be positive if the $D_{3/2}$ level lies above the $D_{5/2}$ level (negative fine structure interval). On the contrary, it should be negative in the case of a positive fine structure interval, so that the experiment yields as an incidental result the sign of the fine structure constant in the Na nD levels [7.75]. Figure 7.24 shows the Fourier spectra of the quantum beat pattern observed in level $10D$ for three different electric field values. The positive Stark shift is an evidence for the negative fine structure constant of the level. Similar results have been obtained in levels $11D$ and $12D$. From the variation of the Stark shift with the electric field the tensor polarizabilities of these states have been obtained. They have been found to be enormous, about seven orders of magnitude larger than the polarizabilities of Na $3S$ and $3P$ states. The tremendous enhancement observed between low-lying and highly excited states is due to the increase of the atomic radius (proportional to n^2) and to the decrease of the energy splittings between neighboring levels (proportional to n^{-3}). A very simple hydrogenic model, corrected for the quantum defects of the D states, accounts very well for the experimental results and shows that the atomic polarizability should increase roughly as n^7.

The very large values of the observed polarizabilities indicate that these states should be very sensitive to all kinds of electric perturbation and collisional processes, as demonstrated in the experiment described in the next subsection.

7.5.6 Collision Studies

Quantum beat spectroscopy could also be used to investigate various types of collision processes in atomic excited states. As mentioned in Subsection 7.2.3, the relaxation times of population, orientation and alignment of the excited states could be simultaneously deduced from the observation of the quantum beat signal decay in the presence of collisions. For the time being, such studies have not been systematically performed, although some qualitative results may be found in the literature [7.76].

The effect of collisions on the transient experimental decay of atomic or molecular excited state populations, on the other hand, has been investigated in several types of experiments. As mentioned already in Subsection 7.5.1, many self-quenching cross sections of molecular excited states have in particular been determined. We will not describe these kinds of experiments. We will just briefly discuss in this subsection an experiment performed once more in the Na nD states by GALLAGHER et al. [7.77], which illustrates an interesting effect occurring in atomic highly excited states.

Using again a setup similar to the one described in the previous sections, GALLAGHER et al. have studied the decay rates of Na nD state populations (for $n=5$ to $n=10$), as a function of foreign gas pressure in the milliTorr range. Collisions with Ar, Ne, He have been investigated. In all cases, a lengthening of the effective lifetimes of the D states has been observed, which is interpreted as collisional angular momentum mixing of the D states with $l>2$ states. All the nl states with the same n and $l \geq 2$ are indeed separated by much less than thermal energy (see Fig. 7.11), so that the $D \rightarrow l>2$ collisions are essentially elastic and should have a very large cross section. When an atom is transferred by a collision process to a higher angular momentum state, it radiates much less than in the D states essentially because the allowed radiative decay channels for $l>2$ fall in the infrared. At sufficiently high pressure (mTorr) the mixing time is short compared with the radiative lifetimes of any of the states, so that the effective lifetime observed in these conditions is the statistically averaged lifetime over all the accessible states, which is longer than the nD state lifetime. The theoretical value for this statistical average is proportional to $n^{4.5}$, whereas the lifetime variations measured by GALLAGHER et al. give a $n^{4.43}$ dependence of the lengthened lifetime.

7.6 Comment About the Ultimate Resolution of a Time-Resolved Fluorescence Experiment: The Time-Delayed Techniques

It is clear that the resolution of a quantum beat experiment is basically limited, as in double resonance and level-crossing spectroscopy, by the atomic or molecular natural linewidth. Since the detected atoms live in the excited state an average time $1/\Gamma$, an energy splitting will be resolved by the method only if it is larger than Γ. For example, if one makes the Fourier transform of the single frequency beat signal

$$S(\omega_0, t) = [A + B \cos \omega_0 t] \, e^{-\Gamma t}, \tag{7.44}$$

one will get a spectrum which exhibits a zero frequency peak and a peak around ω_0 frequency which are separated only if $\omega_0 > \Gamma$. Following a classical idea already developed in other types of experiments [7.78], it may be tempting to try to improve this resolution by observing only atoms which have survived in the sample more than the average time $1/\Gamma$. The major drawback of such an experiment is of course the big loss in signal resulting from the fact that only a small fraction of the initially excited atoms is detected. However in some special cases, corresponding to strongly allowed easily saturable transitions, the high density of initially excited atoms may be very important, and the loss in signal due to the delayed detection may be overcome to some extent by signal averaging procedures. Some experiments along these lines have been proposed [7.79] and performed using either classical spectral lamps [7.79, 80] or lasers [7.64, 65, 81, 82] and we will discuss in this section the principle and the results of these experiments.

7.6.1 Quantum Beats Restricted to Long-Lived Atoms

The principle of a time-delayed quantum beat experiment is very simple [7.83]. Instead of detecting radiation from all the atoms, one observes only the light emitted by atoms which have lived in the excited state more than a minimum time t_1. By performing the Fourier analysis of the beats detected from t_1 to $t = \infty$, one obviously gets a spectrum with resonances exhibiting a narrow central peak whose width is inversely proportional to the delay time t_1. These narrow resonances have also important undulations in the wings, which result from the Fourier transform of the sharp detection step function "switched on" at time t_1. These undulations may be suppressed by a convenient apodisation[1] procedure incorporated in the Fourier transform computation. In fact, the narrowing effect due to delayed observation has not been directly demonstrated by using the quantum beat technique, but with the "time-delayed level-crossing method" which is closely related to quantum beats and which we shall now describe.

7.6.2 Time-Delayed Level-Crossing Spectroscopy

The method applies when the Bohr frequency intervals between the decaying levels may be varied by sweeping an external field (magnetic field for example). Let us discuss the theory in the simple case of a single Bohr frequency ω_0 (two-level excited state). Instead of observing the

---

[1] Apodisation is generally understood to be a technique to suppress side maxima or side lobes in the angular pattern of radiation from a finite aperture by making the intensity distribution across the aperture gradual. Here we consider apodisation in time domain in order to suppress side peaks in the wings of a central line of the spectrum.

quantum beat signal $S(\omega_0, t) = (A + B \cos \omega_0 t) e^{-\Gamma t}$ for fixed ω_0 and varied times, one fixes a time interval $t_1 - t_2$ and studies the signal $S(\omega_0, t_1, t_2)$ integrated in this interval as a function of ω_0. This partially integrated quantum beat signal, given by the relation

$$S(\omega_0, t_1, t_2) = \int_{t_1}^{t_2} (A + B \cos \omega_0 t) e^{-\Gamma t} dt \tag{7.45}$$

may be readily calculated. For example, when $t_2 = \infty$, one finds the very simple result

$$S(\omega_0, t_1, \infty) = e^{-\Gamma t_1} \left(\frac{A}{\Gamma} + \frac{B\Gamma}{\Gamma^2 + \omega_0^2} \cos \omega_0 t_1 - \frac{B\omega_0}{\Gamma^2 + \omega_0^2} \sin \omega_0 t_1 \right). \tag{7.46}$$

More general expressions of the signal in which the A and B constants are explicitly calculated from the pulse excitation parameters may be found in the literature [7.64, 79, 81, 84], but the simple formula of (7.46) will be sufficient for our discussion.

Studied as a function of ω_0, the delayed signal $S(\omega_0, t_1, \infty)$ exhibits a central peak around $\omega_0 = 0$ with a width which for $t_1 > 1/\Gamma$ is inversely proportional to t_1. On each side of this narrow central peak, one gets several wiggles with decreasing amplitudes. These undulations are related to the fact that the detection has a step function sensitivity. When ω_0 varies, the phase of the beat modulation at time t_1 changes periodically which entails an oscillation in the $S(\omega_0, t_1, \infty)$ curve. For zero time delay t_1, the quantum beat signal is completely integrated and the expression of (7.46) reduces to the well-known level-crossing signal. One gets a Lorentzian shaped curve around $\omega_0 = 0$ with a width Γ and no sideband wiggles. Compared to the conventional level-crossing signal [$t_1 = 0$ in (7.46)], the time-delayed signal [$t_1 \gg 1/\Gamma$ in (7.46)] has some attractive features. The narrower width of the central peak allows one, for example, to separate two level-crossing peaks unresolved in the natural linewidth. The sideband wiggles may in some cases be useful in order to determine with precision the center of the line with the help of a computer fitting. If they are undesirable, these wiggles may be suppressed by apodisation [7.79].

Figure 7.25 shows typical delayed crossing signals observed by SCHENCK et al. [7.64] in the $6^1 P_1$ resonant state of Ba. The signals correspond to the zero-field level crossing (Hanle effect) and are observed for various delay times $t_1 = \Delta t$ after the laser pulse. From the shape of these curves, the lifetime and g-factor of the level have been obtained. The g-factor of $^1 P_1$ level in Ca has been measured in the same way.

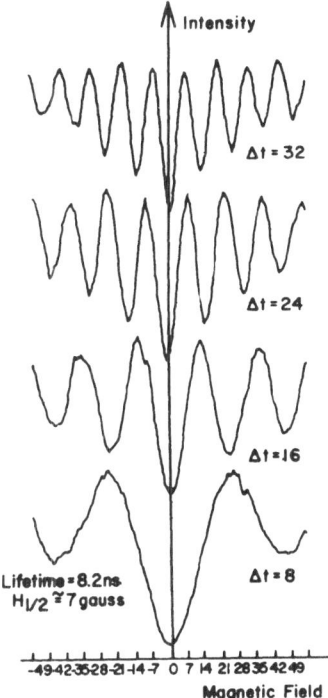

Fig. 7.25. Delayed zero field level crossing signal observed in the $6\,^1P_1$ resonant state of Ba. Δt represents (in ns) various delay times t_1 after the laser pulse (from [7.64])

A similar level-crossing experiment has recently been performed independently by DEECH et al. [7.80] and by FIGGER and WALTHER [7.81] on the $3\,^2P_{3/2}$ excited state of Na. The first group used a classical spectral lamp excitation and the second an NLPD laser. Delaying the signal by several lifetimes has allowed both groups to resolve from each other level-crossing resonances which could not at all be resolved in a conventional level-crossing experiment. An improvement by a factor 2 to 4 over the precision of previous experiments has been obtained in the determination of the a and b hyperfine structure parameters of this Na level. This slight improvement over standard level-crossing techniques is attributed by DEECH et al. [7.80] to the fact that the positions of the crossings are determined with less reliance on the theory of the line profile than is required for unresolved curves. It seems however that the Na case is an especially favorable one for this kind of experiment, since the oscillator strength of the corresponding transition is very large and the level-crossing structures barely unresolved in the excited state. It would be fallacious to consider that the technique may be generally applied to defeat the natural linewidth and to pave the way to ultrahigh-resolution spectroscopy. As in all similar types of experiments, the gain in ultimate resolution is always rather limited and difficult to obtain. Whether it is more advantageous to delay the detection in order to resolve the structure or to keep all the information and determine the line centers by a computer fitting of an unresolved resonance is a difficult question whose answer may be different in each specific case. Besides the application to the $3P_{3/2}$ level of Na, to be complete one should mention similar time-resolved experiments attempted by SCHENCK and PILLOFF [7.65] to measure the very small unresolved hyperfine structure of the $4D_{3/2}$ level of Na and by FELDMAN and METCALF [7.82] to determine with great accuracy the fine structure of the 3^3P level of He.

7.7 Conclusion

The experimental results reported in this chapter clearly demonstrate that time-resolved fluorescence and quantum beat detection using lasers constitute an expanding field in high-resolution spectroscopy. These methods have already yielded numerous and valuable information about lifetimes and energy structures of excited states in atoms and molecules, which were difficult or impossible to study otherwise. In particular, quantum beats appear to be very promising for the study of highly excited atomic Rydberg states, for ion excited state spectroscopy and for molecular spectroscopy.

Various variant techniques could certainly be developed to further broaden the scope of possible experiments. Instead of observing the time dependence of the fluorescence light emitted by an impulsively excited level, one could alternatively study the absorption of a probe laser beam exciting the atomic system from this level to an upper level or to a continuum state. The absorption of this probe beam, if it is properly polarized, should also exhibit a quantum beat behavior which could be detected for example as a modulation of the photoionization current (if the probe beam is ionizing the atom). If one uses as a probe a very short laser pulse, the beats could also be observed as a periodic variation of the total rate of produced ions when the delay between the exciting and the probing pulse is varied. Such methods would be very valuable for the investigation of long-lived metastable states which do not exhibit fluorescence. Probing by a picosecond delayed laser pulse would also allow one to obtain a very high time resolution, in the GHz range, comparable to the one of beam-foil type of experiments, without having to rely on fast detection systems. Experiments along these lines are actually planned in several laboratories [7.85–87].

Acknowledgements. The author thanks Professor COHEN-TANNOUDJI, Dr. C. FABRE and W. M. FAIRBANK for helpful discussions and critical reading of the manuscript. He wishes also to thank all those who have kindly made available to him preprints of their work prior to publication.

References

7.1 W. HANLE: Z. Physik **30**, 93 (1924)
7.2 F. D. COLEGROVE, P. A. FRANKEN, R. R. LEWIS, R. H. SANDS: Phys. Rev. Letters **3**, 420 (1959)
7.3 E. B. ALEXANDROV: Opt. Spectrosc. **14**, 233 (1963)
7.4 A. CORNEY, G. W. SERIES: Proc. Phys. Soc. (London) **83**, 213 (1964)
7.5 M. I. PODGORETSKII, O. A. KHRUSTALEV: Sov. Physics Uspekhi **6**, 682 (1964)
7.6 E. B. ALEXANDROV: Sov. Physics Uspekhi **15**, 436 (1972)
7.7 G. W. SERIES: *Physics of the one and two electron atoms*, ed. by F. BOPP, H. KLEIN-POPPEN (North Holland Publishing Co, Amsterdam 1969) pp. 268—295
7.8 G. BREIT: Rev. Mod. Phys. **5**, 91 (1933)
7.9 A. CORNEY, G. W. SERIES: Proc. Phys. Soc. (London) **83**, 207 (1964)
7.10 E. B. ALEXANDROV: Opt. Spectrosc. **17**, 957 (1964)
7.11 J. N. DODD, R. D. KAUL, D. M. WARRINGTON: Proc. Phys. Soc. (London) **84**, 176 (1964)
7.12 J. N. DODD, W. J. SANDLE, D. ZISSERMANN: Proc. Phys. Soc. (London) **92**, 497 (1967)
7.13 T. HADEISHI, W. A. NIERENBERG: Phys. Rev. Letters **14**, 891 (1965)
7.14 H. J. ANDRÄ: Phys. Rev. Letters **25**, 325 (1970)
7.15 U. FANO, J. M. MACEK: Rev. Mod. Phys. **45**, 553 (1973)
7.16 H. J. ANDRÄ: Physica Scripta **9**, 257 (1974)
7.17 W. W. CHOW, M. O. SCULLY, J. O. STONER: Phys. Rev. A **11**, 1380 (1975)
7.18 A. COMPAAN, L. Q. LAMBERT, I. D. ABELLA: Phys. Rev. A **8**, 1641 (1973)
7.19 P. F. LIAO, P. HU, R. LEIGH, S. R. HARTMANN: Phys. Rev. A **9**, 332 (1974)
7.20 R. L. SHOEMAKER, F. A. HOPF: Phys. Rev. Letters **33**, 1527 (1974)
7.21 R. M. HERMAN, H. GROTCH, R. KORNBLITH, J. M. EBERLY: Phys. Rev. A **11**, 1389 (1975)
7.22 C. COHEN-TANNOUDJI: Ann. Physique **7**, 423 and 469 (1962)
7.23 W. HAPPER: Rev. Mod. Phys. **44**, 169 (1972)
7.24 R. G. BREWER, R. L. SHOEMAKER: Phys. Rev. A **6**, 2001 (1972)
7.25 M. DUMONT: Thesis, unpublished (Paris 1971)
7.26 M. DUCLOY: Thesis, unpublished (Paris 1973)
7.27 M. GROSS: Thèse de 3e cycle, unpublished (Paris 1975) and M. GROSS, M. SILVERMAN, S. HAROCHE: to be published
7.28 C. COHEN-TANNOUDJI: *Atomic Physics* 4, ed. by G. ZU PUTLITZ, E. W. WEBER, A. WINNACKER (Plenum Press, New York, London 1975) pp. 589–614
7.29 C. COHEN-TANNOUDJI: Frontiers in Laser Spectroscopy, ed. by R. BALIAN, S. HAROCHE, S. LIBERMAN (North Holland Publishing Co, Amsterdam 1976)
7.30 P. AVAN and C. COHEN-TANNOUDJI: to be published
7.31 A. OMONT: J. Phys. (Paris) **26**, 26 and 576 (1965)
7.32 A. BEN REUVEN: Phys. Rev. **145**, 7 (1966); ibid. **141**, 34 (1966)
7.33 P. FRANKEN: Phys. Rev. **121**, 508 (1961)
7.34 J. BROSSEL, F. BITTER: Phys. Rev. **86**, 308 (1952)
7.35 J. N. DODD, W. N. FOX, G. W. SERIES, M. J. TAYLOR: Proc. Phys. Soc. **74**, 789 (1959)
7.36 W. HAPPER, in Atomic Physics 4, loc. cit., pp. 651–682

7.37 J. A. MYER, C. L. JOHNSON, E. KIERSTEAD, R. D. SHARMA, I. ITZKAN: Appl. Phys. Letters 16, 3 (1970)

7.38 T. W. HÄNSCH: Appl. Opt. 11, 895 (1972)

7.39 C. V. SHANK, E. P. IPPEN: In *Topics in Applied Physics*, Vol. 1, Dye Lasers, ed. by F. P. SCHÄFER (Springer Berlin, Heidelberg, New York 1973) pp. 121–143

7.40 H. J. ANDRÄ, in Atomic Physics 4, loc. cit., pp. 635–649

7.41 T. A. ERDMANN, H. FIGGER, H. WALTHER: Opt. Commun. 6, 166 (1972)

7.42 W. GORNIK, D. KAISER, W. LANGE, J. LUTHER, H. H. RADLOFF, H. H. SCHULZ: Appl. Phys. 1, 285 (1973)

7.43 D. KAISER: Phys. Letters 51A, 375 (1975)

7.44 T. F. GALLAGHER, S. A. EDELSTEIN, R. M. HILL: Phys. Rev. A 11, 1504 (1975)

7.45 W. GORNIK, D. KAISER, W. LANGE, J. LUTHER, K. MEIER, H. H. RADLOFF, H. H. SCHULZ: Phys. Letters A 45, 219 (1973)

7.46 H. FIGGER, K. SIOMOS, H. WALTHER: Z. Physik 270, 371 (1974)

7.47 K. SIOMOS, H. FIGGER, H. WALTHER: Z. Physik 272, 355 (1974)

7.48 J. HELDT, H. FIGGER, K. SIOMOS, H. WALTHER: Astron. Astrophys. 39, 371 (1975)

7.49 H. J. ANDRÄ, A. GAUPP and W. WITTMANN: Phys. Rev. Letters 31, 501 (1973)

7.50 H. HARDE, G. GUTHÖRLEIN: Phys. Rev. A 10, 1488 (1974)

7.51 H. J. ANDRÄ, H. J. PLÖHN, W. WITTMANN, A. GAUPP, J. O. STONER, M. GAILLARD: to be published (1975)

7.52 M. GAILLARD, H. J. ANDRÄ, A. GAUPP, W. WITTMANN, H. J. PLÖHN, J. O. STONER: Phys. Rev. A, 12, 987 (1975)

7.53 G. A. CAPELLE, H. P. BROIDA: J. Chem. Phys. 58, 4212 (1973)

7.54 J. A. PAISNER, R. WALLENSTEIN: J. Chem. Phys. 61, 4317 (1974)
     J. A. PAISNER: Thesis, Stanford University, unpublished (1974)

7.55 M. BROYER, J. VIGUÉ, J. C. LEHMANN: J. Chem. Phys. 63, 5428 (1975)

7.56 K. C. SHOTTON, G. D. CHAPMAN: J. Chemical Phys. 56, 1012 (1972)

7.57 J. C. KELLER, M. BROYER, J. C. LEHMANN: Compt. Rend. B 277, 369 (1973)

7.58 G. W. HOLLEMAN, J. I. STEINFELD: Chem. Phys. Letters 12, 431 (1971)

7.59 D. J. DAGDIGIAN, H. W. CRUSE, R. N. ZARE: J. Chem. Phys. 60, 2330 (1974)

7.60 W. M. JACKSON: J. Chem. Phys. 61, 4177 (1974)

7.61 K. R. GERMAN: J. Chem. Phys. 62, 2584 (1975)

7.62 J. G. PRUETT, R. N. ZARE: J. Chem. Phys. 62, 2050 (1975)

7.63 W. GORNIK, D. KAISER, W. LANGE, J. LUTHER, H. H. SCHULZ: Opt. Commun. 6, 327 (1972)

7.64 P. SCHENCK, R. C. HILBORN, H. METCALF: Phys. Rev. Letters 31, 189 (1973)

7.65 P. S. SCHENCK, H. S. PILLOFF: Bull. Am. Phys. Soc. 20, 678 (1975)

7.66 R. WALLENSTEIN, J. A. PAISNER, A. L. SCHAWLOW: Phys. Rev. Letters 32, 1333 (1974)

7.67 S. HAROCHE, J. A. PAISNER, A. L. SCHAWLOW: Phys. Rev. Letters 30, 948 (1973);
     S. HAROCHE, J. A. PAISNER: *Laser Physics*, ed. by R. G. BREWER, A. MOORADIAN (Plenum Press, New York 1975) pp. 445–455

7.68 J. A. PAISNER: Thesis, Stanford University, unpublished (1974)

7.69 J. S. DEECH, R. LUYPAERT, G. W. SERIES: J. Phys. B 8, 1406 (1975)

7.70 H. J. ANDRÄ: Private communication

7.71 S. HAROCHE, M. GROSS, M. SILVERMAN: Phys. Rev. Letters 33, 1063 (1974)

7.72 C. FABRE, M. GROSS, S. HAROCHE: Opt. Commun. 13, 393 (1975)

7.73 K. W. MEISSNER, K. F. LUFT: Ann. Phys. (Leipzig) 29, 968 (1937)

7.74 T. W. HÄNSCH, K. C. HARVEY, G. MEISEL, A. L. SCHAWLOW: Opt. Commun. 11, 50 (1974)
     F. BIRABEN, B. CAGNAC, G. GRYNBERG: Phys. Letters 48A, 469 (1974)
     M. D. LEVENSON, M. M. SALOUR: Phys. Letters 48A, 331 (1974)

7.75 C. FABRE, S. HAROCHE: Opt. Commun. 15, 254 (1975)

7.76 W. Lange, J. Luther, A. Steudel, in *Advances in Atomic and Molecular Physics*, Vol. 10 (Academic Press Inc., New York, San Francisco, London 1974) pp. 187–190

7.77 T. F. Gallagher, S. A. Edelstein, R. M. Hill: Phys. Rev. Letters **35**, 644 (1975)

7.78 I. J. Ma, G. zu Putlitz, G. Schütte: Physica **33**, 282 (1965)

7.79 G. Copley, B. P. Kibble, G. W. Series: J. Phys. B **1**, 724 (1968)

7.80 J. S. Deech, P. Hannaford, G. W. Series: J. Phys. B **7**, 1131 (1974)

7.81 H. Figger, H. Walther: Z. Phys. **267**, 1 (1974)

7.82 M. Feldman, H. Metcalf: Bull. Am. Phys. Soc. **20**, 678 (1975)

7.83 G. W. Series: In *Proc. of the Intern. Symp. in Very High Resolution Spectroscopy*, ed. by R. A. Smith (Wiley, New York 1975)

7.84 R. C. Hilborn, R. L. de Zafra: J. Opt. Soc. Am. **62**, 1492 (1972)

7.85 M. Gaillard: Private communication

7.86 C. Fabre: Private communication

7.87 T. W. Ducas, M. G. Littman, M. L. Zimmerman: Phys. Rev. Letters **35**, 1752 (1975)

8. Doppler-Free Two-Photon Absorption Spectroscopy

N. BLOEMBERGEN and M. D. LEVENSON

With 15 Figures

8.1 Historical Introduction

With the formulation of quantum mechanics by DIRAC [8.1] and others in the late twenties, it became clear that the interactions of electromagnetic radiation and matter also comprise processes in which several photons are involved. An early example is provided by Raman scattering. In this process one photon is absorbed, another is emitted and the difference in photon energy is taken up by a transition in the material system from an initial state $|i\rangle$ to a final state $|f\rangle$. The symbolic Hamiltonian describing this event is proportional to $a_1 a_2^\dagger c_i c_f^\dagger$, where a_1 represents an annihilation operator for photons in mode 1, $a_2^\dagger$ represents the creation operator of a photon in mode 2, while c_i takes an electron out of state $|i\rangle$, and $c_f^\dagger$ puts one electron in state $|f\rangle$. The closely related two-photon absorption process is described by a term in the Hamiltonian proportional to $a_1 a_2 c_i c_f^\dagger$. The theory for this latter process was developed by GOEPPERT-MAYER in her PhD thesis at the University of Göttingen [8.2]. The energy balance for the two processes is shown at the top of Fig. 8.1. While the Raman effect was discovered in 1927, the two-photon absorption process was first demonstrated [8.3] in 1961, about thirty years after the theoretical paper of GOEPPERT-MAYER. This difference in the experimental development can be explained by the fact that the Raman scattering involves the spontaneous emission of a Stokes-shifted photon. The spontaneously scattered light intensity at ω_2 is proportional to the intensity of the incident field at ω_1. The power absorbed in the two-photon absorption is, however, proportional to the square of the incident intensity. The demonstration of two-photon absorption requires a high light intensity, and had to await the advent of lasers. Furthermore, the frequency of the laser must be adjustable if the two-photon transition occurs between sharp energy levels and no accidental coincidences between laser frequency and energy separations exist. It is therefore understandable that two-photon spectroscopy could really develop fully only after high power tunable dye lasers became available [8.4].

Spontaneous Raman scattering required neither extremely high intensity nor tuning of the incident light. It is not a nonlinear process in the sense of the semiclassical description. The stimulated Raman effect, however, is a nonlinear process in this sense and may be described in terms of a polarization, which is a nonlinear (cubic) function of the applied field magnitudes. Its demonstration was made possible only after high intensity lasers became available.

The current interest in two-photon absorption spectroscopy is based on the following characteristics:

1) The initial and final states have the same parity.

2) The final state may have an excitation energy in the far UV, while the incident light beam has a frequency in the near UV or blue part of spectrum.

3) It is possible to eliminate momentum transfer between the electromagnetic field and the atom or molecule, and consequently to eliminate Doppler broadening.

The first two characteristics make it possible to reach different states that cannot be reached from the same initial state in a one-photon process. The last characteristic assures that the highly excited states can be investigated with high spectroscopic resolution.

The momentum transfer relations are indicated in the bottom part of Fig. 8.1. In general, the Doppler broadening in angular frequency is $\Delta\omega = \Delta k \cdot v$, where $\hbar \Delta k$ is the change in momentum of the electromagnetic field and v is the atomic velocity.

The fact that the Doppler broadening is less for forward Raman scattering than for backward scattering has been firmly established for hydrogen gas. The absence of Doppler broadening, if two photons of equal energy and opposite momentum are absorbed, was first analyzed by CHEBOTAYEV and coworkers [8.5] in 1970, and experimentally demonstrated by CAGNAC et al. [8.6] and by LEVENSON and BLOEMBERGEN [8.7] in 1974. Numerous results have subsequently been published and Doppler-free two-photon spectroscopy promises to be an important new spectroscopic tool. It is different from other nonlinear techniques that eliminate Doppler broadening such as saturation "Lamb-dip" spectroscopy, discussed elsewhere in this volume. In that method a small segment of the inhomogeneous Doppler distribution of resonant frequencies is selected to contribute to the signal. In two-photon absorption without momentum transfer all atoms or molecules contribute to the signal. A simple way to see this is to consider the case of an atom at rest with the frequency v of the light wave adjusted so that $2hv$ corresponds exactly to the energy difference between two sharp energy levels with the same parity. Now consider an atom moving with a velocity component v parallel to one of the light beams. The apparent frequency of this beam is down-shifted by an amount $-(v/c)v$. The apparent frequency of the light beam, propagating in the opposite direction, is up-shifted by the

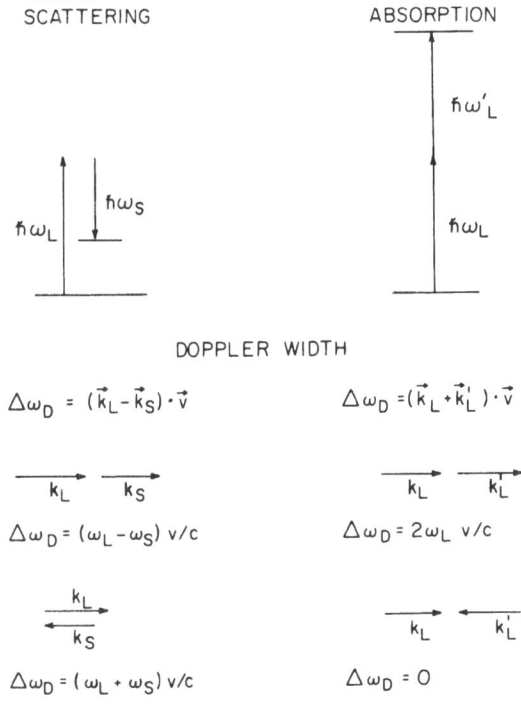

Fig. 8.1. Comparison of the energy and momentum relations for Raman and two-photon absorption processes

same amount, $+(v/c)v$. Thus, the two linear Doppler shifts cancel each other exactly, and for any atom the resonant condition is fulfilled, if quadratic Doppler shifts are ignored.

Similar considerations apply to higher-order processes [8.8]. In Fig. 8.2 the energy and momentum configuration are sketched for Doppler-free three-photon absorption, and a hyper-Raman process in which two photons are absorbed and another one is emitted. In this manner states of the opposite parity could also be investigated in high resolution.

It is, of course, also possible to eliminate the Doppler shift for both one-photon and two-photon spectroscopy [8.9] by utilizing well-collimated atomic and molecular beams moving at right angles to the light beam. Very small difference frequencies can, of course, be detected by the method of coherent quantum beats. This method does not require a very monochromatic source. Combination of optical pumping and radiofrequency spectroscopy is another very powerful method.

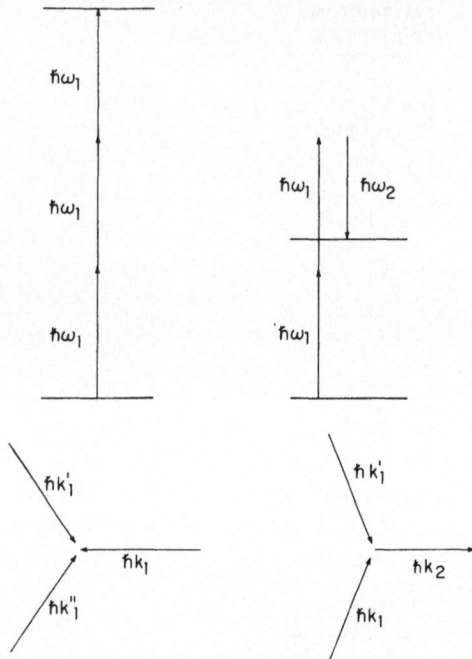

Fig. 8.2. Energy and momentum relations for two three-photon processes with zero momentum transfer

In the present chapter a review will be given of two-photon absorption spectroscopy in vapors. The theoretical description is presented in Section 8.2. The experimental method will be described in Section 8.3, and new spectroscopic results obtained with this technique will be reviewed in Section 8.4. A summary of the advantages and disadvantages of the technique as well as an outlook for its future significance concludes the chapter.

8.2 Theory of Two-Photon Absorption

The evolution of a material system under the influence of interaction with electromagnetic fields is described by the Liouville equation for the total density matrix,

$$\dot{\varrho} = -i\hbar^{-1}[\mathscr{H}, \varrho] \tag{8.1}$$

where the Hamiltonian consists of four parts

$$\mathscr{H} = \mathscr{H}_0 + \mathscr{H}_{\text{field}} + \mathscr{H}_{\text{int}} + \mathscr{H}_{\text{damping}}. \tag{8.2}$$

Here $\mathscr{H}_0$ is the Hamiltonian of the atom or molecule, leading to a set of energy eigenstates $|n\rangle$ with energy W_n. The effect of dc applied magnetic or electric fields may be incorporated in $\mathscr{H}_0$. The field Hamiltonian consists of a number of harmonic oscillators, one for each eigenmode of the electromagnetic field. The interaction Hamiltonian for a one-electron system may be written as

$$\mathscr{H}_{\text{int}} = -\frac{e}{2c}(\boldsymbol{p} \cdot \boldsymbol{A} + \boldsymbol{A} \cdot \boldsymbol{p}) + \frac{e^2}{2mc^2}\boldsymbol{A} \cdot \boldsymbol{A} \tag{8.3a}$$

where A is the vector potential of the electromagnetic field and $e(<0)$ is the electronic charge. An equivalent representation is the multipole expansion

$$\mathscr{H}_{\text{int}} = -e\boldsymbol{r} \cdot \boldsymbol{E} - \tfrac{1}{2}e\boldsymbol{Q} : \nabla\boldsymbol{E} - \boldsymbol{m} \cdot \boldsymbol{H} + \dots . \tag{8.3b}$$

The first term is the electric dipole moment interaction, the second term the electric quadrupole interaction, with the quadrupole tensor defined by

$$Q_{ij} = r_i r_j - \tfrac{1}{3}r^2\delta_{ij} . \tag{8.3c}$$

The third term represents the magnetic dipole interaction, etc. In many important applications in optics only the electric dipole moment term needs to be retained.

The damping Hamiltonian takes account of the fact that the material system may interact weakly with a large number of degrees of freedom. The spontaneous emission into the quasi-continuum of eigenmodes of the electromagnetic field and the influence of collisions with other atoms or molecules in a gas, or phonons in a crystal, may be taken into account statistically by simple phenomenological damping terms.

The expectation value of the electric polarization is given by

$$\boldsymbol{P} = N \operatorname{Tr}(e\boldsymbol{r}\varrho) \tag{8.4}$$

where N is the number of one-electron systems per unit volume. The set of equations must be made self-consistent, by requiring that P and E satisfy Maxwell's wave equation.

$$\nabla \times \nabla \times \boldsymbol{E} + \frac{1}{c^2}\frac{\partial^2\boldsymbol{E}}{\partial t^2} = -\frac{1}{c^2}\frac{\partial^2\boldsymbol{P}}{\partial t^2} . \tag{8.5}$$

The set of Eqs. (8.1–5) is, of course, too general to provide much physical insight. It describes all optical phenomena, including lasers and all kinds of nonlinear interactions. To proceed further, the situation is usually restricted to a very small number of modes of the electromagnetic field (except for the damping by spontaneous emission) and a very small number (often two or three) of energy levels of the material system. These are selected on the basis that they make a dominant contribution, because they are at or near resonance with the selected electromagnetic modes.

For the purpose of two-photon absorption, attention may be restricted to two modes of the electromagnetic field, and these fields will be described classically. The electric field is given by

$$E = \tfrac{1}{2}(\hat{e}_1 E_1 e^{ik_1 \cdot r - i\omega_1 t} + \hat{e}_2 E_2 e^{ik_2 \cdot r - i\omega_2 t}) + \text{c.c.} \tag{8.6}$$

where c.c. means complex conjugate. This expression contains the polarization directions, amplitudes, wave vectors and frequencies of the two waves. Some, or all, of the quantities may be taken equal to each other. If they are all equal, one has the degenerate problem of two-photon absorption from a single wave.

In the semiclassical description, where the field quantities are considered classically, the density matrix reduces to one for the material system alone. Stationary solutions for the polarization given by (8.4) can be obtained in terms of ascending powers of the field amplitudes E_1 and E_2 by iterative perturbation procedures [8.10]. This procedure has the advantage that all energy levels of the material system can be kept in the calculation, and it permits establishment of the connection between two-photon absorption and other nonlinear optical processes. This approach will be followed in the next subsections. The final Subsection 8.2.6, will briefly describe some characteristics of solutions restricted to three-level systems but with no restriction as to the field amplitudes.

8.2.1 Nonlinear Susceptibilities

In systems with inversion symmetry, such as atomic or molecular vapors, on which the interest is centered in this chapter, the electric polarization can contain only odd powers of the electric field amplitudes. The lowest non-vanishing nonlinear polarization is a term cubic in the applied field amplitudes. For the sake of generality, the excitation of a third electromagnetic mode with amplitude E_3, and frequency ω_3 is temporarily admitted. A polarization at the sum frequency $\omega_4 = \omega_1 + \omega_2 + \omega_3$ will be created if the system is initially in the ground state $|g\rangle$. Retaining only the electric dipole term in the interaction Hamiltonian, steady state perturbation theory yields the result ([8.10]),

$$\hat{e}_4 \cdot P(\omega_4) = \tfrac{1}{2}\chi^{(3)} E_1 E_2 E_3 e^{i(k_1 + k_2 + k_3) \cdot r - i\omega_4 t} + \text{c.c.} \tag{8.7}$$

with

$$\chi^{(3)} = \frac{1}{4} Ne^4\hbar^{-3} \sum_{nn'n''} (r \cdot \hat{e}_1)_{gn} (r \cdot \hat{e}_2)_{nn'} (r \cdot \hat{e}_3)_{n'n''} (r \cdot \hat{e}_4)_{n''g}$$

$$\times \frac{1}{(\omega_{ng} - \omega_1 - i\Gamma_{ng}/2)(\omega_{n'g} - \omega_1 - \omega_2 - i\Gamma_{n'g}/2)(\omega_{n''g} - \omega_4 - i\Gamma_{n''g}/2)}$$

$$+ \text{ permutations of indices 1 through 4 .} \tag{8.8}$$

From the structure of $\chi^{(3)}$, which is defined as a scalar quantity in (8.8) it is evident that one can define a fourth-rank tensor susceptibility, whose components are given by choosing the unit vectors $\hat{e}_1$ through $\hat{e}_4$ along the three axes of a Cartesian coordinate system,

$$\chi^{(3)} = \hat{e}_1 \hat{e}_2 : \chi^{(3)}_{\text{tensor}} : \hat{e}_3 \hat{e}_4 .$$

The damping constant Γ_{ng} represents the full width at half maximum of the transition $|g\rangle \rightarrow |n\rangle$. If this is the natural width, it is determined by the lifetime against spontaneous emission of the state $|n\rangle$. The 23 other permutations of the indices 1 through 4 arise from assuming different sequences in which the photon absorptions at ω_1, ω_2, and ω_3 and photon emission at ω_4 take place. This polarization, when substituted in Maxwell's equation, will obviously create a wave at the sum frequency ω_4. The intensity in this wave will be proportional to $|\chi^{(3)}|^2$. The perturbation expression is valid when

$$\hbar|\omega_{gn} - \omega - i\Gamma_{gn}/2| \gg |e(r \cdot \hat{e})E| . \tag{8.9}$$

It is always valid off resonance, and even at resonance, provided that the damping in energy units is larger than the magnitude of the interaction Hamiltonian. Each of the terms in (8.8) contains the product of three denominators. Resonances may occur when either one of the applied frequencies or some linear combinations of these, such as $\omega_1 + \omega_2$, corresponds to the difference in energy between a pair of levels of the material system, $\hbar\omega_{ng} = W_n - W_g$.

It is clear from (8.6) that the real physical fields contain both positive and negative frequency components. The frequencies ω_1, ω_2, and ω_3 may be chosen positive or negative, or equal to each other. Depending on these choices and depending on whether some resonant conditions are met or not, different physical nonlinear processes are described. For two-photon absorption one takes $\omega_2 = -\omega_3$ and consequently $\omega_4 = \omega_1$. Furthermore, the resonant condition $\omega_{n'g} = \omega_1 + \omega_2$ is satisfied, i.e., the excited state $|n'\rangle = |f\rangle$ can be reached by absorption of two photons from the ground state $|g\rangle$. There are only two such resonant terms, depending on the order of ω_1 and ω_2. The summation over n' is reduced to a single final state $|f\rangle$. The double summation over n and n'' may be written as the square of a single summation. Note that $\omega_{ng} - \omega_1 - i\Gamma_{ng} \approx \omega_{ng} - \omega_1 = \omega_{ng} - \omega_4$. For two-photon absorption $\chi^{(3)} = i\chi''^{(3)}$ becomes a pure imaginary quantity

$$\chi''^{(3)}(-\omega_1, \omega_1, \omega_2, -\omega_2)$$
$$= \frac{1}{2} N e^4 \hbar^{-3} \Gamma_{gf}^{-1} \left| \sum_n \frac{(r \cdot \hat{e}_1)_{gn}(r \cdot \hat{e}_2)_{nf}}{\omega_{ng} - \omega_1} + \sum_n \frac{(r \cdot \hat{e}_2)_{gn}(r \cdot \hat{e}_1)_{nf}}{\omega_{ng} - \omega_2} \right|^2 .$$

$$(8.10)$$

The nonlinear polarization at frequency ω_1 is 90° out of phase with the field at ω_1 and proportional to the intensity at ω_2

$$P(\omega_1) = i\chi''^{(3)} E_1 |E_2|^2 .$$

The power absorbed from the beam at ω_1 is proportional to $|E_1|^2 |E_2|^2$ and to imaginary part of $\chi^{(3)}$. The same is true for the power absorbed from the beam at ω_2. The total power absorbed by a unit volume of the material system in these two-photon transitions is $(1/2)(\omega_1 + \omega_2)\chi''^{(3)} |E_1|^2 |E_2|^2$. Dividing this expression by the energy $\hbar(\omega_1 + \omega_2)$ involved in each transition, and by the number of atoms per unit volume, one finds the transition rate for two-photon absorbtion per atom at resonance,

$$w_{f \leftarrow g} = e^4 \hbar^{-4} \Gamma_{gf}^{-1} \left| \sum_n \frac{(r \cdot \hat{e}_1)_{gn}(r \cdot \hat{e}_2)_{nf}}{\omega_{ng} - \omega_1} + \sum_n \frac{(r \cdot \hat{e}_2)_{gn}(r \cdot \hat{e}_1)_{nf}}{\omega_{ng} - \omega_2} \right|^2$$
$$\times \frac{1}{4} |E_1|^2 |E_2|^2 .$$

$$(8.11)$$

This expression had already been obtained by MAYER from Fermi's golden rule [8.2]. The quantity between the vertical bars is proportional to the second-order matrix element connecting the ground state and final state. The density of final states at resonance is equal to $4\Gamma_{gf}^{-1}$. It is well to remember at this point that $|E_1|$ and $|E_2|$ represent the real physical field amplitudes, which are twice the Fourier amplitudes of the positive frequency components as shown in (8.6).

Other nonlinear processes describable by the general expression for the third-order nonlinear susceptibility given by (8.8) include third harmonic generation, obtained by taking $\omega_1 = \omega_2 = \omega_3$ and $\omega_4 = 3\omega_1$, and generation of the combination frequency $\omega_4 = 2\omega_1 + \omega_2$, by taking $\omega_1 = \omega_3 \neq \omega_2$. The intensity generated in these parametric processes is proportional to the square absolute value $|\chi^{(3)}|^2$. Both these nonlinear processes were

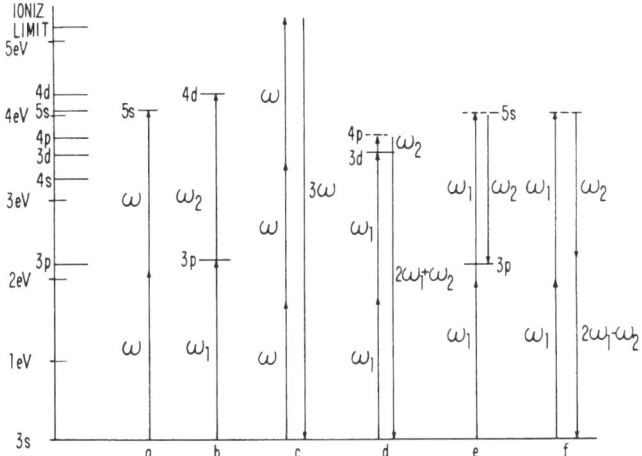

Fig. 8.3a–f. Nonlinear processes in the Na-atom. (a) Two-photon absorption, (b) Two-photon absorption with resonance of intermediate level, (c) Third harmonic generation, (d) Sum-frequency three-wave mixing, (e) Hyper-Raman three-photon process, (f) Parametric mixing with generation of the combination frequency $2\omega_1 - \omega_2$

demonstrated in Na vapor by HARRIS and coworkers [8.11], by SOROKIN et al. [8.12] and BONCH-BRUEVICH et al. [8.13] well before the Doppler-free two-photon absorption was demonstrated in the same material. The general expressions (8.7) and (8.8) are also useful to analyze more precisely what happens to the two-photon transition rate, when ω_1 approaches a single-photon resonant frequency ω_{ng}. In this case the two-photon process becomes mixed with the cascade of two one-photon absorption processes. The distinction between these resides in the role played by the diagonal density matrix element ϱ_{nn} in the calculation of $\chi^{(3)}$. There are rather subtle differences in damping characteristics, which are exhibited by the more detailed solutions in Subsection 8.2.6, if there is a real population in the intermediate state ϱ_{nn}. Analogous discussions have been given by SHEN [8.14] for the distinction between a Raman process, and resonant fluorescence following a one-photon absorption. The different nonlinear processes mentioned here are schematically represented in Fig. 8.3 with some energy levels of the Na atom sketched in. The cases a–d correspond to experiments that have actually been carried out in Na vapor.

8.2.2 Doppler Width

The results in the preceding section were derived for an atom at rest. Now consider the case that the light beams at ω_1 and ω_2 move in the opposite direction, $\boldsymbol{k}_1 = k_1 \hat{\boldsymbol{x}}$ and $\boldsymbol{k}_2 = -k_2 \hat{\boldsymbol{x}}$. The apparent light frequencies seen by an atom moving with a velocity component v_x along the direction of the first light beam are $\omega_1 - k_1 v_x$ and $\omega_2 + k_2 v_x$, respectively. The transition rate in the neighborhood for two-photon resonance may now be written by the following replacement of (8.11),

$$w_{f \leftarrow g} = \frac{1}{4} e^4 \hbar^{-4} |E_1|^2 |E_2|^2 \frac{\Gamma_{gf}}{[\omega_{fg} + (k_1 - k_2)v_x - \omega_1 - \omega_2]^2 + (\Gamma_{gf}/2)^2}$$

$$\times \left| \sum_n \frac{(r \cdot \hat{e}_1)_{gn}(r \cdot \hat{e}_2)_{nf}}{\omega_{ng} + k_1 v_x - \omega_1} + \frac{(r \cdot \hat{e}_2)_{gn}(r \cdot \hat{e}_1)_{nf}}{\omega_{ng} - k_2 v_x - \omega_2} \right|^2 . \tag{8.12}$$

The Doppler corrections in the off-resonant denominator of the second-order matrix element are usually negligible. The Doppler profile of the two-photon absorption line is obtained by integrating the Lorentzian distribution over the Gaussian distribution of the thermal velocity component v_x. The magnitude of the Doppler broadening is clearly $(k_1 - k_2)v_{th}$. If the two frequencies ω_1 and ω_2 are sufficiently close together, this Doppler broadening is negligible compared to the homogeneous width $\Gamma_{fg}/2$.

The special case $k_1 = k_2$ requires separate attention. The Doppler broadening vanishes exactly if one photon is absorbed from each of the beams. In this degenerate case the possibility exists, however, to absorb two photons out of the first beam, or two photons out of the second beam alone. The total transition rate is with $\omega_1 = \omega_2$, $k_1 = -k_2$,

$$w_{f \leftarrow g} = \frac{1}{4} e^4 \hbar^{-4} |E_1|^2 |E_2|^2 \frac{\Gamma_{gf}}{(\omega_{fg} - \omega_1 - \omega_2)^2 + (\Gamma_{gf}/2)^2}$$

$$\times \left| \sum_n \frac{(r \cdot \hat{e}_1)_{gn}(r \cdot \hat{e}_2)_{nf} + (r \cdot \hat{e}_2)_{gn}(r \cdot \hat{e}_1)_{nf}}{\omega_{ng} - \omega_1} \right|^2$$

$$+ \frac{\frac{1}{4} e^4 \hbar^{-4} |E_1|^4 \Gamma_{gf}}{(\omega_{fg} - 2\omega_1 + 2k_1 v_x)^2 + (\Gamma_{gf}/2)^2} \left| \sum_n \frac{(r \cdot \hat{e}_1)_{gn}(r \cdot \hat{e}_1)_{nf}}{\omega_{ng} - \omega_1} \right|^2$$

$$+ \frac{\frac{1}{4} e^4 \hbar^{-4} |E_2|^4 \Gamma_{gf}}{(\omega_{fg} - 2\omega_1 - 2k_1 v_x)^2 + (\Gamma_{gf}/2)^2} \left| \sum_n \frac{(r \cdot \hat{e}_2)_{gn}(r \cdot \hat{e}_2)_{nf}}{\omega_{ng} - \omega_1} \right|^2 .$$

$$\tag{8.13}$$

The line profile now consists, in general, of a Doppler-broadened profile from the last two terms in (8.13), on which the Doppler-free spectral line from the first term is superimposed. The equivalent of (8.13) may be found in a paper by CAGNAC et al. [8.8]; it was first given by CHEBOTAYEV et al. [8.5] for the special case of equal amplitudes and polarization vectors, $E_1 = E_2$, $\hat{e}_1 = \hat{e}_2$. The conditions for the validity of (8.12) and (8.13) are

1) The states $|g\rangle$ and $|f\rangle$ are non-degenerate. It is often straightforward to extend the discussion to degenerate cases with the aid of selection rules.

2) The intermediate states $|n\rangle$ are all off-resonance, $|\omega_{ng} - \omega_1| \gg |k_1 v_x|$ and $|\omega_{ng} - \omega_1| \gg \Gamma$.

3) The perturbation condition given by (8.9) is satisfied. The theory can be extended to cases where conditions 2 and 3 are not obeyed, as will be discussed in Subsection 8.2.6.

4) Processes in which a photon is scattered from the light beam with wave vector k_1 to the light beam with wave vector k_2 can be ignored. If the two frequencies are equal or nearly equal, this is not correct for a very small fraction of the molecules with velocity component $(k_1 + k_2)v_x = \omega_1 - \omega_2$. If the frequencies are sufficiently different $|\omega_1 - \omega_2| > |2k v_x|$, it is assumed that no Raman level exists, $\hbar(\omega_1 - \omega_2)$ above the ground level.

5) Quadratic Doppler shifts are ignored. They have the magnitude of $-(1/2)(v^2/c^2)\omega_{fg}$, where v is the atomic thermal speed. They give rise to a displacement of the center of gravity and an asymmetric line profile, which has been discussed in detail by CHEBOTAYEV and BAKLANOV [8.15] for the $1S \rightarrow 2S$ two-photon transition in atomic hydrogen. This is one of the few cases where the second-order Doppler correction which has an order of magnitude of one part in 10^{11} may become important in optics.

It is useful to present the result for Doppler-free two-photon absorbtion, represented by the first term in (8.13) with $\omega_1 + \omega_2$ in terms of an absorption cross section. The transition rate is divided by the number of photons per cm^2 per s traveling in the direction of k_1, $cn|E_1|^2/8\pi\hbar\omega_1$. The two-photon absorption cross section will be proportional to the power flux density in the opposite beam $P/A = cn|E_2|^2/8\pi$, and may be expressed as

$$\sigma^{(2)}(\omega) = \frac{16\pi^2 \Gamma_{fg}\omega e^4 \hbar^{-3}/c^2}{(\omega_{fg} - 2\omega)^2 + (\Gamma_{gf}/2)^2} \left| \sum_n \frac{\langle f|z|n\rangle\langle n|z|g\rangle}{\hbar(\omega_{fg} - 2\omega)} \right|^2 \frac{P}{A}$$

where all quantities are in cgs units. The index of refraction has been taken as unity in the vapor, $n = 1$. If the matrix elements $\langle f|z|n\rangle$ and $\langle n|z|g\rangle$ are expressed in units of the Bohr radius $a_0 = h^2/4\pi^2 m e^2$, and the energy denominators $\hbar\omega_{ng}$ are expressed in Rydbergs $R_0 = 2\pi^2 m e^4/h^3$, and (P/A) is expressed in watts/cm^2, the practical result is obtained

$$\sigma^{(2)}(\omega) = 4.597 \times 10^{-34} \left| \sum_n \frac{\langle f|z|n\rangle\langle n|z|g\rangle}{\hbar(\omega_{ng} - 2\omega)} \right|^2$$

$$\frac{\Gamma_{gf}\omega/\pi}{(\omega_{fg} - 2\omega)^2 + (\Gamma_{gf}/2)^2} \left(\frac{P}{A}\right) \text{cm}^2 . \qquad (8.13a)$$

8.2.3 Selection Rules for Two-Photon Transitions

In the electric dipole approximation the interaction Hamiltonian between field and matter acts only on the electron orbital. If one chooses the polarization directions in the usual manner according to σ^-, σ^+, and π, the components of the dipole interaction transform as the irreducible tensor components $T_1^{(q)}$ with $q=-1$ for σ^-, $q=+1$ for σ^+, and $q=0$ for π-polarization. Each time this operator of odd parity is applied, the orbital angular momentum must change by ±1. By applying this operator twice, one sees immediately that the orbital angular momentum in a two-photon process must change by

$$\Delta L=0 \quad \text{or} \quad \pm 2. \tag{8.14}$$

Furthermore,

$$\Delta m_L = q_1 + q_2. \tag{8.15}$$

In the case of extreme Paschen-Back decoupling by a very high external magnetic field, the magnetic quantum numbers of electron and nuclear spin are not changed,

$$\Delta m_s = \Delta m_I = 0 \tag{8.16}$$

because the Hamiltonian does not act on these spin variables.

The simple selection rules, (8.14–16), adapted for the high-field case serve as a useful starting point to discuss the situation that spin-orbit coupling $\lambda L \cdot S$ and hyperfine interactions, such as $A I \cdot S$, cannot be ignored and are larger than the Zeeman splitting in an external field.

As long as the energy denominator may be taken the same for all terms in the multiplet of the intermediate state $|n\rangle$, i.e., for $\hbar(\omega_{ng}-\omega_1) \gg \lambda$ and A, the summation over all substates of the intermediate multiplet with equal weight leads to a scalar operation which cannot change the character of the selection rules. In such a case, these depend only on the properties of the initial and final states. The selection rules given explicitly in the $|L, S, I, m_L, m_I\rangle$ representation may be readily transformed by the rules of the addition of angular momenta to other representations, in which, for example, the total angular momentum F with $F = L + S + I$ and $m_F = m_L + m_s + m_I$ are good quantum numbers. Use of the Wigner-Eckart theorem yields immediately

$$|\Delta F| \leq 2 \quad \text{and} \quad \Delta m_F = q_1 + q_2.$$

Other more specialized conditions may be derived from the known properties of Clebsch-Gordan coefficients. A systematic discussion in terms of group-theoretical arguments may be found in several papers [8.8, 16, 17].

An important special case is the two-photon transition between two atomic S-states. Since the orbital angular momentum vanishes in both the initial and final states we must have $\Delta m_L = 0$. If both light beams have a circular polarization in the same sense, $|q_1 + q_2| = 2$, no transitions can be induced between the two S-states, since (8.15) cannot be satisfied. This forms the basis for eliminating the Doppler broadening background for two light waves with the same frequency traveling in opposite directions. One is given a σ^+, and the other a σ^- sense of polarization with respect to a fixed positive x-axis. Only the process, represented by the first term in (8.13), where one photon is absorbed from each of the beams can occur. Furthermore, spin-orbit coupling in the initial and in the final states is absent. Since the spin configuration cannot change— provided all states in the intermediate p-state multiplet occur with equal weight—we have the zero-field selection rule $\Delta F = 0$, $\Delta m_F = 0$. The high-field selection rule is given by (8.16). There is an intermediate situation in which the smaller hyperfine coupling in the upper S-state is in the Paschen-Back regime, while the electron and nuclear spin in the lower S-state are not yet decoupled. One must then carry out the projections of the spin states of the lower level onto those of the higher level. At any magnetic field strength, the spectral pattern is independent of the direction of the applied field with respect to the polarization (and propagation) directions of the light beams. This is a direct consequence of the orbital isotropy in both initial and final states [8.18].

Another situation of interest is the transition from $3s^2 S_{1/2}$ ground state to $4d^2 D_{5/2,3/2}$ multiplet while near-resonance conditions with an intermediate $3p^2 P_{3/2,1/2}$ doublet exist [8.19]. No dipole matrix element exists between $^2D_{5/2}$ and the $^2P_{1/2}$ states. Thus, the transition from the ground S-state to the $D_{5/2}$-state shows only one resonance as ω_1 is varied while $\hbar(\omega_1 + \omega_2)$ is kept fixed at the separation between the S- and $D_{5/2}$-states. The two-photon transition to the $D_{3/2}$-state will show resonant behavior when $\hbar\omega_1$ corresponds to either resonant line in the $3p^2 P_{3/2,1/2} \rightarrow 3s^2 S_{1/2}$ doublet. It should also be noted that a destructive interference occurs between the two resonant contributions when $\hbar\omega_1$ falls in between the two doublet lines. This is apparent from (8.12). The summation over n may be restricted to the near-resonant pair of P-states. One of the denominators in the second-order matrix element will be positive $\omega_1 < \omega_{sp5/2}$; the other will be negative $\omega_2 > \omega_{sp3/2}$. Numerical ratios of the matrix elements may be evaluated and the appropriate projection on nuclear spin states may be taken for a complete numerical evaluation. At resonance of the intermediate state, the two-photon transition rate is comparable to the rate of two consecutive one-photon transitions. The transition rate may increase by six to eight orders of magnitude [8.19]. If only one intermediate level is of importance, the selection rules depend, of course, on the nature of this near-resonant intermediate level. They may be derived from the selection rules for two successive one-photon processes.

It may happen that one of the light frequencies is adjusted in such a manner that a resonance occurs for an electric-quadrupole-allowed single-photon transition, but all electric-dipole single-photon transitions are far off resonance. In such a situation the electric quadrupole term cannot be ignored as the resonance causes it to acquire the same magnitude as an off-resonance dipole transition. The transition rate for a mixed dipole-electric quadrupole two-photon transition may be written in analogy to (8.12) as

$$
w_{f' \leftarrow g} = \frac{1}{4} e^4 \hbar^{-4} |E_1|^2 |E_2|^2 \frac{\Gamma_{gf'}}{[\omega_{f'g} + (k_1 - k_2)v_x - \omega_1 - \omega_2]^2 + (\Gamma_{f'g}/2)^2}
$$
$$
\times \left| \frac{(\frac{1}{2}Q : ik_1 \hat{e}_1)_{gn'} (r \cdot \hat{e}_2)_{n'f'}}{\omega_{n'g} + k_1 v_x - \omega_1 + i\Gamma_{n'g}/2} \right|^2 . \tag{8.17}
$$

It is assumed that the frequency ω_1 is near resonance with a quadrupole-allowed transition between two levels with the same parity $|n'\rangle$ and $|g\rangle$, and that all other terms are non-resonant and negligible. The final state of this two-photon transition $|f'\rangle$ has, of course, the opposite parity from $|g\rangle$. The appropriate selection rules on the orbital angular momentum are, of course, $\Delta L = \pm 1$ or ± 3. More detailed rules may be derived from the known selection rules for the quadrupole interaction matrix element $[(1/2)eQ : \nabla E_1 \hat{e}_1]_{gn'}$, consisting of the inner product of the quadrupole tensor and the electric field gradient tensor. In (8.17) the gradient operator has been replaced by ik_1 appropriate for a plane wave with wave vector k_1.

8.2.4 Two-Photon Ionization

The final state $|f\rangle$ may lie in the ionization continuum. In this case the continuum state wave is characterized by the wave vector of the photo-electron K. An integration over the continuum of final states in the neighborhood of the resonance for the two-photon process must be carried out. This has been discussed in several review papers [8.16, 20]. Peaks in the two-photon ionization rate occur, when there is a one-photon resonance with a bound intermediate level $|n\rangle$. Lambropoulos et al. [8.21] have pointed out the significance for two-photon ionization peaks caused by a resonant quadrupole resonance via an intermediate state $|n'\rangle$. Armstrong and Wynne [8.22] have observed multi-photon transitions to auto-ionizing states lying in the continuum beyond the ionization potential. Characteristic Fano-type interference effects occur also for the nonlinear optical processes to an auto-ionizing state embedded in a continuum. As the subject matter of this review is restricted to Doppler-free high-resolution two-photon spectroscopy, two-photon ionization will not be pursued further. The integration over a continuum of final momentum states tends to wash out Doppler-free features. This brief section is merely intended to provide some references, which may serve as a starting point for further study of this related topic.

8.2.5 Power-Dependent Stark Shift and Broadening

When a dc electric field is applied to an atom or a molecule, a quadratic Stark splitting commonly occurs, partially lifting the spatial degeneracy. Levels with the same absolute value of spatial quantum number $\pm m_J$, where J indicates an atomic multiplet level or a rotational level of a molecule, remain degenerate. Furthermore, a shift in the center of gravity of an atomic level may occur. A P-state will, for example, be pushed up by second-order electric dipole perturbation from a S-state, or D-state, with a lower energy; it will be pushed down by second-order perturbations from S- and D-states with a higher energy. These Stark shifts and splittings are theoretically well known and have been detected by two-photon Doppler-free spectroscopy [8.23]. It may be remarked parenthetically that quantum beat spectroscopy [8.24] can detect the splitting, but not the shift in the center of gravity of a multiplet.

There will also be a second-order shift between a pair of energy levels connected by an electric dipole interaction with an oscillating electric field. Such a shift may be calculated by standard second-order perturbation theory. The displacement of the ground state will for example, be given by

$$\Delta W_g = -\frac{1}{4}\hbar^{-1}|E_1|^2 \, \Sigma_n \frac{(er \cdot \hat{e}_1)_{gn}(er \cdot \hat{e}_1^*)_{ng}}{\omega_{ng} - \omega_1}. \tag{8.18}$$

When ω_1 is near resonance with ω_{ng}, the magnitude of the light-induced shift can become appreciable. The sign of the shift changes when ω_1 passes from below resonance to above resonance. The same (8.18) of course also describes the dc Stark shift if we take $\omega_1 = 0$ and drop the factor $\frac{1}{4}$, as for the dc case the "anti-resonant" terms, with ω_1 replaced by $-\omega_1$, become equally important. If the ground state $|g\rangle$ is, for example, an atomic S-state, the electric field admixes some P-character. The mixing becomes very pronounced near resonance. At this point the damping in the resonant denominator cannot be ignored and the term $-i\Gamma_{ng}/2$ must be added to the denominator in (8.18). When this is done and the Doppler

shift in frequency is taken into account, the real second-order energy
shift becomes

$$\Delta W_g = -\frac{1}{4} \hbar^{-1} |E_1|^2 |(er \cdot \hat{e}_1)_{gn}|^2 \frac{\omega_{ng} + k_1 v_x - \omega_1}{(\omega_{ng} + k_1 v_x - \omega_1)^2 + (\Gamma_{ng}/2)^2}. \quad (8.19)$$

The corresponding imaginary part signifies an additional damping. If
there is no collisional or other homogeneous broadening of the ground
state, its lifetime would be infinite, as there can be no spontaneous
emission from the ground state. The excited state has a natural lifetime
for spontaneous emission which is then responsible for the natural
width Γ_{ng} of the transition. Due to the admixture of this excited state the
new ground state acquires a "borrowed" damping leading to a power-
dependent broadening

$$\Delta \Gamma_{gg} = +\frac{1}{4} \hbar^{-2} |E_1|^2 |er \cdot \hat{e}_1|^2 \frac{\Gamma_{ng}/2}{(\omega_{ng} + k_1 v_x - \omega_1)^2 + (\Gamma_{ng}/2)^2}. \quad (8.20)$$

When ω_1 is near resonance with ω_{ng} and if $\omega_1 + \omega_2$ is near resonance with
the two-photon transition ω_{fg}, then ω_2 is necessarily near resonance with
ω_{fn}. The corresponding quadratic Stark shift of the final level $|f\rangle$ must
then also be taken into account. It will also receive an additional damping
from admixture of the state $|n\rangle$. This may be important if the pure $|f\rangle$
state is metastable and has an intrinsic long lifetime against spontaneous
emission. An important example is provided by the $1S \to 2S$ two-photon
transition in atomic hydrogen.

The total Stark shift for the two-photon transition thus becomes

$$\Delta \omega_{fg} = -\frac{1}{4} \hbar^{-2} |E_1|^2 |(er \cdot \hat{e}_1)_{ng}|^2 \frac{\omega_{ng} + k_1 v_x - \omega_1}{(\omega_{ng} + k_1 v_x - \omega_1)^2 + (\Gamma_{ng}/2)^2}$$

$$-\frac{1}{4} \hbar^{-2} |E_2|^2 |(er \cdot \hat{e}_2)_{fn}|^2 \frac{\omega_{fn} - k_2 v_x - \omega_2}{(\omega_{fn} - k_2 v_x - \omega_2)^2 + (\Gamma_{fn}/2)^2}. $$

$$(8.21)$$

If $\omega_1 + \omega_2 = \omega_{fg}$, then $\omega_{ng} + k_1 v_x - \omega_1 \approx -(\omega_{fn} - k_2 v_x - \omega_2)$ and the two
contributions will have opposite sign.

The damping parameter for the nearly Doppler-free transition is
increased by an amount

$$\Delta \Gamma_{fg} = +\frac{1}{4} \hbar^{-2} |E_1|^2 |(er \cdot \hat{e}_1)_{ng}|^2 \frac{\Gamma_{ng}}{(\omega_{ng} + k_1 v_x - \omega_1)^2 + (\Gamma_{ng}/2)^2}$$

$$+\frac{1}{4} \hbar^{-2} |E_2|^2 |(er \cdot \hat{e}_2)_{fn}|^2 \frac{\Gamma_{fn}}{(\omega_{fn} - k_2 v_x - \omega_2)^2 + (\Gamma_{fn}/2)^2}. $$

$$(8.22)$$

The second-order light-induced Stark shift is a readily observable effect in two-photon Doppler-free transitions [8.25]. If this transition occurs between a ground state and a metastable level, such a two-photon transition can become exceedingly sharp, if collisional broadening is absent. Under such circumstances the power broadening may also become observable.

The perturbation approach adopted so far is plausible although the validity of using the second-order perturbation first to calculate energy level shifts and broadening, and next to use the same interaction to compute the two-photon transition rate is questionable. Fortunately, this procedure can be justified by a rigorous non-perturbation solution outlined in the next section, which also takes into account saturation broadening. This occurs when the frequency variation of a field also produces important changes in the populations of the energy levels.

8.2.6 Rigorous Solution for Two-Photon Transitions in a Three-Level System

It is possible to obtain solutions of the density matrix equation of motion of a material system which are correct to all powers in the amplitude of one of the applied fields [8.26]. This can be done by transforming to a rotating coordinate system in which this one field becomes time independent. One thus takes account of the saturation effects, the Stark shift and power broadening by one electromagnetic mode completely. The other modes are still treated by a perturbation procedure.

Another case in which all powers of the amplitudes of two or more applied fields may be retained in the solution occurs when each mode of the electromagnetic field can induce a transition only between one pair of energy levels. Physical justification for this assumption exists if each mode is near resonance with only one particular level separation or if polarization selection rules forbid all possible transition but one. Consider, for example, the case of three energy levels and two applied electromagnetic modes as illustrated in Fig. 8.4. The mode at frequency ω_1 induces transitions only between levels $|1\rangle$ and $|3\rangle$; the mode at ω_2 can induce transitions only between levels $|2\rangle$ and $|3\rangle$.

The three-level system is described by the nine elements of the three-by-three density matrix. Three off-diagonal elements are the complex conjugates of the other three transposed off-diagonal elements. The three diagonal elements, proportional to the population in each of the three energy levels, obey a trace relation, if the total number of atoms is kept constant. Thus, there are two independent real equations for the diagonal elements and three complex equations for the off-diagonal elements. The latter may be represented for an atom moving with a

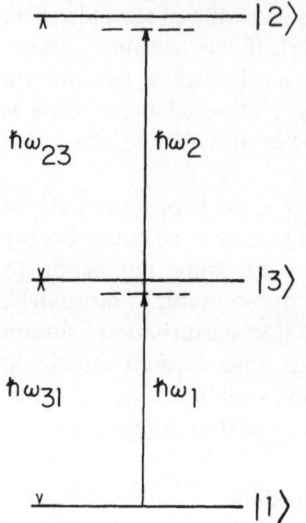

Fig. 8.4. Two-photon absorption in a three-level system near resonance (with notation used in Sect. 8.2.6)

velocity v_x parallel to the beam at ω_1 by

$$
\begin{aligned}
\varrho_{13} &= \tilde{\varrho}_{13} e^{i(\omega_1 - k_1 v_x)t} \\
\varrho_{23} &= \tilde{\varrho}_{23} e^{-i(\omega_2 + k_2 v_x)t} \\
\varrho_{12} &= \tilde{\varrho}_{12} e^{i[\omega_1 + \omega_2 + (k_2 - k_1)v_x]t} \approx \tilde{\varrho}_{12} e^{i(\omega_1 + \omega_2)t} .
\end{aligned}
\tag{8.23}
$$

The amplitudes $\tilde{\varrho}_{13}, \tilde{\varrho}_{23}, \tilde{\varrho}_{12}$ are slowly varying. All rapidly oscillating, non-resonant terms in $\tilde{\varrho}_{13}, \tilde{\varrho}_{23}, \tilde{\varrho}_{12}$ and the diagonal elements are dropped. This truncation procedure is the basic assumption in this approximation. There is only one Fourier component for each matrix element. The equation of motion for the density matrix, (8.1), under the influence of an electric dipole interaction Hamiltonian may then be written out explicitly in the following components:

$$
\dot{\tilde{\varrho}}_{13} + i\tilde{\varrho}_{13}(\Delta - i/T_2) = i\alpha(\varrho_{33} - \varrho_{11}) - i\beta\tilde{\varrho}_{12} ,
\tag{8.24a}
$$

$$
\dot{\tilde{\varrho}}_{23} - i\tilde{\varrho}_{23}(\Delta' + i/T_2) = i\beta(\varrho_{33} - \varrho_{22}) - i\alpha\tilde{\varrho}_{21} ,
\tag{8.24b}
$$

$$
\dot{\tilde{\varrho}}_{12} + i\tilde{\varrho}_{12}(\Delta + \Delta' - i/\tau_2) = i\alpha\tilde{\varrho}_{32} - i\beta\tilde{\varrho}_{13} ,
\tag{8.24c}
$$

$$
\dot{\varrho}_{11} = i\alpha(\tilde{\varrho}_{31} - \tilde{\varrho}_{13}) - (\varrho_{11} - \varrho_{11}^0)/T_1 ,
\tag{8.24d}
$$

$$
\dot{\varrho}_{22} = i\beta(\tilde{\varrho}_{32} - \tilde{\varrho}_{23}) - (\varrho_{22} - \varrho_{22}^0)/T_1 ,
\tag{8.24e}
$$

$$
\dot{\varrho}_{33} = i\alpha(\tilde{\varrho}_{13} - \tilde{\varrho}_{31}) + i\beta(\tilde{\varrho}_{23} - \tilde{\varrho}_{32}) - (\varrho_{33} - \varrho_{33}^0)/T_1 .
\tag{8.24f}
$$

In order to avoid non-essential algebraic complications it has been assumed that the width of the one-photon allowed resonances is the same

$$\Gamma_{31}/2 = \Gamma_{32}/2 = 1/T_2, \quad \text{while} \quad \Gamma_{21}/2 = 1/\tau_2 . \tag{8.25}$$

The relaxation time for population changes in the three levels is assumed to be equal

$$\Gamma_{11} = \Gamma_{22} = \Gamma_{33} = 1/T_1 . \tag{8.26}$$

The populations in thermal equilibrium, in the absence of applied light waves, are given by ϱ_{11}^0, ϱ_{22}^0, and ϱ_{33}^0. For optical frequencies and not too elevated temperatures one may take

$$\varrho_{11}^0 = 1, \varrho_{22}^0 = \varrho_{33}^0 = 0 . \tag{8.27}$$

Furthermore, the following abbreviations have been introduced

$$\alpha = \tfrac{1}{2}(er \cdot \hat{e}_1)_{13} E_1 \hbar^{-1} , \tag{8.28}$$

$$\beta = \tfrac{1}{2}(er \cdot \hat{e}_2)_{32} E_2 \hbar^{-1} , \tag{8.29}$$

$$\Delta = \omega_1 - k_1 v_x - \omega_{31} , \tag{8.30}$$

$$\Delta' = \omega_2 + k_2 v_x - \omega_{23} . \tag{8.31}$$

The set of Eqs. (8.24a–f) is obviously linear in the five independent matrix elements, $\tilde{\varrho}_{13}, \tilde{\varrho}_{23}, \tilde{\varrho}_{13}$ and the population differences $\varrho_{22} - \varrho_{33}$, and $\varrho_{11} - \varrho_{22}$. The steady state solutions are obtained by setting the time derivatives equal to zero. Explicit solutions for the resulting set of five linear algebraic equations have been given by BREWER and HAHN [8.27] in the following form, if one takes $\varrho_{11}^0 = 1$:

$$\varrho_{22} - \varrho_{33} = \frac{-P}{T_1(MP - DQ)}, \tag{8.32a}$$

$$\varrho_{11} - \varrho_{22} = \frac{Q}{T_1(MP - DQ)}, \tag{8.32b}$$

$$\tilde{\varrho}_{12} = a\left(\frac{\varrho_{22} - \varrho_{33}}{\Delta' + i/T_2} + \frac{\varrho_{11} - \varrho_{33}}{\Delta + i/T_2}\right), \tag{8.32c}$$

$$\tilde{\varrho}_{23} = \frac{\alpha\tilde{\varrho}_{21} + \beta(\varrho_{22} - \varrho_{33})}{\Delta' + i/T_2}, \tag{8.32d}$$

$$\tilde{\varrho}_{13} = \frac{-\beta\tilde{\varrho}_{12} - \alpha(\varrho_{11} - \varrho_{33})}{\Delta - i/T_2}. \tag{8.32e}$$

The following definitions must be inserted into (8.32a–e) in order to obtain a solution correct to all powers in α and β, i.e., for arbitrarily large field amplitudes,

$$a = \alpha\beta \Big/ \left[(\Delta + \Delta' - i/\tau_2) - \frac{\alpha^2}{\Delta' - i/T_2} - \frac{\beta^2}{\Delta - i/T_2} \right], \tag{8.33a}$$

$$M = a^*b^*e + abe^* - \frac{2\alpha^2/T_2}{\Delta^2 + 1/T_2^2} + \frac{2\beta^2/T_2}{\Delta'^2 + 1/T_2^2}, \tag{8.33b}$$

$$Q = a^*b^*j + abj^* - \left(\frac{2\alpha^2/T_2}{\Delta^2 + 1/T_2^2} + \frac{4\beta^2/T_2}{\Delta'^2 + 1/T_2^2} + 1/T_1 \right), \tag{8.33c}$$

$$e = i\alpha\beta \left(\frac{-1}{\Delta + i/T_2} + \frac{1}{\Delta' + i/T_2} \right), \tag{8.33d}$$

$$b = \frac{1}{\Delta - i/T_2} + \frac{1}{\Delta' - i/T_2}, \tag{8.33e}$$

$$D = - \left(\frac{2\alpha^2/T_2}{\Delta^2 + i/T_2^2} + 1/T_1 \right) + \frac{a^*e}{\Delta + i/T_2} + \frac{ae^*}{\Delta - i/T_2}, \tag{8.33f}$$

$$j = -i\alpha\beta \left(\frac{2}{\Delta' + i/T_2} + \frac{1}{\Delta + i/T_2} \right), \tag{8.33g}$$

$$P = - \frac{2\alpha^2/T_2}{\Delta^2 + 1/T_2^2} + \frac{a^*j}{\Delta + i/T_2} + \frac{aj^*}{\Delta - i/T_2}. \tag{8.33h}$$

The complexity of these solutions is such that little physical insight is gained from them. They should include all saturation effects, Stark shifts and power broadening due to both fields. To illustrate this, focus the attention on (8.33a) and rewrite it in the form

$$a = \frac{\alpha\beta}{\Delta\left(1 - \dfrac{\beta^2}{\Delta^2 + 1/T_2}\right) + \Delta'\left(1 - \dfrac{\alpha^2}{\Delta'^2 + 1/T_2^2}\right) - i\left(\dfrac{1}{\tau_2} + \dfrac{1}{T_2}\left[\dfrac{\beta^2}{\Delta^2 + 1/T_2^2} + \dfrac{\alpha^2}{\Delta'^2 + 1/T_2^2}\right]\right)}. \tag{8.33a'}$$

Note that the expectation value of the polarization at the frequency ω_2 is proportional to ϱ_{23} which according to (8.32c, 32d, 33a') contains a term proportional to $\alpha^2\beta$ and the denominator of (8.33a'). The nonlinear polarization at ω_2 responsible for two-photon absorption is indeed proportional to $E_1^2 E_2$ and its tuning behavior as the frequencies ω_1 and ω_2 are varied is given by the denominator in (8.33a'). The second-order Stark shifts and the power broadening given previously by (8.21) and

(8.22) are thus recovered. Their simple derivation is thus justified by the general solution.

The sets of equations (8.32) and (8.33) must also contain the description of saturation spectroscopy experiments. The three-level scheme with waves applied at two frequencies can, of course, be used equally well to describe Raman effect situations. In this case both waves should propagate in the same forward direction to obtain a minimum Doppler width of Raman transitions. Many different authors [8.28] have used the same basic equations of motion of the form (8.24a–f) to describe a variety of physical situations. Different approximations are made in the solutions, depending on the focus of attention. The most general solutions are too cumbersome to provide the desired insight.

Transient solutions of the same set of equations, for various initial conditions of the density matrix, i.e., for different forms of preparation of the system, have also been widely discussed. They are, for example, useful in describing coherent quantum beats due to initial values of off-diagonal elements of the density matrix between closely spaced states. BREWER and HAHN [8.27] mention the possibility of coherent emission of two oppositely directed light waves following a suitable preparation of the three-level system by two-photon absorption in analogy with observed coherent Raman beats for two light waves traveling in the same direction.

If one of two fields is omitted by taking, for example, $\beta = 0$, the solution reverts to the well-known Bloch solutions for a two-level system with one rotating field component. One re-obtains the steady state saturation as well as the transient effects of optical or optical "induction", optical nutation, etc.

In certain cases special solutions for a four-level system may be obtained by the density matrix method valid for arbitrary amplitudes of the fields [8.29].

When non-resonant responses are important and many energy levels are involved, the perturbation approach must be used. It always has the advantage of simplicity, and yields reasonably accurate predictions even in many resonant situations.

8.3 Experimental Techniques

All experiments to observe two-photon absorption spectra without the complication of Doppler broadening can be divided into three sections as is shown schematically in Fig. 8.5. The first is necessarily a laser system which produces radiation at the desired frequencies with controlled linewidth, power, and beam parameters. There also must be a means of

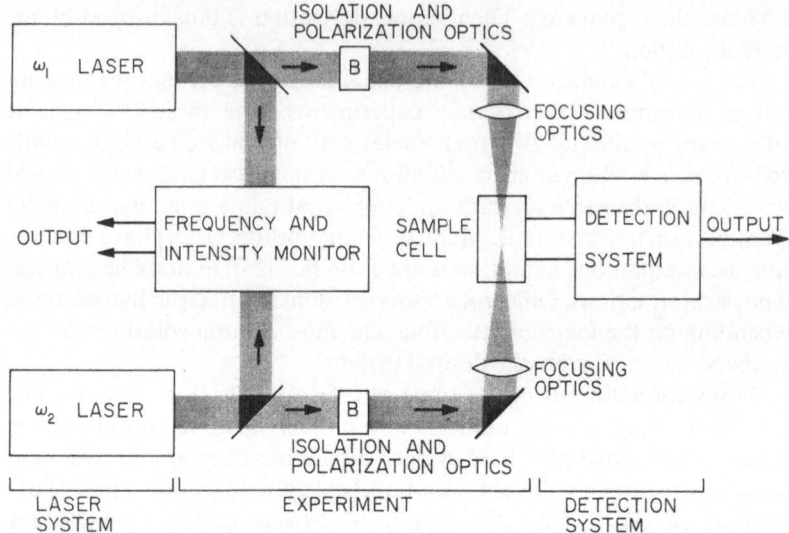

Fig. 8.5. A typical configuration for observing two-photon absorption without Doppler broadening. Photons of one or two frequencies are generated in the laser system, they excite molecules in the experimental region, and the detection system extracts evidence of the excitation from background signals

detecting the simultaneous absorption of multiple quanta by the sample medium and a means of plotting the normalized absorption profile as a function of frequency. Between the laser and detection subsystems there is the experiment proper, consisting of optics to define the photon flux and polarization and a sample cell containing the vapor being studied. Each of the subsystems presents experimental complexities and often technological difficulties; we shall discuss each subsystem in turn.

8.3.1 Laser System for Doppler-Free Multiphoton Absorption

Two-photon transitions without Doppler broadening have presently been observed with ultraviolet, visible and infrared lasers. Clearly, different technologies are required for each spectroscopic region. For the near ultraviolet, visible, and near infrared regions broadly tunable narrowband dye lasers are more than adequate [8.4]. The wavelength range spanned by these devices can be extended into the vacuum ultraviolet by nonlinear mixing and harmonic generation techniques. Between 0.6 and 5 microns, the optical parametric oscillator provides a tunable source with resolution of up to 15 MHz and sufficient power to probe moderately strong transitions [8.30]. For wavelengths beyond 5 microns, a variety of molecular gas lasers produces a large number of useful frequencies [8.31]. The narrow continuous tuning range of each of these lines, however, restricts the number of transitions which can be probed. More widely tunable gas, spin-flip and diode lasers are being developed.

All laser systems suitable for these experiments share certain qualities. All must have narrow linewidths and stable mode frequencies. Single-frequency operation is desirable, but not necessary. For multifrequency lasers, the observed absorption is generally reduced by a factor inversely proportional to the number of frequency components. The inherent linewidths of the spectroscopic structures being resolved are a few megahertz or less. The Doppler width of a visible line is typically one gigahertz. A useful laser must thus have a linewidth and frequency jitter less than roughly 100 MHz, and preferable as small as 1 MHz. The lasers must have stable low-order transverse mode structures, preferably TEM_{00} so that the output can be efficiently concentrated in the sample region. The peak laser power must be sufficient to produce detectable signals as the result of two-photon absorption, but not so great nor so unstable as to lead to unexpected light shifts and power broadening. Often increased resolution can be obtained in return for decreased peak power. The repetition rate or average power must be sufficient to permit data collection at an acceptably rapid rate, and the entire system must be reliable enough so that the experimeter can concentrate on the physics of the experiment rather than the maintainance of the laser.

These desirable qualities may be summarized in terms of a variety of "figures of merit". One such figure is roughly proportional to the data collection rate in a two-photon experiment

$$R = \frac{(\text{Peak power in watts})}{(\text{Wavelength in } \mu\text{m}) (\text{Number of frequency components})}$$
$$\times \frac{(\text{Average power in watts}) \times 1000}{(\text{linewidth in MHz})^2}$$

where the linewidth refers to a single-frequency component and the wavelength factor in the denominator results from the focusing properties of gaussian beams. Other figures of merit will undoubtedly occur to the reader. The minimum acceptable value of such a parameter will depend upon the characteristics of the rest of the experiment. It is our intention only to provide a reference frame.

8.3.2 Dye Lasers

Both pulsed and cw dye lasers have been employed successfully in Doppler-free two-photon absorption experiments. Technologically, the main problem with these devices is the design of a resonator structure

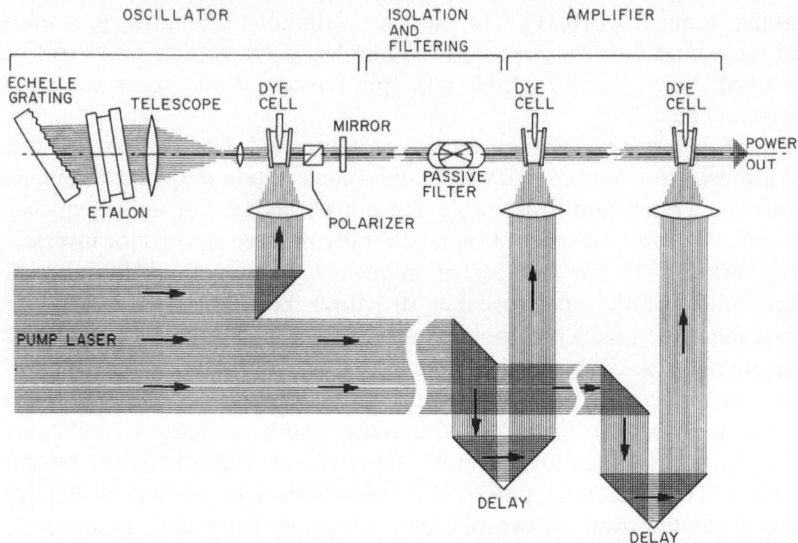

OSCILLATOR   ISOLATION   AMPLIFIER
                             AND
                             FILTERING

ECHELLE
GRATING   TELESCOPE   DYE   DYE   DYE
                             CELL   CELL   CELL
                       MIRROR
                                                                    POWER
                                                                    OUT
                             PASSIVE
                             FILTER
ETALON   POLARIZER

PUMP LASER

                             DELAY

                                      DELAY

Fig. 8.6. An oscillator-amplifier dye laser according to Hänsch [8.32]. The grating, etalon and passive filter may all be tuned synchronously by means of varying the pressure of the gas surrounding each. The passive filters and amplifier stages are optional. Isolation between stages is accomplished in this short pulse laser by separating them by distances longer than can be covered at the speed of light within the pulse length of the exciting laser

capable of producing a narrowband output at a stable frequency. To some extent all narrowband laser applications share these problems, and some engineering solutions—rigid resonator structures, vibration isolation, etc.—are common to all lasers. The broad gain band of a typical dye laser provides some special difficulties, however. Two or more frequency selective elements are necessary to reduce the output linewidth to acceptable levels. All the frequency selective elements must be tuned synchronously to maintain narrowband operation. The high optical pumping intensity needed to invert a liquid dye mixture also leads to difficulties. Windows in the dye cell tend to be damaged by decomposing dye, and the resonator structure cannot be entirely shielded from vibrations due to the dye circulation system. Without rapid and often necessarily turbulent circulation, thermal gradients destroy the optical quality of the gain medium. Remarkably innovative engineering solutions exist now for many of these difficulties. Each type of dye laser has its own set of advantages and weaknesses, but the most flexible design seems to be the Hänsch-type pulsed dye laser, [8.32, 33], the latest version of which appears in Fig. 8.6.

The Hänsch-type dye laser is generally excited by relatively short (< 16 ns) pulses produced by a nitrogen or Q-switched ruby laser. The oscillator cavity contains a transversely pumped dye cell for gain and a telescope to expand the beam that comes from the dye cell so that it illuminates a relatively large area of the etalon-grating combination which provides frequency selection. The light returning from the filtering elements is then refocused into the dye cell by the telescope, is amplified, and exits through a partially transparent mirror. Since the length of the exciting pulse is too short to allow the radiation to make many round trips through the cavity, no longitudinal modes are created. The output has the spectral characteristics of filtered and amplified spontaneous emission. The resonator structure when properly aligned is somewhat unstable. As a result transverse mode structure resembles a TEM_{00} mode with a confocal parameter of typically 6 cm. The repetition rate is limited only by the pumping source.

While the high gain of a laser-pumped dye permits relatively lossy elements to be used as frequency selection components within the oscillator cavity, the minimum linewidth that can be achieved with a single-frequency laser of this sort is roughly 400 MHz. To narrow the spectrum further, the output of the oscillator must be filtered by a passive confocal interferometer. The laser spot on the interferometer mirrors is quite small, and excessive oscillator power will result in damage to the coatings. The passive filter, etalon and grating can be simultaneously tuned to vary the frequency by immersing them in an evacuable chamber and varying the pressure of the surrounding gas [8.34]. The linewidth of the beam emitted by the interferometer may be 10 MHz or less, but the power level that results from the various constraints is too low for multiphoton absorption.

Two stages of amplification are necessary to boost the peak output power back to an acceptable level. Since the amplifiers are pumped by the same short pulse laser that excites the oscillator, the amplifier pumping pulses must be delayed with respect to the oscillator pump. The linewidth of the output which results from the amplifier chain is limited by the uncertainty principle to the inverse of the pulse length. This pulse length in turn is a property of the pump laser.

Commercially available nitrogen lasers produce peak powers of up to 1 MW in pulses 8 ns long (FWHM). The output of an amplified Hänsch-dye laser pumped in this way is typically 40 kW peak with a 6 ns pulse length and a linewidth of 120 MHz. A 60 Hz repetition rate results in an average power of 14 mW producing a value of $R = 82$ at $\lambda = 486$ μm for the previously defined figure of merit.

This design can be somewhat simplified by combining the functions of the intracavity etalon and the passive filter. A plane Fabry-Perot interferometer with a spacing of several centimeters used as an intracavity etalon will allow oscillations on several bands separated by the free spectral range of the etalon. The linewidths of the individual bands are limited by the length of the pumping pulse just as is the output of the oscillator-amplifier design. Output powers of 20 kW can be obtained from an oscillator in this multiline mode, and if the total width of the spectrum which is to be resolved is less than half of the free spectral range of the etalon, multiline operation does not degrade the experiment. Tuning within the band pass of the grating is accomplished by simply tilting the etalon. Any pair of frequency components which sum to the resonant frequency can contribute to the two-photon absorption; however, the total absorption cross section depends upon the phase relationships among the frequency components. If the relative phase is random the various two-photon absorption processes add incoherently in a quantum mechanical sense, and the absorption is less than it would be in the single-frequency case by a factor inversely proportional to the number of frequency components. If, however, a single frequency had somehow been selected from the comb of output modes, the cross section would have been reduced still further.

If the relative phases of the output frequency components are a constant, the modes add together to produce sharp pulses separated in time by the inverse of the free spectral range of the interferometer. This mode-locking produces a net enhancement of the two-photon absorption over that expected in the single-frequency case. If succeeding mode-locked pulses are made to overlap within the sample region, and if the mode-locking is perfect, the enhancement in the average cross section is proportional to the number of modes. While techniques exist for ensuring mode-locking, they have proved difficult to apply to narrow-band lasers of this sort [8.35]. The uncertainty and variability in the amplitudes and phase relation of the frequency components result in an uncertainty in the measured absorption cross sections and provide a noise level on the two-photon absorption signal when this sort of simplified Hänsch laser is employed.

The high gain of a laser-pumped dye oscillator combined with the short risetime of a nitrogen laser pump minimizes the delay between the pump pulse and dye laser output. Thus there is no problem in temporally synchronizing the outputs of two dye lasers pumped by the same nitrogen laser, and experiments in which a material simultaneously absorbs two quanta of different energies can be endeavored with lasers of the Hänsch design [8.36]. Resonant enhancement of two-photon cross sections is generally not needed as adequate signal can be readily obtained in most cases.

The very high output powers obtained in the oscillator-amplifier version of the Hänsch laser can be used to produce narrowband tunable ultraviolet either by sum or second harmonic generation in nonlinear crystals [8.33] or by third-order mixing in atomic vapors [8.11, 12].

Many variations of the basic Hänsch design have been made to accommodate the peculiarities of particular dyes and pump sources.

True single-mode operation of a laser of this type can be obtained if a pump source with pulse length longer than about 50 ns (such as a xenon laser) is employed along with several judiciously chosen intracavity etalons. The resulting 15 MHz linewidth is useful in a number of high-resolution experiments where the output of cw dye lasers is insufficient to excite the transition.

While the pulse energy of flash lamp pumped dye lasers is considerable, and the pulse lengths are long enough to permit good resolution, nearly insoluble engineering problems have limited the best resolution presently obtained to roughly 500 MHz. Relatively low repetition rates also reduce the speed at which data can be collected with such apparatus.

The high average power of a cw dye laser partially compensates for its low peak power and makes this device an attractive alternative to the Hänsch laser in some two-photon absorption experiments. In addition, a number of published laser designs and two commercial units have linewidths and frequency stabilities in the 0.2 to 15 MHz range [8.37–39]. Since the frequencies and output powers of these lasers can be stabilized by standard servo techniques, cw dye lasers are preferred over pulsed dye lasers for ultra-high precision applications [8.38].

Many of the most successful design concepts are incorporated in the two commercially produced narrowband cw dye lasers, the Coherent Radiation 599 and the Spectra Physics 580A diagrammed in Fig. 8.7. Both employ a free-flowing jet of dye solution as the gain medium and a three-mirror cavity configuration. Narrowband operation is achieved with piezoelectrically tunable intracavity etalons, and the output frequency is scanned continuously over limited ranges by simultaneously tuning the etalon and cavity length. Output powers in narrowband operation range up to 300 mW.

Problems of photochemical instability of the coumarin and oxazine dyes have limited the operating range of these dye lasers to the red to green section of the spectrum. In a cw dye laser, narrowband operation is achieved by a delicate balancing of broadband gain and frequency specific loss. The optics used to ensure narrowband operation must be designed for a specific and rather narrowly defined gain parameter. If the gain should change as a result of dye decomposition, the boundary between narrowband laser operation and no laser operation is readily crossed.

With cw lasers, the problems of synchronization and power stability disappear, generally to be replaced by the problem of low signal level, although many two-photon transitions have a large enough cross section to be detected by cw dye lasers. With values of the figure of merit R ranging from 0.1 to 2 for the various laboratory and commercial lasers of this type, cw dye lasers put much more of a burden on the detection system than does the Hänsch laser.

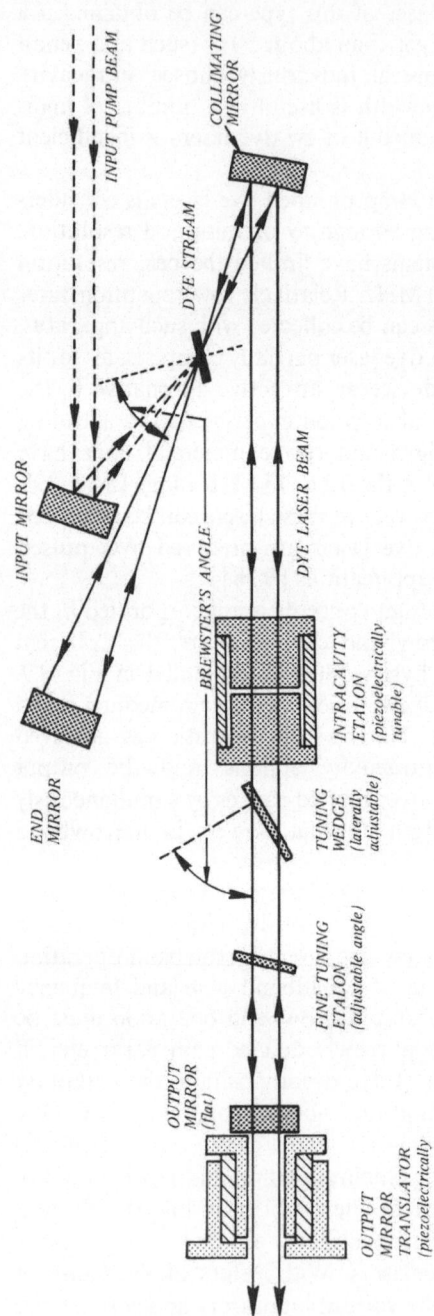

Fig. 8.7. A typical single-mode dye laser configuration. The tuning wedge selects a band of oscillation a few wave numbers wide within the broad tuning range of the dye. The fine tuning etalon narrows the spectrum further and the piezoelectrically tunable interacavity etalon selects a single longitudinal mode. The frequency of that mode can be swept by simultaneously varying the plate spacing of the intracavity etalon and the length of the cavity by means of a piezoelectric translator on the output mirror (after Spectra Physics Corp.)

8.3.3 Optical Parametric Oscillators

As the output wavelength increases beyond 650 nm, dye lasers become increasingly difficult to use. Although proven laser dyes exist that produce wavelengths out to one micron, the dye laser is a less convenient source of near infrared light than is the optical parametric oscillator [8.30]. Parametric oscillation has been demonstrated in a number of laboratories at wavelengths ranging from 0.4 μ to 11 μ. The large, perfect crystals of exotic materials needed for these devices often cannot be easily obtained.

Excellent temperature control on the nonlinear crystal is required for reproduceable tuning, and the band over which oscillation occurs is often a wavenumber or so. The cavity configurations generally used in an optical parametric oscillator do not lend themselves to easy single longitudinal mode operation. However, the overall cavity is often so short that the frequency interval between longitudinal modes is greater than the width of spectrum being resolved. The multifrequency nature of the output of such a device seems to produce fewer difficulties than encountered in using the simplified Hänsch laser discussed previously. The mode frequencies are readily tuned by translating the output mirror piezo-electrically. Resolution is again limited only by the uncertainty principle [8.40].

One commercially produced optical parametric oscillator—the Chromatix 1020—has been found suitable for Doppler-free two-photon spectroscopy in the near infrared. The Chromatix oscillator uses $LiNbO_3$ as the oscillator crystal and produces a total peak power of 120 W in twenty longitudinal modes. This is an inherently pulsed device as the large pump powers needed to achieve threshold can damage the oscillator crystal if maintained continuously. The repetition rate can be quite high, however, and thus the average power can be as much as 3 mW. With a proved linewidth and frequency stability of 15 MHz for the individual modes, this parametric oscillator has a two-photon figure of merit of $R = 0.1$.

As yet no two-photon absorption experiments have been attempted using two quanta of different energies produced by independent parametric oscillators. While the buildup time for the oscillation can be appreciable, other experiments have shown that it is possible to synchronize two such tunable sources by pumping them both with the same laser pulse.

8.3.4 Molecular Gas Lasers

Two-photon transitions beyond the range spanned by organic dye lasers and optical parametric oscillators must presently be probed with fixed frequency lasers. That situation will change with further engineering development of presently extant laser sources, but in the meantime we are fortunate that a number of interesting transitions fall near the gain bands of highly developed molecular gas lasers.

These molecules, notably CO_2, N_2O, and CO, produce powerful cw laser oscillation on a great number of vibrational-rotational transitions when excited by a glow discharge [8.31]. The desired laser line may be selected by an intracavity diffraction grating, and single longitudinal mode oscillation is obtained in this middle infrared region by choosing an overall cavity length short enough that only one such mode frequency falls within the gain band of the molecules. Tuning, if desired, can be accomplished by translating the output mirror [8.41].

Watts of power are available on a cw basis, and in a mechanically stable resonator, the output frequency will vary by less than 1 MHz. It is common, however, to lock the frequency of such a laser to the Lamb dip in the center of its gain band by servoing the output mirror position to the power level. The frequency stability thus achieved can be transferred to another laser operating at an arbitrary point within the gain band by frequency offset locking [8.42]. Since the absolute frequencies of many of these laser transitions are known to considerable accuracy, the result is a system capable of measuring the absolute frequency of a two-photon transition [8.42].

The output power of these molecular gas lasers is also remarkably stable, with an rms ripple of less than one part in 10^6. Feedback techniques also exist for stabilizing the output further, at least for amplitude fluctuations of relatively low frequency. The output power of these lasers is so stable that the lack of sensitive quantum detectors in the middle infrared has led experimenters to detect multiphoton absorptions by monitoring the consequent decrease in the power transmitted through the sample [8.43].

The figure of merit defined previously for visible lasers also applies to these devices. Assuming 1 W of output power and 20 kHz linewidth, the value for R is 4000 at $\lambda = 9.4$ μm. Because of detection difficulties in the far infrared, this high value of R is barely adequate to permit detection of the strongest expected two-photon absorptions due to vibrational-rotational transitions. It can be improved further (at a considerable cost in power stability, however) in laser systems employing a transversely excited pulsed discharge in addition to the cw gain medium (i.e., a hybrid cw—TEA laser). The linewidth of such a device would, however, be limited by the uncertainty principle to a few MHz [8.44].

The figures of merit estimated for various types of laser sources utilized in Doppler-free two-photon absorption experiments are summarized in Table 8.1. Choice of appropriate apparatus necessarily involves other factors, such as detectors.

Table 8.1. Figures of merit for various types of lasers for use in two-photon Doppler spectroscopy

Type	Wavelength (nm)	Figure of merit
Hänsch laser (oscillator + amplifier)	380–760	82
Doubled Hänsch laser	230–360	10^{-3}
Simplified Hänsch laser	420–650	2
Commercial cw dye laser	500–646	2
Parametric oscillator	0.6– 3.5 μ	0.1
Typical molecular laser	5 –11 μ	4000

8.3.5 Detection Techniques

The most common detection scheme has been to collect light emitted by the sample as a result of fluorescent decay of the states excited by multiphoton absorption. At least some of this decay fluorescence should appear at much shorter wavelengths than the incident laser and thus be readily separable from elastic scattering. Since sensitive and efficient quantum detectors have long existed in the visible and ultraviolet regions, this is the most commonly employed technique in dye laser experiments. Recently developed infrared detectors having a detectivity of $d^* = 2 \times 10^{10}$ cm Hz$^{1/2}$/watt could be exploited to extend this technique to the 10 μm region.

The sensitivity of this fluorescence detection technique depends on several factors including the characteristics of the decay channels of the states excited by multiphoton absorption. For maximum signal, the decay photons must be collected over as large a solid angle as possible. Generally 10% of the quanta radiated can be collected and filtered to remove stray scattering. With quantum efficiencies in the range of 10% for the cathode surfaces of typical photomultipliers, one can expect a photo-electron event for every 100 or so fluorescent decays. Quantum detectors, of course, produce signals in the absence of light, but many years of development have reduced this dark current to a few counts per second or less for cooled photomultipliers with bialkali cathodes. So-called photon counting electronics developed for light scattering will clearly detect a light level ten counts per second above this background. If the incident cw laser intensity is 30 mW and if 10% of the absorption events result in photons of the desired wavelength, an attenuation of this beam by one part in 10^{11} due to multiquantum events can be detected.

Had a direct absorption measurement been attempted instead, the desired signal would have been hopelessly buried in laser fluctuations and detector noise.

When the laser source is repetitively pulsed, the fluorescent decay photons necessarily appear within a few microseconds of the laser pulse. In this case the output of the photodetector can be averaged by a gated integrator system which accepts signals during limited time intervals. This technique eliminates most of the photomultiplier background without the need for cooling the detector, and eliminates as well spurious signals due to overhead lights, etc. The minimum detectable signal is then on the order of one photoelectron per laser pulse. If the energy per pulse is 100 μJ, attenuations of the primary beam of 10^{-11} may also be readily detected by this technique.

Various phenomena may degrade the signal-to-noise ratio obtained by these methods. At the low light levels involved, all materials fluoresce somewhat when excited by a laser and care must be taken to shield the photodetectors from such spurious radiation. If the states excited by multiquantum absorption decay by the emission of narrow fluorescence lines, an interference filter may provide sufficient wavelength selection without significantly reducing the solid angle over which quanta are collected. Conventional monochromators have better resolution than interference filters, but their throughput is often inadequate. Still, the tandem monochromators developed for Raman scattering experiments will attenuate unwanted frequency components by eight orders of magnitude.

Undesirable fluorescence may result from linear absorption in the sample itself. Many alkali metal vapors, for example, form dimers capable of absorbing the frequencies needed for two quantum absorption in the atoms [8.45]. Many of the excited molecules will then decay by fluorescing at wavelengths longer than the incident wavelength but some molecules may absorb a second quantum and dissociate into atoms, one of which may be sufficiently excited to radiate at an atomic line of wavelength shorter than the incident radiation. True two-photon absorption can also be observed in trace contaminants as the result of accidental resonant enhancement.

If the detected fluorescence corresponds to a resonance line, self-absorption can be a problem. Often this phenomenon will result in the bulk of the fluorescence signal occurring on the resonance line of longest wavelength in violation of the expected branching ratios. Truly metastable states, like the hydrogen 2S-state, may be directly excited by multiquantum absorption. Generally these states decay eventually by a route involving the emission of a photon, although the emission may be triggered by a collision or a purposely applied dc electric field.

While the technology of quantum detection is quite advanced, situations do occur when it cannot be used. One can imagine cases where the dominant decay route for an excited state does not involve the radiation of any detectable quantum. In these situations, detection techniques developed for other purposes can be applied to Doppler-free multi-quantum spectroscopy.

One well-developed technique of this sort relies upon photoionization of an excited species. Photoionization is quite likely to occur in multi-quantum absorption experiments in any case. The ion produced can be detected with high efficiency in space charge ionization diode, or by a secondary emission electron multiplier [8.46, 47]. The required electrodes must be designed into the optical absorption cell, but the outputs of these detection devices have many of the same characteristics as quantum detectors.

All other techniques having failed, it is indeed possible to detect the decrease in transmission of an incident laser beam due to Doppler-free two-photon absorption [8.43]. Since the expected attenuation will be small, at least one laser with extreme amplitude stability is essential. Even so, phase-sensitive-detection techniques are necessary to extract the small transmission dip from detector and laser noise. Fortunately, techniques of this sort have been well developed by experimenters working in saturated absorption spectroscopy [8.48].

Basically the idea is to monitor the intensity of a stable laser beam transmitted through a sample cell while the absorption coefficient of the cell is modulated with a known frequency and phase. The modulation may be accomplished by pulsing or chopping the output of a second counterpropagating laser beam. Thus multiquantum absorption will occur when both beams are incident whenever the sum of the photon energies of the two waves equals a transition energy. It is also possible to modulate the frequency of the second or pump laser by dithering one mirror. For transitions between vibrational or rotational levels of a molecule, a third modulation technique is more convenient. In order to achieve coincidence between a molecular transition frequency and the fixed frequency of a gas laser, the molecular transition must be tuned with a dc Stark field. It is then relatively easy to superimpose a small ac component upon the tuning field and modulate the transition frequency of the molecule. The laser outputs and frequencies can then be highly stabilized by servo techniques, and yet the modulation required for synchronous detection is retained [8.48].

Sufficient sensitivity and linearity are provided for detection of this effect in the infrared by a number of detectors. In the 10 μm region, a helium-cooled copper-doped germanium detector or a mercury-cadmium-telluride detector is probably preferred. If the frequency of a laser or of the transition is modulated, the detected ac signal will depend upon the amplitude of the frequency modulation. Excessive modulation will distort the experimental lineshape, making it broader than true lineshape of the transition. The effect is not serious if only the center frequency of the transition is desired, otherwise a correction factor must be computed.

When the frequency of a laser is varied, the output intensity changes as well in most cases. Since the multiphoton absorption signal depends upon a product of laser intensities, a true measurement of the relative cross sections of features resolved by the Doppler-free technique necessarily requires that the absorption signal be normalized to the correct product of laser intensities. In the case of pulsed lasers where the intensity variation is most severe, this is best accomplished on a shot-by-shot basis with an on-line computer. This same computer could then average the resulting ratios to reduce random noise. Alternatively the absorption signal and laser intensities can be separately averaged by gated integrators or photon counters and the average of the signal divided by the product of the average laser intensities either digitally or by means of operational amplifiers [8.7]. The problem is less severe when cw laser sources are employed, but even then some sort of normalization is desirable.

8.3.6 Two-Photon Absorption Cell Configurations

At the intersection of the laser and detection systems stands the actual experiment. It is there that two or more laser beams are brought together in such a way that the momenta of the interacting photons cancel and it is there that the results of the interaction with the sample vapor are monitored. The apparatus of the experimental region can be deceptively simple, some lenses and mirrors, a cell to hold the vapor, and polarization rotators, but care taken in the design of this simple apparatus is absolutely essential.

A typical configuration with which to observe two-quantum absorption without Doppler broadening appears in Fig. 8.5. A laser beam at frequency ω_1 is incident from the top. The desired polarization condition is selected by a glan-laser prism and the appropriate retardation plates. The diffraction-limited beam is then focused with a lens in such a way that the minimum spot size occurs inside the sample cell. There is no particular advantage to sharp focusing in a two-photon experiment. The size of the observed effect due to two-quantum absorption will be proportional to the square of the laser power per unit area and to the number of atoms or molecules in the focal region. The length of this focal region, often called the Rayleigh range, depends upon the size of the minimum focal spot. For gaussian beams the formula is

$$Z_r = \pi w_0^2 / \lambda$$

where w_0 is the $1/e^2$ radius of the beam at focus [8.49]. The volume of the focal region is thus roughly $V = 2\sqrt{2}\pi Z_r w_0^2$ and is proportional to the number of atoms excited. The expression for the total signal then becomes,

$$S = \varrho V P^2 / (\pi w_0^2)^2$$

where ϱ is the number density of the sample vapor and P is the rms laser power. Note that the factors proportional to the radius of the focal spot cancel. Thus one will obtain optimum signal whenever the laser beam is focused in such a way that the Rayleigh range is less than the effective sample length. In practice more than adequate signal can often be obtained when this condition is not fulfilled, and the only effect of focusing even this tightly is to introduce light shifts and power broadening.

A second incident laser beam with proper polarization and frequency ω_2 must be focused into the cell from the bottom. It should be nearly collinear with the first, and focused in exactly the same way, to obtain a residual Doppler broadening of magnitude

$$\Delta\omega_d = (\omega_1 - \omega_2)(2kT/mc^2 \ln 2)^{1/2} .$$

Residual Doppler broadening may also occur when two beams of exactly the same frequency are employed in a geometry where the wavevectors are not exactly antiparallel. Such situations are not necessarily the result of experimental sloppiness. When the optimum focusing and collinearity conditions are reached, frequency components may feed back into the laser cavity from the sample region and destabilize the oscillation frequency. A variety of techniques exists for eliminating this effect, but many have undesirable side effects. Isolation techniques based upon the reflection properties of circularly polarized light, for example, restrict the accessible polarization conditions that can be employed. Shutters and short-pulsed lasers lead to uncertainly principle broadening. Passive absorbers necessarily reduce the signal intensities. Isolators based upon the principle of Faraday rotation will work, but presently are not widely available [8.50]. The simplest solution to this feedback problem is to utilize beams that are slightly non-collinear and isolate the laser from the returning beam by spatial filtering. The residual Doppler effect expected will be

$$\Delta\omega_0 = 2\omega_1(2kT/mc^2 \ln 2)^{1/2} \sin\theta$$

where θ is the deviation from collinearity of the laser beams. To achieve this near collinearity in practice one may reflect and refocus the laser beam transmitted through the sample cell by means of a properly chosen lens and a prism retroreflector rather than with a concave mirror.

The geometrical configuration just described will permit the observation of Doppler-free absorption of any even number of quanta. The geometries suitable for experiments involving odd numbers of quanta are more complex. The signals produced by all of these higher-order effects will be enhanced by tight focusing of the interacting beams, and the expected cross sections are small enough in general to require all the sophistication one can muster.

The sample cell itself can be quite simple, or may be more sophisticated, employing the recently developed heat pipe technology and perhaps containing part of the detection apparatus. Some reactive vapors are best handled as atomic or molecular beams, but the Doppler-free technique removes all the collimation requirements previously needed for high-resolution spectroscopy. Thus the "beam" can be quite rudimentary [8.23]. One requirement upon the cell is that the optical windows be of adequate quality, that they not degrade the focal properties of the beam, or scatter it extensively or produce spurious signals as a result of fluorescence, etc. If the cell must be heated, care should be taken that the heating elements not produce magnetic fields sufficient to perturb the energies of the highly excited states being probed.

When nearly collinear beams of frequencies ω_1 and ω_2 incident from opposite directions overlap with sufficient intensity within the sample cell, a signal due to two-photon absorption will appear whenever $h\omega_1 + h\omega_2 = \Delta E$ is the energy separation of two levels of the same parity. The width of this resonance can be quite narrow, and that is, of course, the point of the technique. However, the resonance can be quite easily missed if the sum of the laser frequencies cannot be made to equal the resonant frequency with sufficient precision. Some laser systems do intrinsically possess sufficiently precise resettability to allow an experimenter to set them to the exact transition frequency, but the widely tunable dye lasers and parametric oscillators do not. Thus a well calibrated monochromator with considerable resolution is a necessary, if subsidiary, piece of equipment in a Doppler-free two-photon absorption experiment. The best grating instruments when carefully used can determine a laser's frequency only to within 3 GHz. With Doppler-free resonances as narrow as 10 MHz, a patient and systematic search procedure becomes necessary to find the resonance within the band determined by the monochromator.

When $\omega_1 = \omega_2$ and a single laser frequency is used, a Doppler-broadened component will appear in the two-photon absorption spectrum. This rather broad peak may be strong enough to facilitate the search procedure, but it may inhibit the experiment proper. In the case of transitions between atomic S states, the broadened component may be eliminated by choosing a circular polarization condition in which photons in the interacting beams have opposite spin projections on the propagation axis. Thus the beam entering the cell from the top must be circularly polarized by a quarter-wave plate at B, and the spin of the returning photons can be reversed by forcing them to propagate twice through another quarter-wave plate placed between the cell and the return mirror or corner cube. Atomic S to S transitions cannot absorb two photons with the same spin projection, and this property can be employed to identify such transitions among other lines due either to S-D transitions or contaminant molecules.

The Doppler-broadened component should have half of the integrated intensity of the narrow resonances in a properly aligned experiment, even without polarization selection. If this nonselective excitation presents a problem in the case of S-D transitions or molecular lines, circular polarization will not help, but another strategem for removing it exists. This technique involves using two laser frequencies which sum to the desired transition, but which cannot separately excite it because

$$|\Delta E/\hbar - 2\omega_2| \gg \omega_2(kT/mc^2)^{1/2} \quad \text{and} \quad |\Delta E/\hbar - 2\omega_1| \gg \omega_1(kT/mc^2)^{1/2} .$$

The complexity of a laser system which produces two distinct frequencies is warranted in experiments which require a nearly resonant intermediate state to enhance an otherwise weak two-photon cross section, or to observe features characteristic of the intermediate resonance. Small cross sections occur in a variety of contexts in both atomic and molecular systems. In particular, transitions among vibrational levels, transitions involving quadrupole matrix elements, and spin forbidden transitions are generally too weak to observe without an intermediate state with energy close to that of one photon. However, a multifrequency laser system allows one to tune one of the two absorbed frequencies within a few Doppler widths of an intermediate energy level, and thereby enhance the cross section for multiquantum absorption by a factor of 10^8 [8.19]. The possibility of such resonant enhancement also encourages the search for absorption phenomena involving three or more photons and from which Doppler broadening may be eliminated.

The cross section for a resonantly enhanced two-quantum transition can be fully comparable to that of an allowed single-quantum absorption. With presently available lasers, one can produce significant populations of selected highly excited states by this technique and study the chemical and physical properties of such highly excited species. Properly done, such nonlinear excitation need not result in populating undesired intermediate energy levels or levels near the desired final state. Thus direct nonlinear excitation may in some cases be superior to the stepwise excitation techniques presently employed to reach high-lying energy levels [8.19].

Another technique which enhances both the absorption signal and the resolution is to place the nonlinearly absorbing medium inside a confocal interferometer cavity. The optical intensity inside such a cavity is enhanced over that incident from the outside by a factor of $T/(T+A)^2$ where T is the transmission coefficient of the partially reflecting mirrors and A is the single-pass loss due to absorption in the cavity. This increase in the field intensity occurs only when the frequency of the incident light coincides with a resonance of the interferometer cavity within the latter's resolution; otherwise, the field intensity inside the interferometer is reduced below that of the incident beam by a factor of T.

This frequency selectivity can be used to improve the resolution of a nonlinear spectroscopy experiment beyond the limit set by the laser system. If the sample cell is equipped with low loss windows, an interferometer resolution of better than 20 MHz can be readily achieved. The free spectral range of this cavity (FSR = $C/4L$) should be chosen larger than the expected linewidth of the incident beam in order to avoid spurious secondary resonances. Thus a laser system with a resolution of 150 MHz requires an interferometer of length 50 cm or less. The diameter of the TEM_{00} mode of this cavity at the entrance mirror will be 0.5 mm, and a damage resistant reflective coating should survive many 50 kW incident pulses. To scan over a narrow resonance using this interferometer scheme, one must translate one mirror of the resonator containing the sample. If the length of the scan exceeds the linewidth of the laser, the tuning elements of the laser must be scanned synchronously with the interferometer.

While a confocal interferometer cavity has the advantage that all of the transverse modes are degenerate, it has the disadvantage that the minimum spot size within it is comparatively large [8.49]. Thus the focal region essentially fills the interferometer, making it difficult to collect signals emitted from the entire region of maximum excitation. A more nearly spherical resonator geometry will have a tighter focus at its center for the same overall cavity length, but will present more alignment difficulties. To avoid instabilities, any interferometer must necessarily be isolated from the laser oscillator.

The ultimate precision and resolution of a multiquantum absorption experiment is limited by a number of phenomena. Residual Doppler effect, and the uncertainty principle have already been mentioned. To these should be added transit time effects familiar from saturated absorption, collision broadening, and a group of phenomena collectively termed nonlinear power broadening. These latter effects are especially troubling since the line between adequate signal and excessive broadening can be quite narrow in nonlinear experiments.

The most familiar source of power broadening is saturation, which occurs whenever the incident laser intensities are sufficiently strong to equalize the populations of the two states connected by the nonlinear transition. This effect broadens lineshapes symmetrically in a manner analogous to the linear case. Less familiar in optical spectroscopy is the light shift, which results from a perturbation of the energy levels themselves as a result of the intense optical fields. Equations (8.13, 18, 19) show that this effect is proportional to the intensity, while the multiquantum absorption signal is proportional to the square of the intensity. Thus light shifts can never be entirely eliminated. While this effect merely shifts the energy levels when the optical fields are uniform in space and time, it results in an asymmetric broadening in the more common case where the laser intensities vary. Whenever the fields are weak, the shift and the absorption are both small, but when the fields are strong, both the unidirectional shift and the nonlinear absorption signal increase. There are also intrinsic line-broadening phenomena which result from the interaction of large uniform optical fields with atomic and molecular systems which must be included in any calculation of the nonlinear power-broadening cross section.

These effects which degenerate the resolution and precision of a nonlinear spectroscopy experiment are best handled by increasing the diameter of the region where the optical fields interact with the sample. The reduction in signal due to the decreased intensity is thus partially compensated by an increase in the sample volume. Experiments must be performed, however, with the incident intensity attenuated by various amounts in order to ensure that the nonlinear power broadening is truly under control.

A variety of interferometric and heterodyne techniques has been developed to determine the absolute frequencies and scanning rates of tunable laser systems. The accuracy of this calibration step may determine the precision of the entire experiment. The frequency calibration should be performed simultaneously with the main experiment in order to avoid relying upon a potentially unreproduceable internal standard. When an on-line computer is available, the frequency information can often be recorded on a pulse by pulse basis. Otherwise, some sort of average frequency scale must be included in the experimental data along with a measurement of the nonlinear absorption signal and the laser intensities.

The experimental techniques reviewed here may sound rather involved, but there has been a wide variety of high resolution nonlinear spectroscopy experiments reported since the first demonstration of the effect. Most of the sophisticated apparatus developed for previous high-resolution techniques is directly applicable to Doppler-free multi-quantum absorption, and the technological difficulties encountered are no worse. A new class of experimental investigations is thus opened to the laser spectroscopist.

8.4 Experimental Results

8.4.1 Early Experimental Confirmation of Doppler-Free Two-Photon Absorption

Doppler-free two-photon absorption was first demonstrated in sodium vapor. Of the three earliest experiments, two used the technique discussed here [8.6, 7] to resolve the hyperfine splitting of the $3S \rightarrow 5S$ two-photon transition, and the third resolved the fine structure of the $3S - 4D$ transition by crossing the laser beam with a well-collimated atomic beam [8.9]. The cross sections of both of these transitions for two equal quanta are large as the $3P$ levels of sodium lie nearly half way. Since the $3P$ term dominates the sum in (8.11), the unknown signs of other terms affect the size of the cross section rather little. Thus the early experimenters could accurately estimate the expected signals, as all pertinent matrix elements are well known [8.11]. The quantum energies are not so near the $3S - 3P$ transition, however, that resonant effects alter the two-photon selection rules. All three experiments detected multiquantum absorption events by observing fluorescence due to the decay cascade of the excited state, and all employed rhodamine 6G dye laser sources.

The Na^{23} nucleus has spin $I = 3/2$, and it interacts with the spin of the unpaired electron by means of a magnetic hyperfine Hamiltonian $\mathcal{H} = A\mathbf{I} \cdot \mathbf{S}$ which applies to S-states. The splitting of the $F = 2$ and $F = 1$ hyperfine levels of an S-state is thus $2A_{nS}/h$ in frequency units. The selection rules for the transition result in two absorption lines separated by $\Delta v = (A_{3S} - A_{nS})/h$ where Δv is the required shift of the laser frequency to move from one hyperfine component to the other. While the ground state hyperfine splitting is well known for sodium, the splittings of the $5S$ and $6S$ states were first measured using the Doppler-free two-photon technique.

The phenomenology thus predicted for the $3S \rightarrow 5S$ two-quantum transitions agrees with the experimental results in Fig. 8.8. When the vapor is excited by a linearly polarized traveling wave, only a Doppler-broadened profile is seen. When a second linearly polarized beam propagating in the opposite direction overlaps the first, two narrow

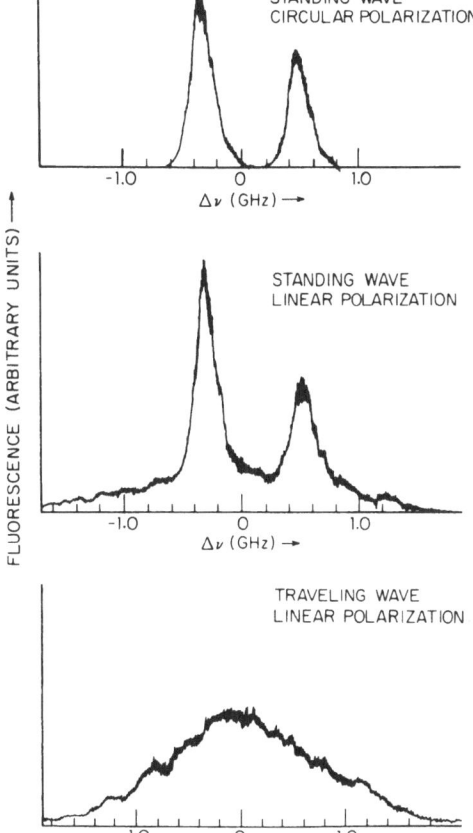

Fig. 8.8. Two-photon absorption signal on the $3S \rightarrow 5S$ transition of atomic Na^{23}. The experimental traces record the observed resonance fluorescence intensity at 330 nm ($4P$-$3S$ transition), following the two-photon absorption

peaks appear superimposed upon the former component. If a single $\lambda/4$ plate is employed to give both beams the same photon spin projection, the fluorescence signal entirely disappears. If the spin of one of the beams is reversed with a second $\lambda/4$ plate, the unbroadened components reappear, but the Doppler background is suppressed.

The ratio of the intensity of the $F = 2$ to that of the $F = 1$ line is $5:3$, from the statistical weights of the F-states in the $3S$ ground level. From the separation of the doublet, one can calculate a value of $A_{5S}/h = 78 \pm 5$ MHz for the hyperfine interaction constant in the $5S$-state. This number is in good agreement with theoretical predictions ([8.51]) and with subsequent data from a cascade pumping experiment. The linewidths of the narrow components in Fig. 8.8 result from the uncertainty

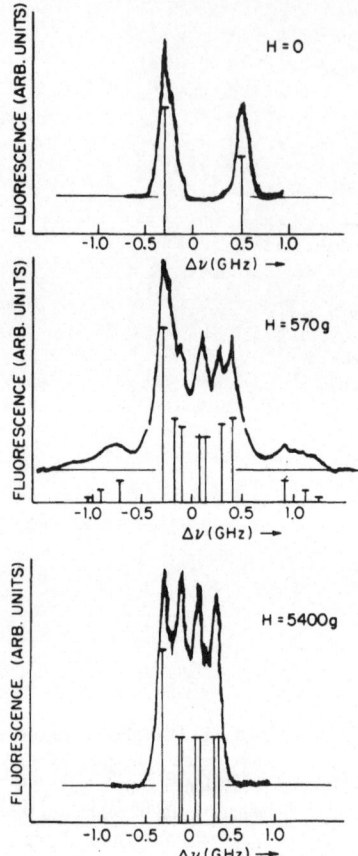

Fig. 8.9. Zeeman effect of the two-photon $3S \rightarrow 5S$ transition in the Na^{23} atom. The verticals indicate the theoretically calculated line positions and their relative strength (after Bloembergen et al. [8.18])

principle broadening of the 4 ns long laser pulse; the natural linewidths are much less.

The Zeeman effect is readily observable and the Zeeman splittings of the atomic hyperfine multiplets are, of course, well known. The effect merits, nevertheless, some attention because of the unusual characteristics for the particular symmetry of the $S \rightarrow S$ transition [8.18]. These are illustrated in Fig. 8.9. The nuclear and electron g-factors in the initial and final states are the same. Because of the selection rules $\Delta m_S = \Delta m_I = 0$, the spectral lines become asymptotically independent of the magnetic field strength in the high-field limit. The spectrum then consists of four equally spaced spectral lines with an individual separation $(1/2)(A_{35} - A_{55})h^{-1}$, as shown at the bottom of Fig. 8.9. The spectrum for intermediate fields, in which one or neither of hyperfine states is in the decoupled Paschen-Back regime, consists of 13 components whose position and intensity may be calculated exactly, as the eigenvalue problem factorizes in a number of two-by-two determinants. An intermediate pattern is shown in the middle of Fig. 8.9. It should be noted that the spectrum at any given field strength is invariant for rotation of the magnetic field with respect to the light beam polarization and propagation directions. These unusual features are characteristic of $S \to S$ two-photon electric dipole transitions.

While the early atomic beam experiment [8.9] employed a cw dye laser with an intrinsic resolution of 10 MHz, the achieved experimental resolution was comparable to that in Fig. 8.8 and 9. The tight focusing necessary to obtain an optimum signal in the thin collimated atomic beam introduced a large residual Doppler effect in that experiment. A much higher resolution was, however, soon achieved for the $3S - 4D$ transition by HÄNSCH et al. [8.52]. They used a cw dye laser and counterpropagating beams in an Na vapor cell. The Doppler-broadened background, while not suppressed by any selection rule, is quite negligible to the four narrow components, shown in the lower trace of Fig. 8.10. The splitting between the components with the same value of J corresponds to the hyperfine splitting of the $3S$ ground state. The splitting between the two components on the left (or on the right) corresponds to the fine structure splitting of the $4D$-state.

The phenomena of resonant enhancement and ac Stark effect were soon demonstrated by BJORKHOLM and LIAO in a series of experiments involving two cw dye lasers [8.19, 25]. The frequency of one laser was adjusted to bring it close to the $3S - 3P$ doublet in sodium while the other laser was adjusted so that the sum frequency scanned over the $3S - 4D$ two-photon absorption quartet. The difference in frequency between the two counterpropagating beams introduced a residual Doppler effect of less than 90 MHz.

As the frequency of the first laser approached resonance with the $3S - 3P_{3/2}$ absorption line, the cross sections for the two-photon transitions to the $4D_{5/2}$ and $4D_{3/2}$ levels were enhanced by a factor of 10^8. In order to avoid level shifts and other spurious effects, the beams were expanded and highly attenuated near the resonance condition. Nevertheless, the red-orange decay fluorescence from the excited D-state was clearly observable in the darkened laboratory. Figure 8.11 shows that the cross section for the $3S - 4D_{3/2}$ transitions goes through a sharp minimum and is again enhanced as the laser frequencies are tuned to bring the $3P_{1/2}$ level into resonance. Since no dipole matrix element connects the $3P_{1/2}$ and $4D_{5/2}$ states, the cross section for two-photon transitions to the latter state has only a single resonance. The solid lines in Fig. 8.11 are calculated from (8.12) and adjusted for the residual Doppler effect which varies as the laser frequencies become more and

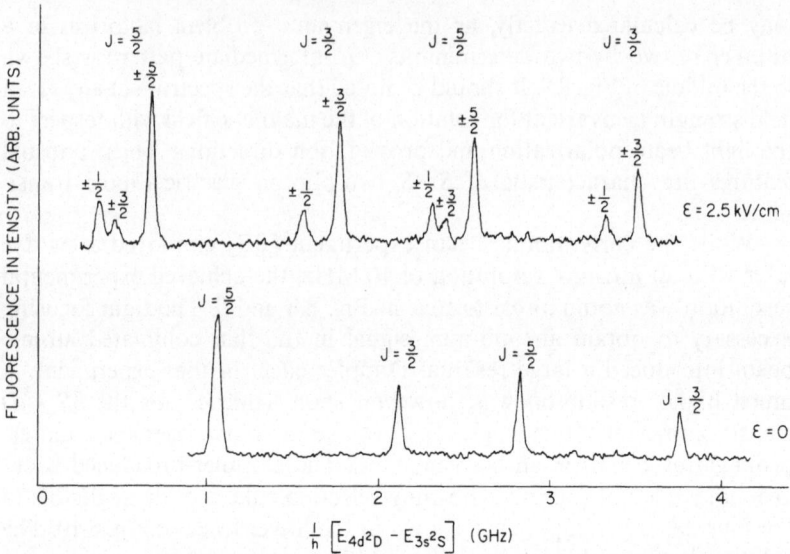

Fig. 8.10. Two-photon transition $3S \rightarrow 4d$ for the Na^{23} atom (bottom), and the Stark shifts and splittings produced by a dc electric field (top). They are quadratic in the dc field strength (after Harvey et al. [8.23])

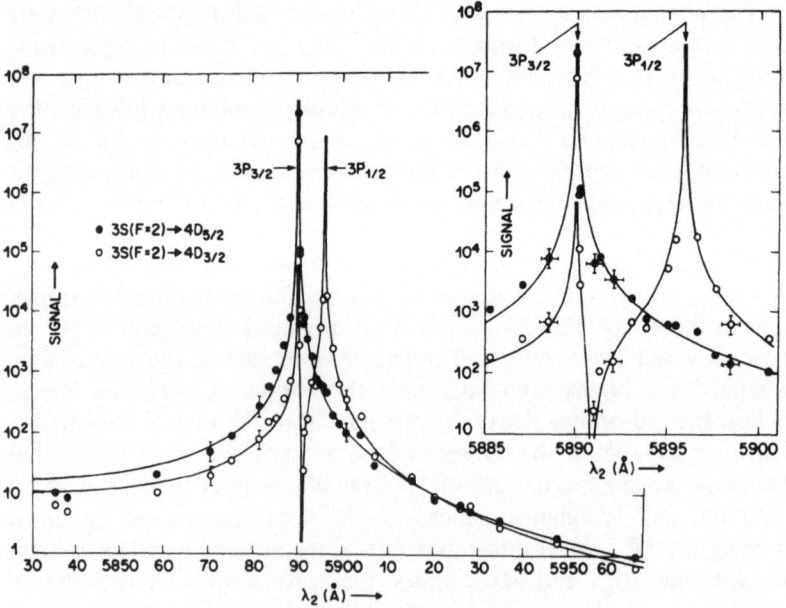

Fig. 8.11. Resonant enhancement of the two-photon absorption rate; $h(\omega_1 + \omega_2)$ is fixed at the $3S\,(F=2) - 4D_{3/2}$ or $3S\,(F=2) - 4D_{5/2}$ transition, while $h\omega_1$ is tuned through the (one-photon) yellow Na-doublet (after Bjorkholm and Liao [8.19])

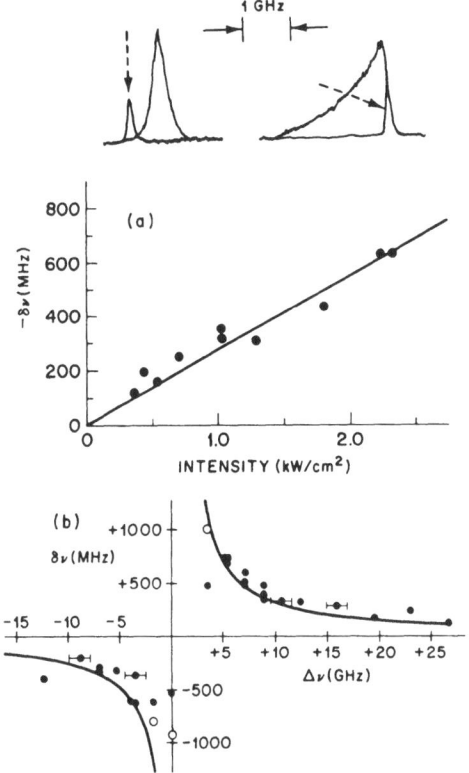

Fig. 8.12. Power-dependent Stark shift as a function of frequency ω_1, near the intermediate (one-photon) resonance (after LIAO and BJORKHOLM [8.24]). Top: Oscillographs of the $3S(F=2) \rightarrow 4D_{5\,2}$ two-photon absorption line. The frequency ω_1 increases to the right. The lines indicated by the arrows were obtained with both laser beam at low power level (< 3 mW). The shifted line on the left is obtained when only the power at 589 nm is increased to 26 mW. The broadened and shifted line on the right is obtained when only the power in the strongly focused beam at 569 nm is increased to 30 mW. Middle: The linear dependence on the induced shift with power (quadratic in field amplitude). Bottom: The increase and change in sign of the shift, as one of the frequencies is tuned through the $3S(F=2) \rightarrow 3P_{3/2}$ intermediate state resonance

more unequal. On the wavelength scale of the insert the Doppler width of the $3S - 3P$ resonance lines is less than 1/100 the diameter of the solid circles.

The power-dependent shifts of both the $3S$ ($F=2$) and the $4D_{5/2}$ atomic energy levels were observed with the laser frequencies tuned near the $3P_{3/2}$ resonance [8.25]. With gaussian beams, the optical intensities cannot be uniform over the sample, and thus atoms at different positions will experience different shifts. However, the laser just below the $3S - 3P_{3/2}$ transition at 589 nm was focused rather loosely so that near its center the intensity was rather uniform. The second beam at 569 nm was focused tightly in that region so that atoms absorbing two quanta would necessarily experience a uniform 589 nm illumination. The results appear in Fig. 8.12. At low intensities the two-quantum absorption line is undisturbed. When the 589 nm intensity is increased, the two-photon transition is shifted to higher frequencies, indicating a depression in energy of the $3S$ ($F=2$) state. The line is also somewhat broadened by other effects described by (8.20). When the intensity of the more tightly focused beam is increased, the two-photon transition is shifted to lower frequencies and asymmetrically broadened as a result of the nonuniform intensity of the field creating the downward shift of the $4D_{5/2}$ state. Shifts of 1000 MHz were observed with intensities under $3\,\text{kW/cm}^2$. The sign of both shifts would change, if the laser at 589 μm would be tuned just above the $3S-3P_{3/2}$ transition.

8.4.2 Alkali Atom Two-Photon Spectroscopy

The success of the initial experiments stimulated many investigators to study the characteristics of alkali atom states which could be excited by two-photon absorption. The Doppler-free transitions are so narrow that many spectroscopic features, which are not resolvable in a Doppler-broadened profile, can be measured. These include:
a) Fine structure splittings of excited d-states,
b) Hyperfine splittings of excited s-states,
c) Isotope shifts,
d) Zeeman splittings,
e) Second-order Stark splittings and shifts,
f) Collision-induced broadening and shifts.
Some of these new results are tabulated in Table 8.2.

The hyperfine splitting of the $3S\rightarrow5S$ transition was remeasured with greater precision with a cw dye laser [8.53]. The Zeeman splitting of the $3S\rightarrow4D$ level was also investigated [8.54]. The quadratic Stark shift of the $3S$-state and the quadratic Stark shift and splittings of the $4D$-state were also investigated in detail with a cw dye laser [8.23]. The results of the application of a dc electric field are shown in the upper part of Fig. 8.10.

Table 8.2. Some spectroscopic results obtained with Doppler-free two-photon absorption

Atom	Level	HFS MHz	FS MHz	Isotope shift	Pressure shift MHz/Torr (Ne–Na)	Pressure broadening MHz/Torr (Ne–Na)	Polarizabilities MHz/(kV/cm)² α_0	α_2
Na	5S	76±5					5.2±3	
	4D		1028±3	−7±1		32±5	155.3±1.7	−38.5±7 $D_{3/2}$
							156.1±1.3	−53.2±5 $D_{5/2}$
	6S	39±3						
	5D		618±12					
Rb	6D			166±40				
	8S			145±40				
	9S			160±40				

It should be noted that radiospectroscopy or coherent quantum beat techniques, in conjunction with cascade pumping, only yield Stark splittings, but not shifts. SCHAWLOW et al. [8.23] were able to deduce the scalar and tensor polarizabilities of the D-states, and to separate the contributions from neighboring p- and f-levels to the quadratic Stark effect. From it they obtained the oscillator strengths $f_{4d \to 5p} = 0.274\,(29)$ and $f_{4d \to 4f} = 0.018\,85\,(24)$.

The cw dye lasers available in 1974 could not excite higher levels in sodium. Studies of the fine structure splitting of the $5D$ level and the hyperfine interaction in the $6S$ state were made with pulsed laser techniques [8.55]. Transitions to higher lying nS levels, with n up to 20, have been observed. The small splittings in these higher lying states can, however, be determined more accurately with radiospectroscopy [8.56, 57] and coherent quantum beat techniques [8.24], in conjunction with cascade optical pumping.

FORTSON and ROBERTS have observed Doppler-free two-photon transitions in rubidium vapor with an optical parametric oscillator as a source [8.40]. The $6D_{5/2}$, $8S$ and $9S$ levels were excited in both the Rb^{85} and Rb^{87} isotopes. While the observed hyperfine splittings were previously measured with better accuracy by an optically pumped radio-frequency spectroscopy technique, neither this method nor quantum beat spectroscopy determines the isotope shifts which are tabulated in Table 8.2.

A more extensive survey of the $5S \to nD$ transitions in rubidium has been reported by STOICHEFF [8.58]. They used a cw dye laser in an experiment similar to the sodium experiment of HÄNSCH [8.52] to resolve the fine structure splitting of the states from $11D$ up to $30D$ in Rb^{85} and Rb^{87} with a resolution of 20 MHz. In rubidium the fine structure is not inverted in these highly excited levels.

The observed fine structure intervals failed to scale with the inverse cube of either the principal quantum number or the effective quantum number. This interesting result implies that arguments based on simple scaling laws do not apply to the Rydberg states of even the heavy alkalis.

The effects of line broadening and displacement have been observed for the $3S - 4D$ transitions as a function of the partial pressure of gaseous admixture of Ne to the Na vapor by CAGNAC and coworkers [8.59]. The effect of collisions with Ne atoms is the same for all four transitions. The increase in total width for Ne at $20°$C is 32 ± 5 MHz/Torr, while the frequency shift is -7 ± 1 MHz/Torr. Since two photons are absorbed in the transitions, the dye laser frequency tuning is changed by half of the above amounts.

The same authors [8.60] also observed the transfer of energy between the $4D_{5/2}$ and $4D_{3/2}$ levels induced by the collisions between Ne and Na atoms. One of D levels is excited, for example the $4d^2 D_{5/2}$. One observes the fluorescence at 568.2 nm from the $4d^2 D_{3/2}$ to the $3p^2 P_{1/2}$ level. This level cannot be reached directly from the $D_{5/2}$ state according to electric dipole transition selection rules. The fluorescence at 568.8 nm gives the decay of the joint population in both the $D_{5/2}$ and $D_{3/2}$ levels. In this manner the authors were able to derive an effective cross section for energy transfer

$$\sigma_{5/2 \to 3/2} = (2.7 \pm 0.9) \times 10^{-14} \, \text{cm}^2$$

in Ne–Na collisions.

8.4.3 Atomic Hydrogen

The $1S \to 2S$ transition in atomic hydrogen is among the most important in physics, and it can be resolved by the Doppler-free two-photon technique. The experiment is a tremendous challenge. To excite the $2S$ level, two quanta of wavelength 243 nm are required. Photons of that energy cannot be produced directly in any present laser, and are generated [8.61] by second harmonic generation in lithium formate of a dye laser beam at 486 nm. The $2S$ state is truly metastable with a lifetime of 0.14 s. That lifetime implies that the $1S \to 2S$ transition frequency can be measured to one part in 10^{17} in a sufficiently careful experiment. Atoms in the $2S$ state can be induced to decay faster via fluorescence at the Lyman α line by collisions with other atoms or by an applied electric field.

HÄNSCH and coworkers have set out to tackle this spectroscopic problem [8.33]. They have succeeded in detecting the Doppler-free two-photon resonance in atomic hydrogen, and in resolving the hyperfine doublet for the $F = 1 \to 1$ and $F = 0 \to 0$ transitions. They also have compared the frequency of the $1S - 2S$ transition in hydrogen and deuterium to that of the Balmer β line at 486 nm in order to measure the Lamb shift of the $1S$ state. Their experimental setup is shown in Fig. 8.13 and some early results in Fig. 8.14.

The fine structure of the one-photon Balmer β line can be resolved using the higher resolution of the saturation spectroscopy technique. Since this transition has an energy almost exactly 1/4 of that the $1S \to 2S$ line, the resolved components can be used as an internal frequency reference for the laser whose second harmonic excites the two-photon transition. Present results for the $1S$ hydrogen and deuterium Lamb shifts determined in this way are within 1/2% of the theoretical values with a 1.2% error bar [8.61].

While the precision achieved in these measurements of the $1S$ Lamb shift is less than that obtained for the $2S$ Lamb shift determined by microwave techniques, the result is impressive nonetheless. If the

MEASUREMENT OF THE HYDROGEN IS LAMB SHIFT

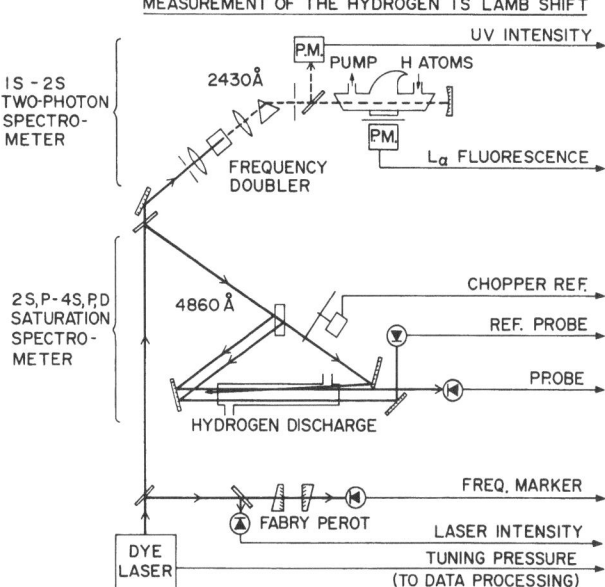

Fig. 8.13. Diagram of experimental apparatus to measure 1S–2S two-photon transition in hydrogen (after HÄNSCH et al. [8.33])

theoretical value of the 1S Lamb shift were assumed correct, a Doppler-free two-photon absorption experiment on the $1S - 2S$ transition in hydrogen with the present resolution would improve the accuracy of the Rydberg by another order of magnitude. Ultimately lasers with less linewidth and frequency jitter will allow such measurements to approach the theoretical maximum precision.

8.4.4 Molecules and Complex Atoms

Excited electronic states with the same parity as the ground state have first been studied by two-photon absorption in molecules without the Doppler-free feature [8.62]. The first Doppler-free resonance of this sort was observed in Na_2 by SCHAWLOW et al. [8.63]. The concentration of this molecule in the cell of atomic sodium was quite small. Still, the incident laser intensity fell close enough to some intermediate state in the molecule to enhance the two-photon absorption cross sections of a few lines to the point where they could be detected. The familiar splitting due to the ground state hyperfine interaction in atomic sodium was entirely absent.

Further studies of this sort are continuing, with emphasis on simple molecules like NO and molecules with high symmetry like C_6H_6. The

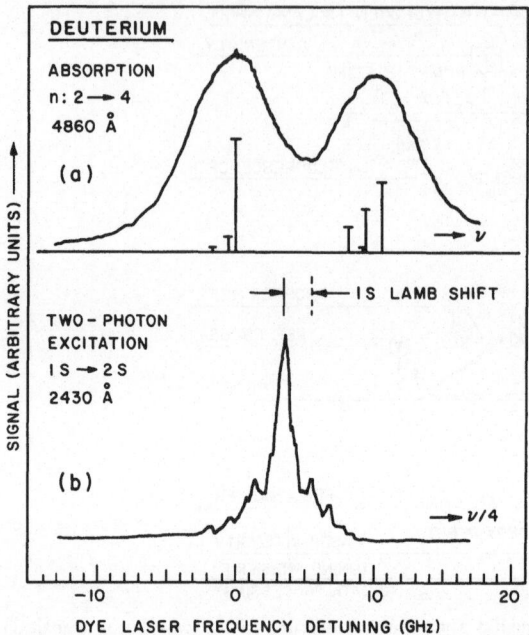

Fig. 8.14. (a) Absorption profile of the deuterium Balmer-β line with theoretical fine structure. (b) Simultaneously recorded two-photon resonance of deuterium $1s$–$2s$ (after Hänsch et al. [8.33])

many closely spaced levels which must be resolved require unusually narrow laser linewidths and powerful data analysis techniques [8.61].

Multiquantum transitions between vibrational and rotational levels of a molecule have been observed in the infrared. While the cross sections for $\Delta v = 1$ transitions are quite small because of parity selection rules, those of $\Delta v = 2$ transitions are acceptably large. The signal level may also be enhanced if an allowed $\Delta v = 1$ transition frequency is nearly equal to the frequency of one of the incident laser beams. The first experiment of this sort was reported by Bischel et al. in methyl fluoride [8.43]. The infrared transition from the vibrational ground state to the doubly excited v_3 vibration was observed. Use was made of the near coincidence of the $P(14)$ line of a CO_2 laser at 9.6 μm with the $R(1, 1)\,0 \rightarrow v_3$ fundamental band of the CH_3F molecules, and the $P(30)$ line of an oppositely directed CO_2 laser beam, also near 9.6 μm, with the $R(2, 1)\,v_3 \rightarrow 2v_3$ hot band. The small amount of continuous tuning needed could be provided by a dc Stark field applied to the CH_3F cell. The experimental arrangement was described in Section 8.2 and a schematic of the apparatus and pertinent molecular energy levels is given in Fig. 8.15. It is clear that there

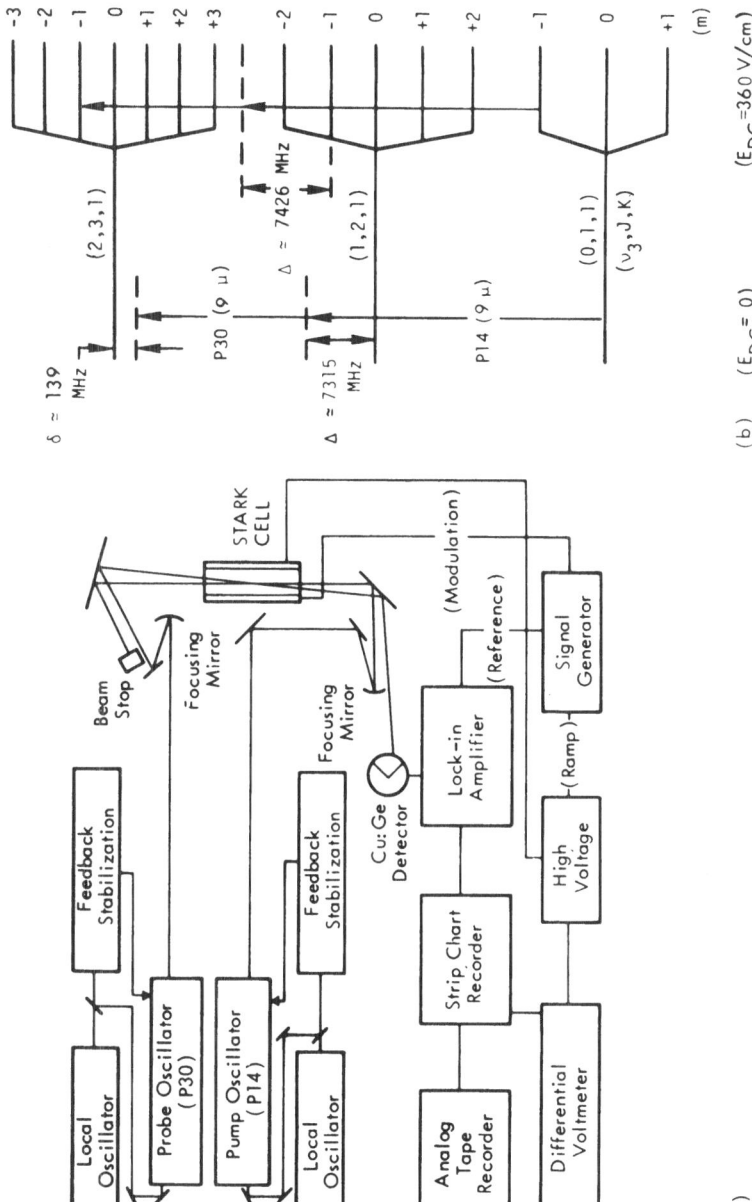

(a)

(b) $(E_{DC} = 0)$ $(E_{DC} = 360 \text{ V/cm})$

Fig. 8.15. (a) Schematic diagram for the apparatus to detect two-photon infrared absorption in CH_3F. (b) Schematic diagram of the Stark-split energy levels in the ν_3 band. The two-photon transition for the labeled energy levels is indicated with both infrared electric fields parallel to the dc field (after BISCHEL et al. [8.43]) is near resonance with the intermediate level represented by the first excited vibrational state. There is no Doppler background, because the frequency difference between the two opposing infrared beams is sufficiently large that two-photon absorption from a single beam is off-resonance.

BISCHEL et al. were able to measure the linewidth and line shift of the Doppler-free line as a function of pressure. They found that the collisions between two methyl fluoride molecules introduce a line broadening of 41.3 ± 1.0 MHz/Torr, and a positive ("blue") frequency shift of 2.1 ± 0.1 MHz/Torr. Addition of helium gas led to a broadening coefficient of 5.0 ± 0.5 MHz/Torr for helium–CH_3F collisions. Accurate Stark shifts and selection rules could also be determined.

This same group also discovered strong two-photon absorption signals in NH_3 [8.64]. The transition between the $(v_2^\pi, J, K) = (0^-, 5, 4)$ state and the $(1^+, 5, 4)$ state of the v_2 band falls near the $P(34)$ line of the $10.6\,\mu m$ CO_2 laser band while the $(1^+, 5, 4) \rightarrow (2^-, 5, 4)$ line is nearly coincident with the $P(18)$ CO_2 line. Thus the $(0^-, 5, 4) \rightarrow (2^-, 5, 4)$ two-photon transition can be excited by these laser frequencies when sufficient tuning of the molecular resonance is achieved with the dc Stark effect. The strong signals obtained (40 times that observed in CH_3F) permitted detailed studies of the effects of collisions and of high optical intensities.

BISCHEL et al. measured pressure broadening coefficients for NH_3 due to self-broadening and due to the foreign gases He, Ne, Xe, H_2, and D_2. In the case of Ne, a pressure shift in the center frequency was also observed. The difference in the laser frequencies leads to a residual Doppler effect at low pressures. At high pressures, the observed linewidth is entirely due to collisions, and in between the phenomenon of Dicke narrowing [8.65] might be expected. A 70 kHz narrowing of the resonance was in fact observed at an Ne pressure of 80 mTorr.

The frequency of the two-photon transition was also measured accurately as $1876.991493 \pm (3)$ cm^{-1} at low intensities. At high intensities the optical level shifts discussed previously lead to a measurable change in the center frequency of the resonance. By separately varying the intensities of the applied fields, these authors measured the transition dipole moment for the $v_2 = 1^+ \rightarrow 2^-$ vibrational hot band. The obtained value of $\mu(1^+ \rightarrow 2^-) = 0.27 \pm 0.5D$ is larger than that of the analogous transition from the ground state. An estimate was also made in the course of this calculation of the transition matrix element between the inversion levels in the excited vibrational state to be $\mu(2^- \leftrightarrow 2^+) = 0.83 \pm 0.08D$. These results demonstrate the value of the optical level shift in the accurate measurement of transition matrix elements, especially matrix elements between excited states. This method avoids the difficulties in measuring small absorptions and estimating the populations of the states involved. With more widely tunable infrared lasers, many studies of this type will be undertaken.

Doppler-free two-photon studies of complex atoms (such as the noble gases) have been hindered somewhat by the large energy differences between the ground and first states. BIRABEN et al. circumvented this difficulty by populating the $3s(3/2)J=2$ metastable level in neon by means of an rf discharge [8.66]. Fluorescence resulting from the decay of atoms excited into the $4d'(5/2)J=2$ state by two-photon absorption could be observed whenever the rf field was briefly extinguished. The isotope shifts, $E(^{22}\text{Ne}) - E(^{20}\text{Ne}) = 2780.0 \pm 2.5 \text{ MHz}$ and $E(^{21}\text{Ne}) - E(^{20}\text{Ne}) = 1452 \pm 6 \text{ MHz}$, were measured as were the hyperfine constants for the upper state: $A = -235.9 \pm 0.6 \text{ MHz}$, $B = 11 \pm 9 \text{ MHz}$, using a single-mode cw dye laser. The great technological interest in separating the isotopes of remarkably complex atoms should stimulate more studies of this sort in the months and years ahead.

8.4.5 Conclusion

The results, obtained since the first successful experiments on Doppler-free two-photon spectroscopy were first announced early in 1974, are sufficiently promising to forecast a rapid further growth of this field, which in some respects complements other high-resolution techniques described elsewhere in this volume. Several new results in atomic spectroscopy have been obtained. Numerous highly excited levels of many atoms and molecules can be investigated in high resolution. Chemical applications in the field of molecular spectroscopy appear particularly promising, and their exploitation has scarcely begun. Doppler-free processes involving more than two photons have not yet been carried out, but are definitely feasible. Levels near and beyond the ionization limit may be investigated with this technique. In fact, three or multiphoton ionization via a two-photon Doppler-free resonance with a bound excited state may prove to be a valuable detection technique to establish the two-photon resonance condition. The high resolution permits the establishment of isotopic shifts. The possible usefulness of two-photon spectroscopy in isotope separation schemes has been discussed by KELLEY et al. [8.67–69].

References

8.1   P. A. M. DIRAC: *The Principles of Quantum Mechanics*, 1st edition (1930), 2nd edition (1958) (Clarendon Press, Oxford)

8.2   M. GOEPPERT-MAYER: Ann. Physik **9**, 273 (1931)

8.3   W. KAISER, C. G. B. GARRETT: Phys. Rev. Letters **7**, 229 (1961)

8.4   See, for example: *Topics in Applied Physics*, Vol. 1: Dye Lasers, ed. by F. P. SCHÄFER (Springer Berlin, Heidelberg, New York 1973)

8.5   L. S. VASILENKO, V. P. CHEBOTAYEV, A. V. SHISHAEV: JETP Letters 12, 113 (1970)
8.6   B. CAGNAC, G. GRYNBERG, F. BIRABEN: Phys. Rev. Letters 32, 643 (1974)
8.7   M. D. LEVENSON, N. BLOEMBERGEN: Phys. Rev. Letters 32, 645 (1974)
8.8   B. CAGNAC, G. GRYNBERG, F. BIRABEN: J. Phys. (Paris) 34, 56 (1973)
8.9   D. PRITCHARD, J. APT, T. W. DUCAS: Phys. Rev. Letters 32, 641 (1974)
8.10 See, for example, N. BLOEMBERGEN: *Nonlinear Optics* (W. A. Benjamin Inc., New York 1965)
8.11 R. B. MILES, S. E. HARRIS: IEEE J. QE-9, 470 (1973);
      J. F. YOUNG, G. C. BJORKLUND, A. H. KUNG, R. B. MILES, S. E. HARRIS: Phys. Rev. Letters 27, 155 (1971);
      D. M. BLOOM, J. T. YARDLEY, J. F. YOUNG, S. E. HARRIS: Appl. Phys. Letters 24, 427 (1974)
8.12 P. P. SOROKIN, J. J. WYNNE, J. R. LANKARD: Appl. Phys. Letters 22, 342 (1973);
      R. T. HODGSON, P. P. SOROKIN, J. J. WYNNE: Phys. Rev. Letters 32, 343 (1974)
8.13 A. M. BONCH-BRUEVICH, V. A. KHODOVOI, V. V. KHRONOV: JETP Letters 14, 333 (1971)
8.14 Y. R. SHEN: Phys. Rev. B9, 622 (1974)
8.15 E. V. BAKLANOV, V. P. CHEBOTAYEV: Sov. J. Quant. Electron. 2, 606 (1975) (in Russian)
8.16 J. M. WORLOCK: *Laser Handbook*, vol. 2, ed. by T. ARECCHI, F. SCHULZ-DUBOIS (North-Holland Publishing Co. 1972) p. 1323
8.17 M. INOUE, Y. TOYOZAWA: J. Phys. Soc. Jap. 20, 363 (1965)
8.18 N. BLOEMBERGEN, M. D. LEVENSON, M. M. SALOUR: Phys. Rev. Letters 32, 867 (1974)
8.19 J. E. BJORKHOLM, P. F. LIAO: Phys. Rev. Letters 33, 128 (1974)
8.20 A. GOLD, in: *E. Fermi Course 42*, ed. by R. GLAUBER (Academic Press, New York 1969) p. 397
8.21 P. LAMBROPOULOS, G. DOOLEN, S. P. ROUNTREE: Phys. Rev. Letters 34, 636 (1975)
8.22 J. A. ARMSTRONG, J. J. WYNNE: Phys. Rev. Letters 33, 1183 (1974)
8.23 K. C. HARVEY, R. T. HAWKINS, G. MEISEL, A. L. SCHAWLOW: Phys. Rev. Letters 34, 1073 (1975)
8.24 C. FABRE, M. GROSS, S. HAROCHE: Opt. Commun. 13, 393 (1975)
8.25 P. F. LIAO, J. E. BJORKHOLM: Phys. Rev. Letters 34, 1 (1975)
8.26 N. BLOEMBERGEN, Y. R. SHEN: Phys. Rev. A 37, 133 (1964)
8.27 R. G. BREWER, E. L. HAHN: Phys. Rev. A 11, 1641 (1975)
8.28 The literature on the electromagnetic response of three-level systems is so vast that no attempt will be made here at a comprehensive listing. The reader will find besides Ref. [8.10, 14, and 27] of this chapter numerous other references quoted in the other chapters of the present volume
8.29 P. AVAN, C. COHEN-TANNOUDJI: J. Phys. Letters (Paris) 36L, 85 (1975)
8.30 R. L. BYER, in: *Laser Spectroscopy*, ed. by R. G. BREWER, A. MOORADIAN (Plenum Press, New York 1974) p. 77
8.31 *Handbook of Lasers*, ed. by R. J. PRESSLEY (Chemical Rubber Co., 1971) p. 298 ff. and references therein, specifically: T. Y. CHANG: Opt. Commun. 2, 77 (1970);
      C. K. N. PATEL: Appl. Phys. Letters 6, 12 (1965);
      C. K. N. PATEL: Phys. Rev. 141, 71 (1966)
8.32 T. W. HÄNSCH: Appl. Opt. 11, 895 (1972);
      R. WALLENSTEIN, T. W. HÄNSCH: Opt. Commun. 14, 353 (1975)
8.33 T. W. HÄNSCH, S. A. LEE, R. WALLENSTEIN, C. WIEMAN: Phys. Rev. Letters 34, 307 (1975)
8.34 R. WALLENSTEIN, T. W. HÄNSCH: Appl. Opt. 13, 1625 (1974)
8.35 C. V. SHANK, E. P. IPPEN: Appl. Phys. Letters 24, 373 (1974)
8.36 M. D. LEVENSON, N. BLOEMBERGEN: Phys. Rev. B10, 4447 (1974)
8.37 R. E. GROVE, F. Y. WU, L. A. HACKEL, D. G. YOUMANS, S. EZEKIEL: Appl. Phys. Letters 23, 442 (1973)

8.38 H. WALTHER: High Resolution Spectroscopy with Tuneable Dye Lasers, in: *Laser Spectroscopy* (op. cit.)

8.39 R. L. BARGER, M. S. SOREM, J. L. HALL: Appl. Phys. Letters **22**, 573 (1973)

8.40 D. E. ROBERTS, E. N. FORTSON: Opt. Commun. **14**, 332 (1975)

8.41 C. FREED: IEEE J. QE-3, 203 (1967);
C. FREED: Designs and Experiments Relating to Stable Lasers, in *Proceedings of the Frequency Standards and Metrology Seminar*, University Laval, Quebec, Canada, 226 (1971)

8.42 J. L. HALL: IEEE J. QE-**4**, 638 (1968);
R. L. BARGER, J. L. HALL: Phys. Rev. Letters **22**, 4 (1969)

8.43 W. K. BISCHEL, P. J. KELLY, C. K. RHODES: Phys. Rev. Letters **34**, 300 (1975)

8.44 W. GONDHALEKAR, E. HOLZHAUR, N. R. HECKENBERG: Phys. Letters A **46A**, 229 (1973)

8.45 W. GORNIK, D. KAISER, W. LANGE, J. LUTHER, H. H. RADLOFF, H. H. SCHULTZ: Appl. Phys. **1**, 285 (1973)

8.46 D. POPESCU, M. L. PASCU, C. B. COLLINS, B. W. JOHNSON, I. POPESCU: Rev. A **8**, 1666 (1973);
also G. V. MARR, S. R. WHERRETT: J. Phys. B **5**, 1735 (1972)

8.47 H. HOTOP, T. C. PATTERSON, W. C. LINEBERGER: Phys. Rev. A **8**, 762 (1973)

8.48 R. G. BREWER: Science **178**, 247 (1972), and references therein

8.49 H. KOGELNIK, T. LI: Appl. Opt. **5**, 1550 (1966)

8.50 L. G. DeSHAZER, E. A. MAUNDERS: Rev. Sci. Instr. **38**, 248 (1967)

8.51 A. ROSEN, I. LINDGREN: Phys. Scr. **6**, 109 (1972)

8.52 T. W. HÄNSCH, K. C. HARVEY, G. MEISEL, A. L. SCHAWLOW: Opt. Commun. **11**, 50 (1974)

8.53 F. BIRABEN, B. CAGNAC, G. GRYNBERG: Phys. Letters A **49A**, 71 (1974)

8.54 F. BIRABEN, B. CAGNAC, G. GRYNBERG: Phys. Letters A **48A**, 469 (1974)

8.55 M. D. LEVENSON, M. M. SALOUR: Phys. Letters A **48A**, 331 (1974)

8.56 R. GUPTA, W. HAPPER, L. K. LAM, S. SVANBERG: Phys. Rev. A **8**, 2292 (1973)

8.57 S. SVANBERG, P. TSEKERIS: Phys. Rev. A **11**, 1125 (1975)

8.58 Y. KATO, B. P. STOICHEFF: in *Lecture Notes in Physics*, Vol. 43: Laser Spectroscopy, ed. by S. HAROCHE, J. C. PEBAY-PEYROULA, T. W. HÄNSCH, S. E. HARRIS (Springer Berlin, Heidelberg, New York 1975) p. 452

8.59 F. BIRABEN, B. CAGNAC, G. GRYNBERG: J. Phys. Letters **36**, L41 (1975)

8.60 F. BIRABEN, B. CAGNAC, G. GRYNBERG: Compt. Rend. B **280**, 235 (1975)

8.61 S. A. LEE, R. WALLENSTEIN, T. W. HÄNSCH: Phys. Rev. Letters **35**, 1262 (1975)

8.62 R. M. HOCHSTRASSER, N. H. SONG, J. E. WESSEL: Chem. Phys. Letters **24**, 7 and 171 (1974)

8.63 A. L. SCHAWLOW: private communication

8.64 W. K. BISCHEL, P. J. KELLEY, C. K. RHODES: (to be published). The authors would like to thank these investigators for information of their results prior to publication

8.65 R. H. DICKE: Phys. Rev. **89**, 472 (1953)

8.66 F. BIRABEN, E. GIACOBINO, G. GRYNBERG: Phys. Rev. A **12**, 2444 (1975);
see also B. CAGNAC, in: *Laser Spectroscopy* (op. cit.) p. 165

8.67 P. L. KELLEY, H. KILDAL, H. R. SCHLOSSBERG: Chem. Phys. Letters **27**, 62 (1974)

8.68 V. P. CHEBOTAYEV, A. L. GOLGER, V. S. LETOKHOV: Chem. Phys. **7**, 316 (1975)

8.69 K. SHIMODA: Appl. Phys. **9**, 239 (1976)

Additional References with Titles

R. ABJEAN, M. LERICHE: On the shape of absorption lines in a divergent atomic beam. Opt. Commun. **15**, 121 (1975)

C. K. AU: Two photon absorption spectroscopy of metastable hydrogen like atoms. Phys. Lett. A **51** A, 442 (1975)

S. N. BAGAYEV, L. S. VASILENKO, A. K. DMITRIJEV, V. G. GOL'DORT, M. N. SKVORTSOV, V. P. CHEBOTAYEV: Narrow resonances in radiation spectrum of the He-Ne laser with methane absorber. Appl. Phys. **10**, 231 (1976)

YE. V. BAKLANOV, V. P. CHEBOTAYEV: Profile of the two photon absorption line due to the $1S$–$2S$ transition in atomic hydrogen (frequency standard). Soviet. J. Quant. Electron. **5**, 342–344 (1975); translation of Kvantovaya Electron. Moskva **2**, 606–609 (1975)

YE. V. BAKLANOV, V. P. CHEBOTAYEV, B. YA. DUBETSKY: The resonance of two-photon absorption in separated optical fields. Appl. Phys. **11**, (1976)

YE. V. BAKLANOV, B. YA. DUBETSKY, V. P. CHEBOTAYEV: Non-linear Ramsey resonance in the optical region. Appl. Phys. **9**, 171 (1976)

K. BERGMANN, W. DEMTRÖDER, P. HERING: Laser diagnostic in molecular beams. Appl. Phys. **8**, 65 (1975)

A. F. BERNHARDT: Isotope separation by laser deflection of an atomic beam. Appl. Phys. **9**, 19 (1976)

A. F. BERNHARDT, D. E. DUERRE, J. R. SIMPSON, L. L. WOOD: High resolution spectroscopy using photodeflection. Opt. Commun. **16**, 166 (1976)

F. BIRABEN, G. GRYNBERG, E. GIACOBINO, J. BAUCHE: Investigation of infrared $4d'$ subconfiguration of neon using Doppler-free two-photon spectroscopy

G. M. CARTER, D. E. PRITCHARD, T. W. DUCAS: Steady-state excitation of a sodium beam using a C.W. laser. Appl. Phys. Lett. **27**, 498 (1975)

V. P. CHEBOTAYEV, A. L. GOLGER, V. S. LETOKHOV: Selective two quantum laser excitation of molecules with overlapping absorption lines. Chem. Phys. **7**, 316 (1975)

C. DELSART, J.-C. KELLER: Hyperfine structures in ^{21}Ne using laser-induced absorption line narrowing. Opt. Commun. **16**, 388 (1976)

T. W. DUCAS, R. R. FREEMAN, M. G. LITTMAN, M. L. ZIMMERMAN, D. KLEPPNER: Stark ionization of high-lying Rydberg states of sodium. In: *Lecture Notes in Physics,* Vol. 43: Laser spectroscopy, ed. by S. HAROCHE, J. C. PEBAY-PEYROULA, T. W. HÄNSCH, S. E. HARRIS (Springer Berlin, Heidelberg, New York 1975) p. 462

T. W. DUCAS, M. G. LITTMAN, R. A. FREEMAN, D. KLEPPNER: Stark ionization of high-lying states of sodium. Phys. Rev. Lett. **35**, 366 (1975)

R. B. DUNNING, T. B. COOK, W. P. WEST, R. F. STEBBINGS: Selective removal of either metastable species from a mixed $^3P_{0,2}$ rare-gas metastable beam. Rev. Sci. Instr. **46**, 1072 (1975)

H. T. DUONG, G. HUBER, P. JACQUINOT, P. JUNCAR, R. KLAPISCH, S. LIBERMAN, J. PINARD, C. THIBAULT, J. L. VIALLE: High resolution laser spectroscopy of the D-lines of on-line produced radioactive sodium isotopes. In: *Lecture Notes in Physics*, Vol. 43: Laser spectroscopy, ed. by S. HAROCHE, J. C. PEBAY-PEYROULA, T. W. HÄNSCH, S. E. HARRIS (Springer Berlin, Heidelberg, New York 1975) p. 144

G. L. EESLEY, M. D. LEVENSON: Single mode operation of a repetitively pulsed dye laser. IEEE. J. Quant. Electron. QE-12, 259 (1976)

P. ESHERICK, J. A. ARMSTRONG, R. W. DREYFUS, J. J. WYNNE: Multiphoton ionization spectroscopy of high-lying, even-parity states in calcium. Phys. Rev. Letters 36, 1296 (1976)

S. M. FREUND, M. RÖMHELD, T. OKA: Infrared-radiofrequency two-photon and multi-photon Lamb dips for CH_3F. Phys. Rev. Letters 35, 1497 (1975)

J. A. GELBWACHS, P. F. JONES, J. E. WESSEL: Doppler free two photon electronic absorption of Nitric oxide and Benzene. Appl. Phys. Lett. 27, 551 (1975)

H. J. GERRITSEN, G. NIENHUIS: Multidirectional Doppler pumping. A new method to prepare an atomic beam having a large fraction of excited atoms. Appl. Phys. Lett. 26, 347 (1975)

D. GRISHKOWSKY, M. M. T. LOY: Self-induced adiabatic rapid passage. Phys. Rev. A 12, 1117 (1975)

M. GROSS, C. FABRE, P. PILLET, S. HAROCHE: Observation of near-infrared Dicke narrowing on cascading transitions in atomic sodium. Phys. Rev. Lett. 36, 1035 (1976)

P. GRUNDEVIK, M. GEESTAVSSON, S. SVANBERG: Isotope shift in dysprosium measured by high resolution laser spectroscopy. Phys. Lett. A 56 A, 25 (1976)

G. GRYNBERG, F. BIRABEN, B. CAGNAC: Three-photon Doppler-free spectroscopy – Experimental evidence. Phys. Rev. Lett. (to be published)

L. A. HACKEL, K. H. CASLETON, S. G. KUKOLICH, S. EZEKIEL: Observation of magnetic octupole and scalar spin-spin interacteous in I_2 using laser spectroscopy. Phys. Rev. Lett. 35, 568 (1975)

C. D. HARPER, M. D. LEVENSON: Fine structure splittings of high ^2D states of ^{39}K. Phys. Lett. A 56A, 361–362 (1976)

YU. I. HELLER, A. K. POPOV: Laser-induced narrowing of autoionizing resonances studied by the method of parametric generation. Phys. Lett. 56A, 453 (1976)

S. M. HEIDER, G. O. BRINK: Simple system for producing an atomic beam of alkaline earth metastable. Rev. Sci. Instr. 46, 488 (1975)

G. HUBER, R. KLAPISCH, C. THIBAULT, T. H. DUONG, P. JUNCAR, S. LIBERMAN, J. PINARD, J. L. VIALLE, P. JACQUINOT: Détermination par spectroscopie laser des moments quadrupolaires de noyaux radioactifs de sodium. Compt. Rend. Acad. Sci. Paris B 282 119 (1976)

J. W. C. JOHNS, A. R. W. McKELLAR, T. OKA, M. RÖMHELD: Collision-induced Lamb dips in laser Stark spectroscopy. J. Chem. Phys. 62, 1488 (1975)

H. KOSCHMIEDER, V. RAIBLE: Intense atomic-hydrogen beam source. Rev. Sci. Instr. 46, 536 (1975)

V. S. LETOKHOV, V. G. MINOGIN, B. D. PAVLIK: Cooling of atoms and molecules by resonant laser field. Opt. Commun. (to be published)

V. S. LETOKHOV, B. D. PAVLIK: Spectial line narrowing in a gas by atoms trapped in a standing light waves. Appl. Phys. 9, 229 (1976)

M. D. LEVENSON: Optical beating spectroscopy of atomic polarrizabilities excited by two photon absorption. Phys. Rev. A June 1976 (to be published)

P. F. LIAO, G. C. BJORKHOLM: Polarization rotation induced by resonant two photon dispersion. Phys. Rev. Lett. 36, 584 (1976)

M. G. LITTEMAN, M. L. ZIMMERMAN, T. W. DUCAS, R. R. FREEMAN, D. HEPPNER: Strutcure of sodium Rydberg state in weak to strong electric fields. Phys. Rev. Lett. 36, 788 (1976)

H. MAEDA, K. SHIMODA: Theory of the inverted Lamb dip with a Gaussian beam. J. Appl. Phys. 47, 1069 (1976)

M. MATSUOKA: Doppler free two photon induced coherence and emission. Opt. Commun. 15, 84 (1975)

J. L. PICQUÉ: Non-optical observation of zero-field level crossing effect in a sodium beam. In: Lecture Notes in Physics, Vol. 43: Laser spectroscopy, ed. by S. HAROCHE, J. C.

Pebay-Peyroula, T. W. Hänsch, S. E. Harris (Springer Berlin, Heidelberg, New York 1975) p. 462

J. L. Picqué, J. Pinard: Direct observation of the Autler-Townes effect in the optical range. J. Phys. B, **9**, L 77 (1976)

F. M. Pipkin: New vistas for precision fine structure measurements. Comm. Atom. Molec. Phys. **5**, 45 (1975)

R. Saloma, S. Stenholm: Two photon spectroscopy: Effects of a resonant intermediate state. J. Phys. B **8**, 1795 (1975)

M. Sargent III: Laser saturation grating phenomena. Appl. Phys. **9**, 127 (1976)

M. Sargent III, P. E. Toschek: Unidirectional saturation spectroscopy. II. General lifetimes, interpretations, and analogies. Appl. Phys. **11** (October 1976)

M. Sargent III, P. E. Toschek, H. G. Danielmeyer: Unidirectional saturation spectroscopy. I. Theory and short diode liefetime limit. Appl. Phys. **11** (1976)

E. Shimizu, K. Namba: Highly selective pumping of an excited state by Doppler free multiphoton absorption. Phys. Lett. A **54A**, 179 (1975)

J. J. Snyder, J. L. Hall: A new measurement of the relatevistic Doppler shift. In: *Lecture Notes in Physics*, Vol. 43: Laser spectroscopy, ed. by S. Haroche, J. C. Pebay-Peyroula, T. W. Hänsch, S. E. Harris (Springer Berlin, Heidelberg, New York 1975) p. 6

R. E. Smalley, D. H. Levy, L. Wharton: Molecular-jet spectroscopy. Laser Focus (November 1975) p. 40

B. P. Stoicheff, Y. Kato: High resolution two photon absorption spectroscopy of highly excited *D* states of Rb atoms. J. Opt. Soc. Am. **65**, 1180 (1975)

R. K. Thareja, S. N. Haque: Multiphoton transitions in hydrogen atom. Phys. Lett. A **48A**, 231 (1974)

H. Walther: Atomic fluorescence induced by monochromatic excitation. In: *Lecture Notes in Physics,* Vol. 43: Laser spectroscopy, ed. by S. Haroche, J. C. Pebay-Peyroula, T. W. Hänsch, S. E. Harris (Springer Berlin, Heidelberg, New York 1975) p. 358

C. C. Wang, L. I. Davis, Jr.: Saturation of resonant two photon transitions in thallium vapor. Phys. Rev. Lett. **35**, 650 (1975)

F. Y. Wu, R. E. Grove, S. Ezekiel: Investigation of the spectrum of resonance fluorescence induced by a monochromatic field. Phys. Rev. Lett. **35**, 1426 (1975)

Subject Index

Applied Physics

A monthly journal

Board of Editors	**S. Amelinckx,** Mol · **A. Benninghoven,** Münster
	V. P. Chebotayev, Novosibirsk · **R. Gomer,** Chicago, Ill.
	V. S. Letokhov, Moskau · **H. K. V. Lotsch,** Heidelberg
	H. J. Queisser, Stuttgart · **F. P. Schäfer,** Göttingen
	A. Seeger, Stuttgart · **K. Shimoda,** Tokyo
	T. Tamir, Brooklyn, N.Y. · **W. T. Welford,** London
	H. P. J. Wijn, Eindhoven

Coverage	application-oriented experimental and theoretical physics:

Solid-State Physics	*Quantum Electronics*
Surface Physics	*Laser Spectroscopy*
Chemisorption	*Photophysical Chemistry*
Microwave Acoustics	*Optical Physics*
Electrophysics	*Integrated Optics*

Special Features	**rapid** publication (3–4 months)
	no page charge for **concise** reports
	prepublication of titles and abstracts
	microfiche edition available as well

Languages	Mostly English

Articles	original reports, and short communications
	review and/or tutorial papers

Manuscripts	to Springer-Verlag (Attn. H. Lotsch), P.O. Box 105280
	D-69 Heidelberg 1, F.R. Germany

Place North-American orders with:
Springer-Verlag New York Inc., 175 Fifth Avenue, New York. N.Y. 10010, USA

Springer-Verlag
Berlin Heidelberg New York

R. Beck, W. Englisch, K. Gürs

Table of Laser Lines in Gases and Vapors

IV, 130 pages. 1976
(Springer Series in Optical Sciences, Vol. 2. Editor: D. L. MacAdam)

Laser Spectroscopy

Proceedings of the 2nd International Conference, Mégève, France, June 23—27, 1975
Edited by *S. Haroche, J. C. Pebay-Peyroula, T. W. Hänsch, S. E. Harris*
230 figures, 30 tables. X, 468 pages (5 pages in French). 1975
(Lecture Notes in Physics, Vol. 43)

Forthcoming Titles

Springer Series in Optical Sciences
Editor: D. L. MacAdam

Vol. 3 Tunable Lasers and Applications

Proc. of Loens Conf. (1976). Editors: A. Mooradian, T. Jaeger

Vol. 4 V. S. Letokhov, V. P. Chebotayev
Nonlinear Laser Spectroscopy

Topics in Current Physics

Beam-Foil Spectroscopy S. Bashkin (editor)

S. Bashkin: Introduction. — *S. Bashkin:* Instrumentation. — *I. Martinson:* Wavelengths Measurements and Level Analysis. — *L. Curtis:* Lifetime Measurements. — *I. Sellin:* Autoionizing Levels. — *H. Marrus:* Studies of H-like and He-like Ions of High Z. — *W. Whaling, L. Heroux:* Applications to Astrophysics. — *O. Sinanoglu:* Fundamental Calculation of Level Lifetimes. — *W. Wiese:* Systematic Effects in Z-Dependence of Oscillator Strengths. — *J. Macek, D. J. Burns:* Coherence, Alignment, and Orientation Phenomena

Springer-Verlag Berlin Heidelberg New York